Renewable Energy: Advanced Technologies and Applications

Renewable Energy: Advanced Technologies and Applications

Edited by Ted Weyland

SYRAWOOD
PUBLISHING HOUSE

New York

Published by Syrawood Publishing House,
750 Third Avenue, 9th Floor,
New York, NY 10017, USA
www.syrawoodpublishinghouse.com

Renewable Energy: Advanced Technologies and Applications
Edited by Ted Weyland

International Standard Book Number: 978-1-68286-469-2 (Hardback)

Cataloging-in-Publication Data

Renewable energy : advanced technologies and applications / edited by Ted Weyland.
    p. cm.
Includes bibliographical references and index.
ISBN 978-1-68286-469-2
1. Renewable energy sources. 2. Renewable energy sources--Technological innovations. 3. Energy development.
I. Weyland, Ted.
TJ808 .R46 2017
621.042--dc23

Printed in the United States of America.

TABLE OF CONTENTS

Permissions

List of Contributors

Index

PREFACE

The world is advancing at a fast pace like never before. Therefore, the need is to keep up with the latest developments. This book was an idea that came to fruition when the specialists in the area realized the need to coordinate together and document essential themes in the subject. That's when I was requested to be the editor. Editing this book has been an honour as it brings together diverse authors researching on different streams of the field. The book collates essential materials contributed by veterans in the area which can be utilized by students and researchers alike.

Renewable energy is defined as energy harnessed from natural resources that can be easily replenished. This book on renewable energy takes into account renewable energy technology and green energy practices that reduce emission and energy wastage. The increased use of renewable energy points to a greener future that can be sustained and shared by larger number of people. Topics in this book provide data and information on the present status of the various renewable energy technologies and the harnessing of renewable energy. This book will help new researchers by foregrounding their knowledge in this branch. From theories to research to practical applications, case studies related to all contemporary topics of relevance to this field have been included herein. The book, with its detailed analyses and data, will prove immensely beneficial to professionals and students involved in renewable energy technology at various levels.

Each chapter is a sole-standing publication that reflects each author's interpretation. Thus, the book displays a multi-facetted picture of our current understanding of applications and diverse aspects of the field. I would like to thank the contributors of this book and my family for their endless support.

Editor

A proposed algorithms for tidal in-stream speed model

Hamed H. H. Aly, M. E. El-Hawary

Department of Electrical and Computer Engineering, Dalhousie University, Halifax, Nova Scotia, Canada, B3H 4R2

Email address:
hamed.aly@dal.ca (H. H. H. Aly,), elhawary@dal.ca (M. E. El-Hawary)

Abstract: In this paper we propose four models for tidal current speed and direction magnitude forecasting model. The first model is a Fourier series model based on the least squares method (FLSM), the second model is an artificial neural network (ANN), the third model is a hybrid of FLSM and ANN and the fourth model is a hybrid of ANN and FLSM for monthly forecasting of tidal current speed. These proposed models are ranked in order depending on their performance. These models are validated by using another set of data (tidal current direction). The proposed hybrid model of FLSM and ANN is highly accurate and outperforms. This study was done using data collected from the Bay of Fundy in 2008.

Keywords: Power System Modeling, Tidal Currents, Forecasting, ANN, Fourier Series Based On Least Squares

1. Introduction

Tidal current energy can be converted into electrical power. It is so hard to store the electrical energy for later use and a control system is required to be connected directly to the consumer. It is very important to have advance knowledge of the tidal current energy to manage the production of the electrical power so that it may ensure that this power will be controlled in an efficient way to allow scheduling different electrical energy resources to minimize interruptions. Prior knowledge of future generated electrical energy or the tidal current energy is known as forecasting. Tidal flow causes somewhat predictable energy output patterns. Forecasting marine currents using data gathered for short periods of time is predictable to within 98% accuracy). On the other hand forecasting wind speed requires data gathered over a longer period. The marine resource is easier to integrate in the electrical grid. Forecasting is the first step in dealing with the future generation of the tidal current power systems. The accuracy of the models used for tidal current forecasting is critical as a tidal in-stream forecast with sufficient accuracy can provide a stable and controlled electric power dynamic performance that allows better dispatching of grid resources and this evens out the use of battery storage and will affect the overall cost of electricity. In addition, tidal current prediction is useful in making operations- and planning related decisions such as towing of activities vessels, fisheries and recreational activities and monitoring of oil slick movements [1, 2].

2. Previous Research for Tidal Currents Forecasting

Sir G. H. Darwin [3] is credited with the idea that tidal oscillation of the ocean may be represented as the sum of a number of simple harmonic waves. Subsequently, Doodson [4-5] proposed using least squares estimation to determine the parameters of the harmonic series which has been widely used for tidal forecasting [6].

Artificial neural networks (ANN) have been used to overcome the problem of exclusive and nonlinear relationships. French, Krajewski & Cuykendall proposed an ANN model to predict rainfall intensity [7]. Raman & Sunilkumar proposed a multivariate modeling of water resources time series by using ANN [8]. Dawson and Wilby considered the potential of using ANN for rainfall-runoff modelling and flood forecasting [9]. Coulibaly et al. used a modified ANN for daily resvoir inflow forecasting [10]. Lee and Jeng [11] used an ANN model for tidal level forecasting using short-term tidal records from three harbours in Taiwan. Campolo & Soldati applied ANN for river flood forecasting [12]. Lee [13] used the ANN Back Propagation with descent algorithm to forecast the tidal level for three different tide types, diurnal, semi-diurnal and mixed tides. This model was used for the short and long term forecasting. In [14], Lee, Tsai and Shieh applied the Back Propagation Neural Network (PBN) to predict long term semi-diurnal tidal levels. Based on the model, the different tide types for other two field data of diurnal and mixed types were used to test the performance of a PBN model.

Vijay and Govil [15] used radial basis function ANN networks (RBF) for tidal data prediction of high and low tides of any day of the year depending on the training data of only one month. They concluded that a Fourier series or a polynomial series alone do not give accurate results. They also reported that using Wavelets yielded approximately the same results as ANN but implementing the Wavelet approach required longer execution time. Chen, Wang and Chu [16] proposed a hybrid of wavelet and ANN models for tidal current prediction. The signal in the multi-resolution analysis (MRA) used in wavelet analysis consists of high and low frequency components. Chen et al. eliminated the high frequency components and used the inverse wavelets to rebuild new signals. The input/output data that were used for the training of the ANN depended on the calculation of the tidal constituent time-lags. Adamowski [17] proposed a hybrid of ANN and wavelet and cross wavelet constituent components for short term river flood forecasting that gave accurate results compared to wavelet and cross wavelet constituent components alone.

In [18] genetic algorithms (GA), were used to carry out the prediction task. A preliminary empirical orthogonal function (EOF) analysis was used to compress the spatial variability into a few eigen-modes, so that GA could be applied to the time series of the dominant principal components (PC). Burrage et al. proposed an optimal multi linear regression model for the tidal current forecasting [19].

Harmonic tidal current constituent analysis or numerical hydrodynamic models are traditional models used for tidal current prediction. These models have their own limitations and nonlinear data adaptive approaches are gaining increased acceptance. Numerical hydrodynamic models require large computing resources and huge input information.

In this paper such an approach, known as a Hybrid model, has been employed for the tidal prediction. The novelty of the method is the use of the ANN or FLSM technique to forecast of the resulting principal components from a few observed tidal levels with the use of FLSM or ANN. The proposed model is easy to use and only depends on the input data (speed or direction) without knowing the tides' constituents because we covered all cycles without referring to the the type of the cycle so we used the model for predicting of the speed and the direction using the time as an input. These models are used for more than month (33.67 days).

3. Overview of Some Forecasting Techniques

In this section some most commonly used forecasting techniques will be outlined and the main focus will be on the techniques used in this paper for tidal current forecasting. Many forecasting techniques are used nowadays ranging from Multiple Linear and Nonlinear Regression, Dynamic Techniques, General Exponential Smoothing Technique, Expert System, Fourier Series Model based on the Least Squares Method, Time Series, Wavelet and An Artificial Neural Network.

3.1. Multiple Linear and Nonlinear Regression[20, 21]

Regression is a commonly used technique for modeling. Regression is used to develop a mathematical model which is represented by an equation or a set of equations that represent the system behavior and treat one variable as a function of others. These equations may be linear or nonlinear and can be used to predict a response from the value of a given predictor(s). They can be used to consider more complex relationships than correlation by using more than two variables or combinations of different order equations. This technique is effective in the case of off line forecasting application and is generally unstable for the on-line forecasting application, because it requires many external variables. It is commonly used in experimental tests where a range of fixed predictor levels are set and tests whether there is a significant increase or decrease in the response variable along the gradient of predictor levels.

In multiple linear regression, the most common estimation method is implemented using an equation of the form:

$$E(y_i) = \beta 0 + \sum_{i=1}^{N} \beta_i x_i (t) + r(t)$$

$E(y_i)$ is the forecasted variable at a certain time t (the dependent or response variable), β_0 is the intercept , β_i is the regression coefficient and r(t) is the residual.

The previous equation is a first order model with one predictor but sometimes there will be a need for increasing the order depending on the used model and the data. The least squares method is used for estimating the parameters for the model.

In the nonlinear model at least one of the parameters appears nonlinearly. Generally speaking in a nonlinear model at least one parameter should appear when a first order derivative with respect to that parameter. For example one may write the nonlinear model in the form of $E(y_i) = \exp (ax + bx^2)$

3.2. Expert System Approach[22]

The expert system method depends on statistical analysis of the past data and the knowledge of experts in the field of interest. The forecast model using this technique emulates the knowledge, experience and identifies the rules and the variables used by the experts. This technique is commonly used for the load forecasting.

3.3. Neural Network Structure

An artificial neural network is a mathematical model inspired by the natural neurons interactions. It is based on simulating the function of the human brain which consists of massive neural networks. The brain has the ability to compute, recognize faces, speech and control activities. The brain has a highly parallel computing structure, and the capability for processing the information. The human brain has more than 10 billion interconnected neurons. Each neuron in the human brain is a cell which uses the reactions to receive, process, and transmit information. The networks of nerve fibers are called dendrites which are connected to the cell body or soma, where the cell nucleus is located. The axon is a single long fiber extending from the cell body. This axon branches into strands and substrands, to connect to other neurons through synaptic terminals or synapses. The neurons receive signals through synapses. The neurons start to activate and emit a signal through the axons when they receive strong signals, the potential of the signals reach a threshold, a pulse is sent down the axon and the cell is fired. Figure (1) shows the general structure of the neural network feed forward system [23, 24].

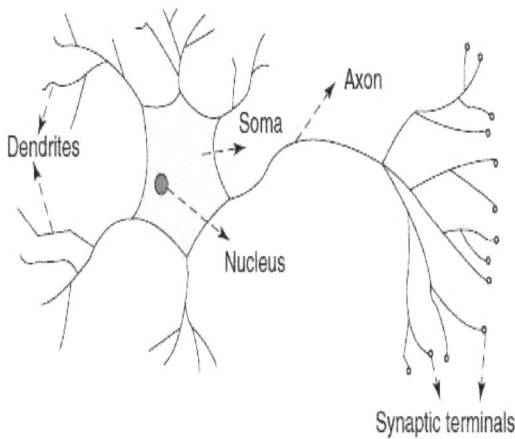

Figure 1. *The mammalian neuron.*

In neural networks, the effects of the synapses are represented by connection weights that modulate the effect of the associated input signals. The transfer function is used to represent the nonlinear characteristic exhibited by neurons. The impulse of the neuron is equal to the weighted sum of the input signals that transformed by the transfer function. By adjusting the weights the artificial neuron starts to learn [24]. An artificial neural network is commonly used for forecasting. As the size of the input data increases, accuracy will increase. The neural network consists of one input layer, one output layer and one or more hidden layers. Each layer consists of a number of neurons. In feed forward networks, the signal flow is coming from input to output and there is no feedback connections. In the recurrent networks there is a feedback connection. There are several neural network structures like recurrent or Elman networks, adaptive resonance theory maps, competitive networks...., which are used according to the properties and requirements of the application.

The neural network has to be configured to produce the desired set of outputs for a certain set of inputs. There are many methods used to configure the ANN. A common way is to set the weights explicitly by using a priori knowledge.

Another simple way is to train the neural network by feeding it teaching patterns and allowing it to change its weights according to some additional learning rule which is easier but requires additional processing time. The output of each neuron can be expressed as a function of the input signals as:

$$Y_j(t) = f\left(\sum_{i=1}^{x} W_{ij} X_i(t) \pm b_i\right) \qquad \text{for } j = 1,\dots H1$$

$$Y_k(t) = f\left(\sum_{j=1}^{H1} W_{jk} Y_j(t) \pm b_i\right) \qquad \text{for } k = 1,\dots H2$$

$$O_r(t) = f\left(\sum_{r=1}^{H2} W_{kr} Y_k(t) \pm b_i\right) \qquad \text{for } r = 1,\dots Z$$

$Y_j(t)$ & $Y_k(t)$ = Quantity computed by the first and second hidden neurons respectively.

$O_r(t)$ = Network output.

X & Z = Number of input and output neurons.

H_1 & H_2 = Number of first and second hidden neurons.

W_{ij}, W_{jk} & W_{kr} = Adjustable weights between input and first hidden layer, the first and second layer and the second and output layer.

b_i = Biases.

f = Transfer function.

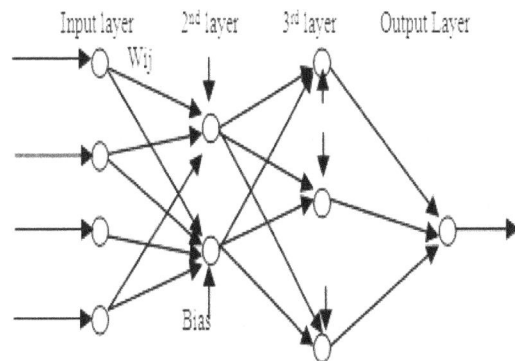

(a) The general structure of the neural network multi layer feed forward system.

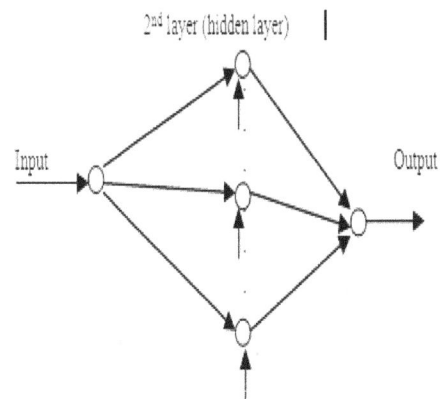

(b) The structure of the model used in this paper.

Figure 2. *Neural network structure.*

There are three types of learning, supervised learning, unsupervised learning, and reinforcement learning. For supervised learning, the input vector is presented at the input nodes together with a set of desired responses, one for each node, at the output layer. A forward path is used, and the errors are calculated which is the difference between the desired and actual data for each node in the output layer. The errors are used to determine weight changes in the network depending on the learning rule. The backpropagation algorithm, the perceptron rule, and the delta rule are typical supervised learning techniques. In unsupervised learning, the output unit is trained to respond to pattern clusters within the input. The system attempts to discover statistically salient features of the input. Here the system develops its own representations from the input. In reinforcement learning the system is taught what to do, and how to map situations to distinguish features of reinforcement learning. Trial and error search and delayed reward all characterize reinforcement learning. In this approach the learner is not informed which actions to take first for solving a certain problem, but it is informed to discover which actions yield the most reward by trying them [23-27]. There are some rules that should be considered while dealing with ANN like the selection of the raw data patterns for training, the topology of the network, and the training algorithm that has faster convergence properties and lower computational time [25].

A neural network is commonly used for forecasting. As the size of the input data increases accuracy will increase. The neural network consists of one input layer which in our case is the time index and one output which is the tidal speed or the tidal direction and one or more hidden layers. Each layer consists of some neurons. The weight matrices, the number of layers, neurons, epochs of training, inputs and the transfer functions affect on the ANN performance. The back propagation algorithm is an efficient method for changing the weights in a feed forward network, with differential activation function units and supervised training, to learn a training set of input/output examples. It depends on gradient descent that adjusts the weights to reduce the system error [28].

Figure (2.a) shows the general structure of the ANN. The system which is used in this paper has only one input and one output as shown in figure (2.b).

3.4. *Fourier Series based on Least Square Model Structure (FLSM) [29-31]*

The estimated data may be defined using Fourier series as:

$$Z_{0estimate}(K) = DC + \sum_{n=1}^{N}(a_i \sin(\omega_i k + \theta_i)) =$$

$$DC + \sum_{i=1}^{N}(a_i \sin(\omega_i k)\cos\theta_i + a_i \cos(\omega_i k)\sin\theta_i))$$

Where: DC= Constant value depending on the data (the average value),

K=discrete time,

a= amplitude,

i=number of harmonics in the wave,

θ= the phase shift. The Fourier series parameters may be

determined using the LSM.

Now let us define the actual data as Z=DC+HX+e(k), e(k)is the error (residuals), then we may apply the least square model as follows to estimate the Fourier series parameters.

$$Xhat= (H^{T}H)^{-1}H^{T}Z ,$$

$$Z_{oestimate}=DC+HX_{hat},$$

$$Z_{innovation}=Z-Z_{oestimate},$$

$$=$$

$$\begin{bmatrix} sin\omega_1 & cos\omega_1 & sin\omega_2 & \dots\dots\dots & sin\omega_i & cos\omega_i \\ sin2\omega_1 & sin2\omega_1 & \dots\dots & \dots\dots & sin2\omega_i & cos2\omega_i \\ sin3\omega_1 & \dots\dots\dots & \dots\dots & \dots\dots & \dots\dots\dots & cos3\omega_i \\ \dots\dots\dots & \dots\dots\dots & \ddots\ddots & \dots\dots\dots & \dots\dots\dots \\ \dots\dots\dots & \dots\dots\dots & \dots\dots\dots & \dots\dots\dots & \dots\dots\dots \\ sinn\omega_1 & cosn\omega_1 & \ddots\ddots & sinn\omega_i & cos8n\omega_i \end{bmatrix}$$

$$, X=\begin{bmatrix} a_1 cos\theta_1 \\ a_1 sin\theta_1 \\ a_2 cos\theta_2 \\ \dots\dots \\ \dots\dots \\ a_i sin\theta_i \end{bmatrix}$$

H is a matrix that has a number of rows equal to the number of the input data and number of columns depending on the number of harmonic used. X is a resulting vector coming from the matrix H and the input data.

In the following section we will use the FLSM, ANN model and a hybrid of ANN and FLSM for the prediction of the tidal speed during a month. We will use the innovation (residuals) data as the input to the ANN in case of a hybrid model. The data that we used in this paper is a commercial data so we used it after multiplying by a factor and shifting it.

4. Proposed Networks Construction of the Tidal Current Speed Forecasting Model

A. ANN model

In this model we use ANN feed forward back propagation for training the proposed model.

B. FLSM Model

In this model we use Fourier series based on the Least Square method for training the proposed model.

C. Hybrid model of FLSM and ANN

This model consists of the tidal current prediction using the FLSM at the beginning then find the error between the exact and the predicted data, and this error is called innovations. Secondly, we try to use these innovations to feed the ANN model as an input. Finally the full model is equal to the FLSM plus the ANN. The flowchart shown in figure (3) describing each step in the hybrid model.

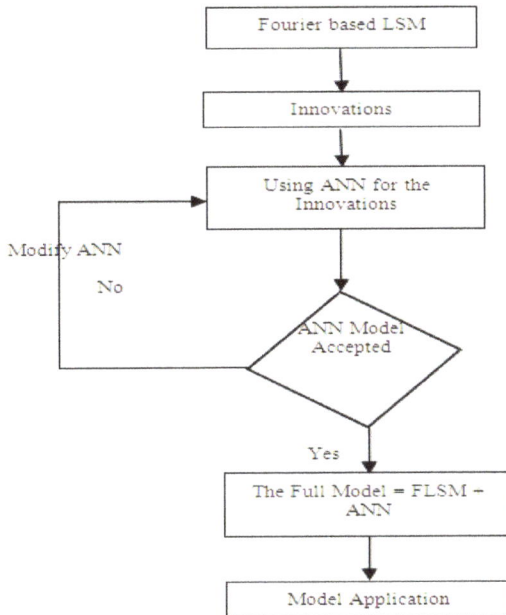

Figure 3. Hybrid model FLSM and ANN flowchart.D. Hybrid model of ANN and FLSM.

This model consists of the tidal current prediction using the ANN at the beginning then find the error between the exact and the predicted data, and this error is called innovations. Secondly, we try to use these innovations to feed the FLSM model as an input. Finally the full model is equal to the ANN plus the FLSM. The flowchart shown in figure (4) describing each step in the hybrid model.

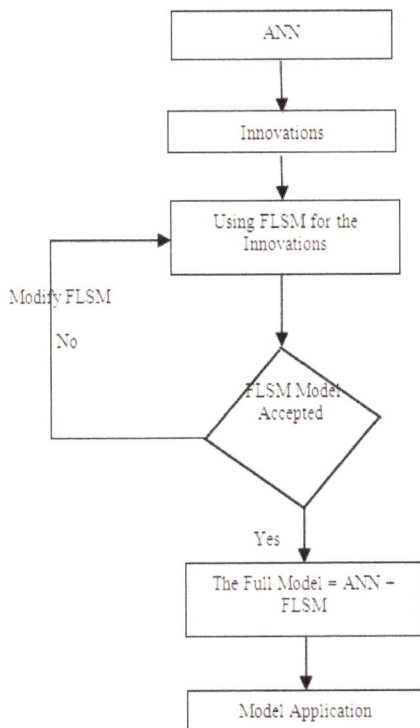

Figure 4. Hybrid model of ANN and FLSM flowchart.

5. Tidal Current Data Identifications (Tidal Speed Models)

5.1. Fourier Series Model based on Least Square Method (FLSM)

In this section we try to use Fourier series based on the least square model for the tidal current prediction. We use the percentage of error (P. E.) for comparing between different proposed methods.

The percentage of error (P. E.) = $((Z_{Actual} - Z_{Predicted})/ Z_{Actual}) *100$.

After using the FLSM we find that the percentage of error is 0.6399% for 70% of the data and 0.817 for the other 30% of the data. This means that the error is depending on the number the data and the time that the data taken as its shape will be different from time to time. Figure (5) shows the exact and the estimated data using FLSM for 70% of the available data and figure (6) shows the other 30%. The time for the collected data is measured after each ten minutes and we try to use a code for the time such that the time 1 is the first ten minutes and the time 2 is the second ten minutes and so on till we reach the end. The time on the second graph was measured also at each ten minutes after ten minutes of the end of the first graph. This means that the time on the second graph starting from the 3001*10 minutes which is after ten minutes from the end of the first graph and this means that the time1 on the second graph is equal to 3001*10 minutes for the first graph. The speed in all graphs is in m/s.

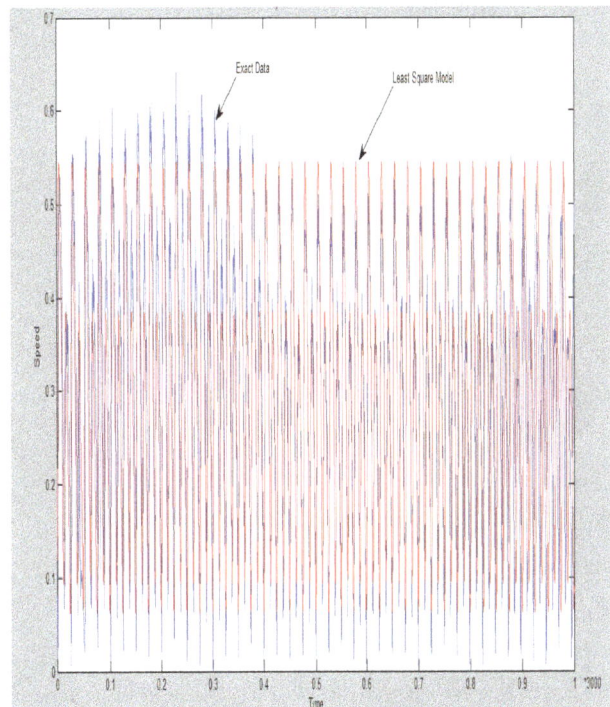

Figure 5. The relation between the speed and the time of the tidal currents after using the FLSM for the exact trained and the estimated data.

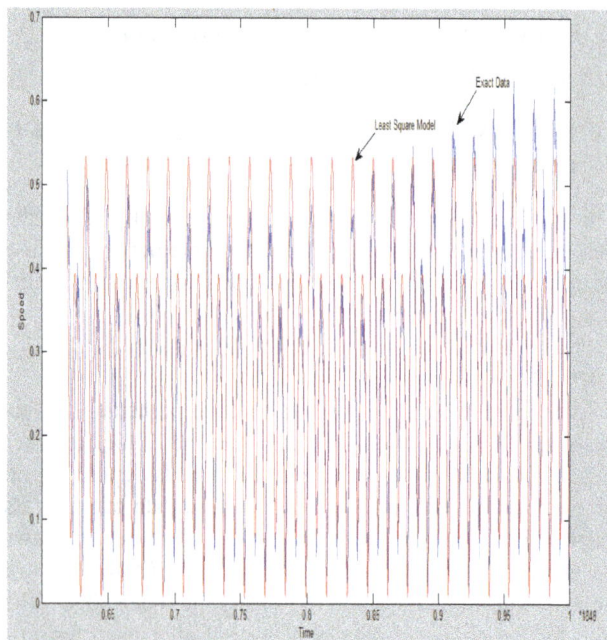

Figure 6. *The relation between the speed and the time of the tidal currents after using the FLSM for the exact untrained and the estimated data.*

5.2. ANN Model

In this section we try to use the ANN for tidal current speed prediction for the same data used in the previous section. We use the exact data without modifications as an input to the ANN. After 20,000 epochs, 225 neurons in the first layer, the mean squared error became 0.000336. The percentage of error of the trained exact data became 0.3903 and for the estimated data became 3.0946. The exact trained data and the estimated ANN data are plotted in figure (7).

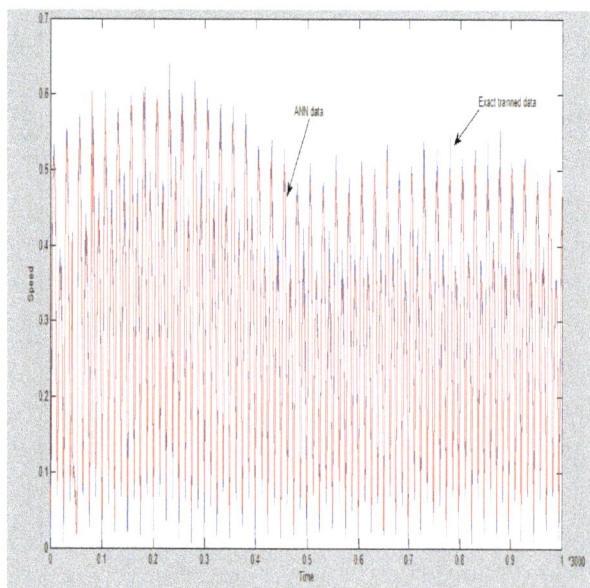

Figure 7. *The relation between the speed and the time for the tidal currents after using the predicted ANN and the exact trained data.*

The percentage of error for the predicted trained data is very small but for the predicted untrained data is still high. The exact untrained data and the predicted ANN data are

shown in figure (8)

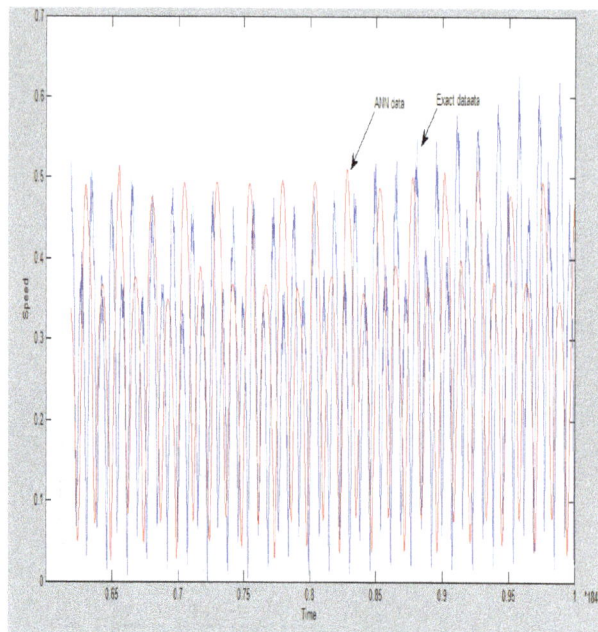

Figure 8. *The relation between the speed and the time for the tidal currents after using the predicted ANN and the exact untrained data.*

5.3. Hybrid model of FLSM and ANN

The models used in the case of FLSM and ANN give a high percentage of error for the predicted not trained data so we will try to use a hybrid of FLSm and ANN. In the hybrid model we try to find the innovation data using FLSM and use this innovation as an input to the ANN. After 20,000 epoch, 225 neurons in the first layer, i=6, the mean squared error became 0.000380. The percentage of error for the innovation of the trained exact data is 0.1304.

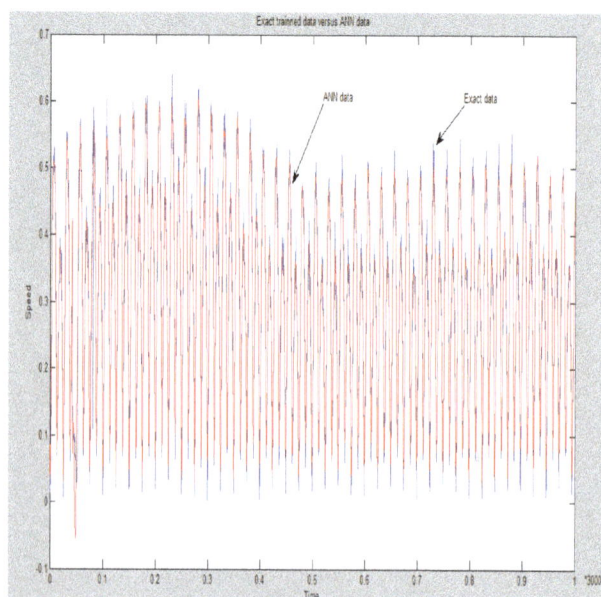

Figure 9. *The relation between the speed and the time for the tidal currents after using the hybrid model of FLSM and ANN and the exact trained data.*

The exact trained data and the predicted from the hybrid model have a percentage of error of 0.3328 and is shown in figure (10). The exact not trained data and the predicted ANN from the hybrid model is shown in figure (10). The predicted ANN from the hybrid model has a percentage of error of 0.4737.

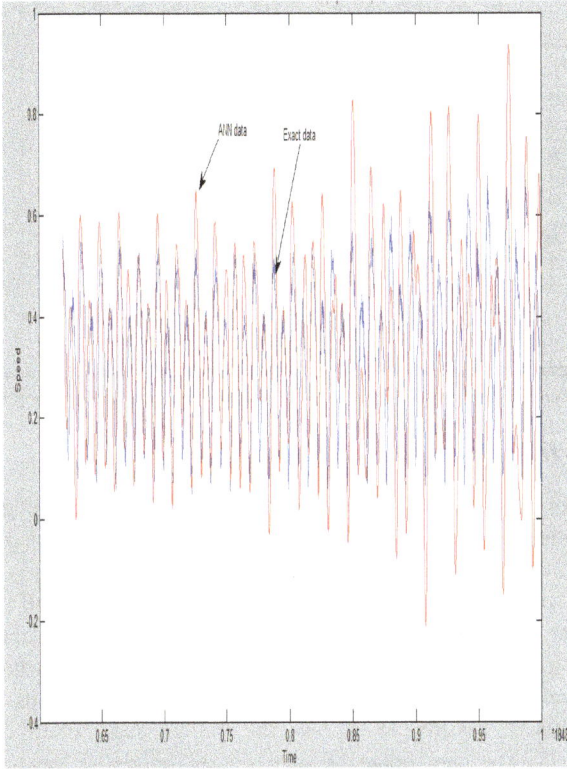

Figure 10. The relation between the speed and the time for the tidal currents after using the hybrid model of FLSm and ANN and the exact untrained data.

5.4. Hybrid model of ANN and FLSM

This model is the same as the previous model except the order. So the data is fed to ANN and then find the innovation data. After that use the innovation data to feed FLSM. The exact trained data and the predicted from the hybrid model has a percentage of error of 0.357 for the trained set of data and 0.672 for the untrained data.

We can summarize the previous analysis in table (1). We conclude that the percentage of error for the hybrid model of FLSM and ANN is the smallest value in the case of the trained or predicted data. So these models can be ranked as follows in terms of high accuracy:

a. Hybrid of FLSM and ANN.
b. Hybrid of ANN and FLSM.
c. FLSM.
d. ANN.

As the number of data points increased and included all changes the error will be decreased. In the next section we try to use the same algorithms for another set of data (the direction magnitude forecasting) to prove the validity of the proposed model.

Table 1. Comparison between different used models for tidal current speed forecasting.

Comparison between different used models

Type of comparison	ANN	FLSM	FLSM + ANN	ANN + FLSM
% Error of trained data (70%)	0.3903	0.6399	0.3328	0.357
% Error of forecasted data (30%)	3.0946	0. 817	0.4737	0.672

6. Validation of the Proposed Models (Tidal Direction Forecasting)

In this section we try to apply the proposed algorithms to forecast the direction magnitude of the tidal current.

6.1. Fourier Series Model based on Least Square Method

We use the FLSM for the tidal direction for the same period and the site that was used for the speed forecasting. Fourier series based on the least square model data and the exact data are drawn in the same graph as shown in figure (11) for 70% of the whole data and in figure (12) for 30% of the whole data. The percentage of the error for 70% of the whole data is equal to 0.7339 and for the other 30% is equal to 0.861. The direction in all graphs is in degree.

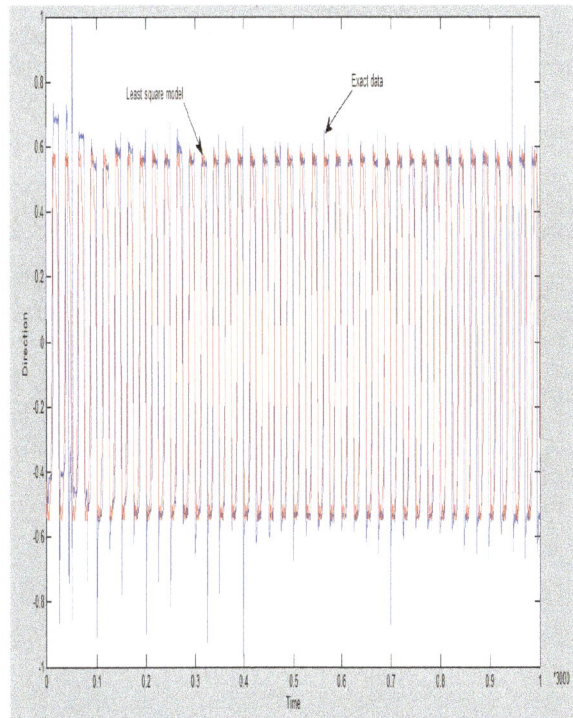

Figure 11. The relation between the direction and the time for the tidal currents after using the FLSM and the exact trained data.

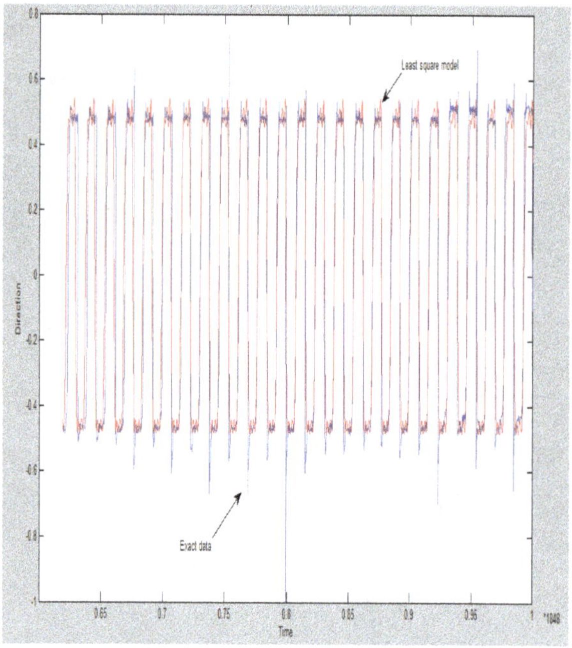

Figure 12. *The relation between the direction and the time of the tidal currents after using the FLSM and the exact data untrained data.*

6.2. ANN Model

We use the direction magnitude data as an input to the ANN without any modifications and training the ANN for 20000 epochs, 225 neurons in the first layer, the mean squared error became 0.000579. The percentage of error for the trained data is 0.5169 and for the estimated data is 3.8496. The exact trained data and the predicted ANN data are plotted in figure (13) and the exact untrained data and the predicted ANN are shown in figure (14). The percentage of error in the untrained predicted data is still high so we use the hybrid of ANN and FLSM.

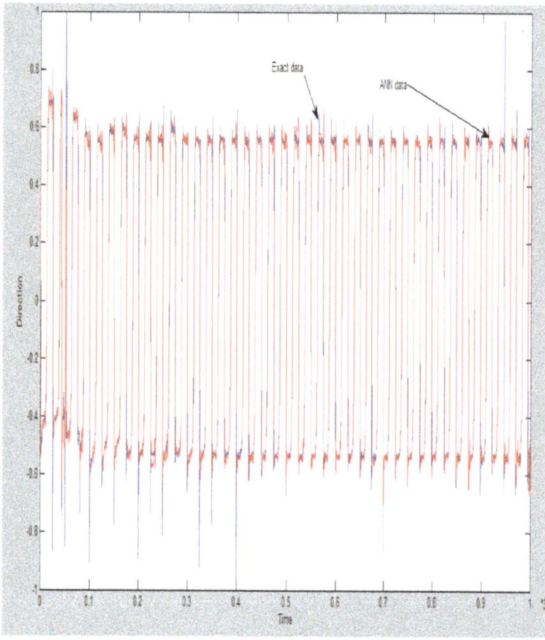

Figure 13. *The relation between the direction and the time for the tidal currents after using the predicted ANN data and the exact trained data.*

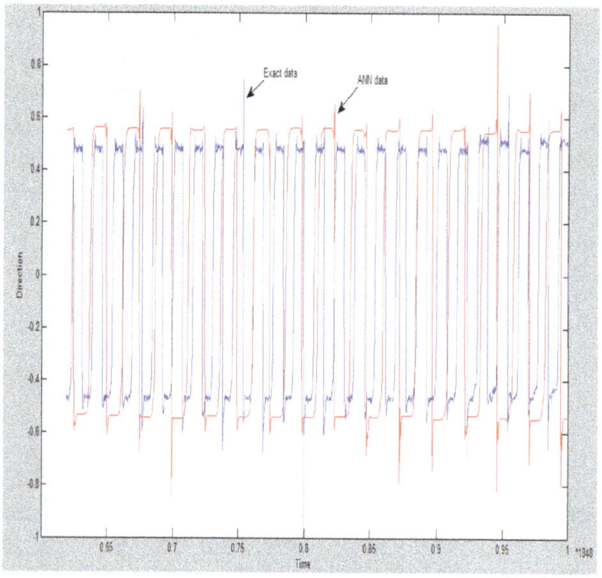

Figure 14. *The relation between the direction and the time for the tidal currents after using the predicted ANN data and the exact untrained data.*

6.3. Hybrid model of FLSM and ANN

As described in the previous section this hybrid model consists of the FLSM and ANN. We use FLSM to predict the tidal current direction magnitude and then calculate the innovations which is the difference between FLSM and the exact data. The innovations are fed to the ANN as an input. We then add the innovations to FLSM to find the whole model. The mean squared error for the innovations is 0.000679. Figure (15) shows the trained data versus the hybrid. The percentage of error for this model is equal to 0.1169 for the trained data. Figure (16) shows the predicted data versus the hybrid data for the untrained period of data. The percentage error for untrained data is equal to 0.4970.

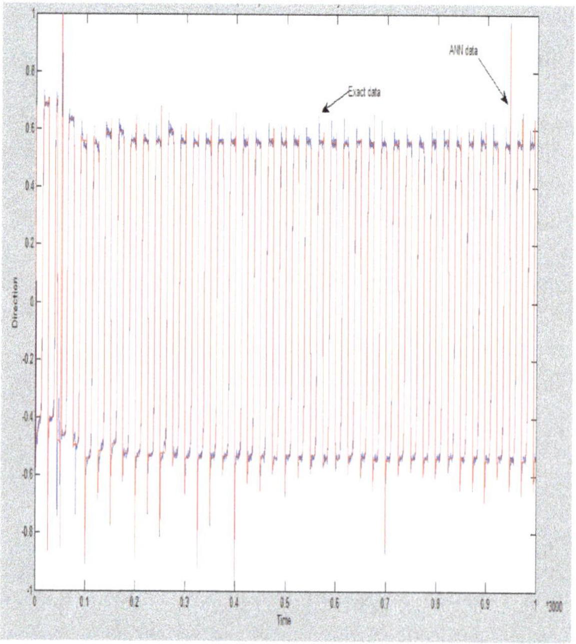

Figure 15. *The relation between the direction and the time for the tidal currents after using the hybrid model and the exact trained data.*

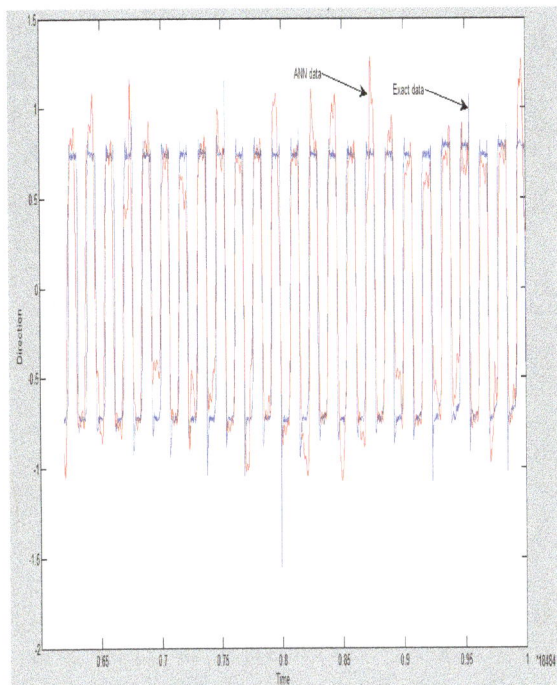

Figure 16. *The relation between the direction and the time for the tidal currents after using the hybrid model and the exact untrained data.*

6.4. Hybrid model of ANN and FLSM

This model again is the same as the previous model except the order. So the data is fed to ANN and then the innovation data is fed to FLSM. The exact trained data and the predicted from the hybrid model has a percentage of error of 0.156 for the trained set of data and 0.675 for the untrained data.

We can summarize the previous analysis in table (2). We find that the hybrid model of FLSM and ANN has the smallest error.

Table 2. *Comparison between different used models for tidal current direction forecasting.*

Comparison between different used models				
Type of comparison	ANN	FLSM	FLSM + ANN	ANN + FLSM
% Error of trained data (70%)	0.5169	0.7339	0.1169	0.156
% Error of forecasted data (30%)	3.9892	0.917	0.4970	0.675

7. Conclusion

Four models are proposed in this paper for tidal current speed and direction magnitude forecasting. Using either FLSM or ANN alone for the prediction of tidal current Speed and direction is not recommended. The models are ranked from the highest accurate to the lowest accurate. The hybrid of FLSM and ANN is the best model for the tidal current speed forecasting model and has a high accu-

racy. The hybrid of ANN and FLSM is the next best model. Then FLSM and ANN at the end. In this paper we use the proposed model for the speed forecasting of the tidal current and the model gives good results. We validated this work by applying the proposed algorithms for another set of data (tidal current direction magnitude) and the simulated results show better performance after using the hybrid model of FLSM and ANN. This is because we add the advantages of FLSM and ANN at the same time to be in one model.

References

[1] "Tidal Stream" Available online (November 2010), http://www.tidalstream.co.uk/html/background.html.

[2] "Marine Current Turbines" Available online (October 2010), http://peswiki.com/index.php/Directory:Marine_Current_Tu rbines_Ltd#How_it_Works.

[3] Darwin, G.H., on an apparatus for facilitating the reduction of tidal observations. Proceedings of the Royal Society, Series A 52, 345–376, 1892; Available at: http://archive.org/details/philtrans02858224.

[4] Doodson, A. T., The Harmonic Development of the Tide-generating Potential, Proc. Roy. Soc., London, pp 305-329, 1923. Available at: http://archive.org/details/philtrans08044568.

[5] Doodson, A.T., The analysis and predictions of tides in shallow water. International Hydrographic Review: Monaco 33, 85–126, 1958.

[6] John, V. "Harmonic tidal current constituents of the Western Arabian Gulf from moored current measurements." Coastal Engineering, Vol. 17,pp. 145–151, 1992.

[7] French, M. N. , Krajewski, W. F. & Cuykendall, R. R. "Rainfall forecasting in space and time using a neural network", Journal of Hydrological Sciences, 1992.

[8] Raman, H. & Sunilkumar, N. "Multivariate modeling of water resources time series using artificial neural networks", Journal of Hydrological Sciences, pp. 145 - 163, 1995.

[9] Christian Dawson & Robert Wilby "An artificial neural network approach to rainfall- runoff modelling", Journal of Hydrological Sciences, pp. 47- 66, 1988.

[10] Coulibaly, P., Anctil, F., and Bobee, B.: Daily reservoir inflow forecasting using artificial neural networks with stopped training approach,Journal of Hydrological Sciences, pp. 244–257, 2000.

[11] T.L. Lee, and D.S. Jeng "Application of Artificial Neural Networks in Tide Forecasting", Journal of Ocean Engineering, Vol. 29, No. 9, pp. 1003-1022, August 2002.

[12] M. Campolo, A. Soldati and P. Andreussi, , "Artificial neural network approach to flood forecasting in the river Arno", Journal of Hydrological Sciences, 2003.

[13] Tsong-Lin Lee "Back-propagation neural network for long-term tidal predictions" Journal of Ocean Engineering, Volume 31, Issue 2, pp. 225-238, February 2004.

[14] Lee, T. L., C. P. Tsai and R. J. Shieh "Applied the Back-propagation Neural Network to Predict Long-Term Tidal Level" Asian Journal of Information Technology, Volume 5, Issue 4, pp. 396-401, 2006.

[15] Ritu Vijay, and Rekha Govil "Tidal Data Analysis using ANN", Journal of World Academy of Science Engineering and Technology, 2006.

[16] Bang-Fuh Chen, Han-Der Wang, Chih-Chun Chu " Wavelet and artificial neural network analyses of tide forecasting and supplement of tides around Taiwan and South China Sea", Journal of Ocean Engineering, Vol. 34, Issue 16, pp. 2161-2175, November 2007.

[17] Jan F. Adamowski "River flow forecasting using wavelet and cross-wavelet transform models" Journal of Hydrological Processes, Volume 22, Number 25, pp. 4877–4891, 2008.

[18] Remya, P.G., Kumar, Raj, and Basu Sujit, "Forecasting Tidal from Tidal levels using genetic algorithm"; Journal of Ocean Engineering, Volume 40, pp. 62-68, February 2012.

[19] Burrage, D.M., C. R. Steinberg and K.P. Black "Predicting long-term currents in the Great Barrier Reef", 11th Australasian Conference on Coastal and Ocean Engineering, Australia, 1993.

[20] William Mendenhall, Terry Sincich A Second Course in Statistics: Regression Analysis, Amazon, 2011.

[21] Golberg, M. , Cho, H. A. "Introduction to Regression Analysis" , Southampton : Wit Press, 2003.

[22] Rahman, S., Bhatnagar, R. "An expert system based algorithm for short term load forecast",IEEE Transaction on Power Systems, Volume 3, Page(s): 392 – 399, 1988.

[23] Hamed H. H. Aly, and M. E. El-Hawary, "A Proposed ANN and FLSM Hybrid Model for Tidal Current Magnitude and Direction Forecasting" accepted at the IEEE Journal of Ocean Engineering, 2013.

[24] S. A. Soliman ; A. M. Al-Kandaria; and M. E. El-Hawary "Time Domain Estimation Techniques for Harmonic Load Models" Journal of Electric Power Components and Systems, Volume 33, Number 10, 2005.

[25] "Artificial Neural Networks" available online (August 2012), http://www.softcomputing.net/ann_chapter.pdf

[26] Zurada, J., "Introduction to Neural Systems", West Publishing, 1992.

[27] Ma L., Khorasani K. "New Training Strategies for Constructive Neural Networks with Application to Regression Problems" Neural Networks, Vol. 17, No. 4, May 2004.

[28] T. Q. D. Khoa, L. M. Phuong, P.T.T.B inh, N. T. H. Lien "Application of Wavelet and Neural Network to Long-Term Load Forecasting" International Conference on Power System Technology , Singapore, 21-24 November 2004.

[29] S. A. Soliman; A. M. Al-Kandaria; and M. E. El-Hawary "Time Domain Estimation Techniques for Harmonic Load Models" Journal of Electric Power Components and Systems, Volume 33, Number 10, 2005.

[30] Atif S. Debs, "Modern Power Systems Control and Operation" Boston, Mass.: Kluwer Academic Publishers, 1988.

[31] Hamed H. Aly "Forecasting, Modeling, and Control of Tidal currents Electrical Energy Systems"PhD thesis, Halifax, Canada. 2012.

Theoretical and experimental treatment of gaseous cementation of iron

Tayeb CHIHI [1], **FATMI Messaoud** [2, 3] *

[1] Laboratory for Elaboration of New Materials and Characterization (LENMC), University of Setif 1, 19000, Algeria
[2] Research Unit on Emerging Materials (RUEM), University of Setif 1, 19000, Algeria
[3] Laboratory of Physics and Mechanics of Metallic Materials (LP3M), University of Setif 1, 19000, Algeria

E-mail address:
fatmimessaoud@yahoo.fr(FATMI M.), Tchihi2001@yahoo.fr(T. CHIHI)

Abstract: Mathematical model is developed for cementation of iron taking into account the diffusion of atomic carbon C through the γ phase. Analytical solutions are obtained assuming constant diffusion coefficients, firstly the analytical method proposed that test to control the process of gaseous cementation, controlled the technological parameters of the cementation such: time (t), temperature (T), initial concentration (Co), potential carbon or atmospheric concentration ($Catm$), and speed of the gas flow (xw), secondly to accelerate the process of the gaseous cementation. Finally the results are quantitatively compared with those obtained experimentally taking into account the micro hardness profile. In addition, it is shown that the layer cemented produced during cementation of iron can be predicted by the numerical simulation.

Keywords: Gaseous Cementation, Iron, Ageing Time, Phase Diagram FeC

1. Introduction

The cementation is a thermo chemicals surface treatments by which the atomic carbon C is introduced into steel work pieces content to carbon between 0.10% to 0.18%C, usually at 850°C 1050°C in an endothermic furnace having for object the enrichment in carbon of a superficial layer, of the piece to treat. This enrichment makes himself by stake in contact with a middle rich on carbon solid, liquid or gas. The process the more utilised is the one of gaseous cementation, that is adjusts in series to the industrial productions. The temperature of cementation was maintained the most constant possible T (850-1050) °C and it in a triple but (to avoid the too fast impoverishment of the cement, to facilitate the regeneration avoiding the stay at the high temperature and to decrease risks of piece distortion). Because the solubility of the carbon in phase austenite, being much order elevated than in phase ferrite, and the coefficient of diffusion increase with the temperature, the cementation is achieved to temperatures superior to the A3 line, of the phase diagram FeC, these temperatures lets stable the austenite, solvent in quantity the carbon [1]. In the metallographic study of carburized steel, the structure of the high-carbon case is usually very important. After high-temperature

exposure in a carbonaceous environment, the carburized steel is then hardened, either by direct quenching from the austenitizing temperature during carburization or by reheating and quenching. The amount and morphology of martensite in this case depends on the carbon content and the transformation kinetics. Other microstructural constituents *ie* (retained austenite, nonmartensitic transformation products, carbides and inclusions) also may be present and influence properties and performance. The depth and amount of carbon diffusion depends on the source of surface carbon and the processing time at temperature. The most common method is gas carburization in an endothermic furnace at temperatures in the range of approximately 850 to 1050 °C. The higher processing temperatures of vacuum or plasma carburization (approximately 1050 °C) also allow higher diffusion rates and the additional benefit of increased solubility of carbon in austenite at a higher temperature. At a higher temperature, the surface of the hot steel part can become saturated to a higher carbon content, which thus increases the diffusivity (and hence the carburizing rate). Once the sample is carburized, the hardening step involves rapid quenching from austenite [2,3]. The depth of the martensitic layer depends of the amount of carbon that can diffuse into the sample.

This depends on the carbon content at the surface and the time-temperature exposure during carburization. Sometimes one adopts that the technical thickness of the layer is the one of which the structure after, tempering decomposes in martensit (95%) + retained austenite, sometimes one calls him efficient thickness, who corresponds to the depth for which the HRC toughness is superior to 55HRC after tempering to low temperature [1]. The thickness of the layer cemented of pieces made of steel is lower or equal to 17% of the thickness total, which vary between 0.5 and 2 mm. In the metallographic study of carburized steel, the structure of the high-carbon case is usually very important. Moderate amounts of retained austenite are proper and unavoidable in the high-carbon case microstructure of carburized steels. However, excessive amounts of retained austenite can lower hardness to unacceptable levels. Excess retained austenite also can lead to grinding problems. The most important cause of excessive amounts of retained austenite is due to the high carbon on the surface. This condition leads martensite start (Ms) temperatures down, causes the formation of plate martensite, and can shift the balance of the temperature range for martensite transformation to well below room temperature. High alloy content also lowers Ms temperatures. The depth and the amount of carbon diffused depend on the carbon source and the processing time at the temperature T.

Common locations of excessive surface carbon concentration are specimen corners at which the austenite is saturated with carbon during the first part of a carburizing cycle. Carbon potential control and quenching methods are means of controlling the austenite content. If retained austenite is unacceptably high, requenching from a lower temperature (distortion) may be considered. Retained austenite also can be reduced by low-temperature treatment in order to allow for more transformation due to the lower Ms caused by high carbon content.

2. The Model

2.1. Cementation

The quantity of the substance distributed (q) that crosses an unit section of the transverse surface by unit of time is proportional to the gradient of the concentration (first Fick's law) [4-7]:

$$q = -D \, \partial C / \partial x$$

Where: C: concentration of the distributing substance, x: coordinate, D: coefficient of proportionality (coefficient of diffusion). The difference between the incoming flux and the retiring flux in the volume 1.dx is equal to the speed of atom accumulation in this volume, therefore we have:

$$\partial C / \partial t = \partial q / \partial x = \partial (D \partial C / \partial x) / \partial x$$

and since D doesn't depend on the concentration C that is given by the Arrhenius's law:

$$D = D_0 \exp(-E / RT)$$

Where D_0: factor of frequency, E: energy of activation kcal/g.atom, R: constant of gases perfect (2cal/g.atom), T: temperature °C. The equation of diffusion becomes to only one dimension: $dC / dt = D d^2 C / dx^2$, this equation can be resolved with applying initial and to limit conditions as following :

$$C(x,0) = C_0 \quad x > 0$$

$$-D \partial C(0,t) / \partial x = \beta' (C_{atm} - C_{ini}) \quad t > 0$$

$$dC(\infty,t) / dx = 0$$

$$C(\infty,t) = C_0$$

Where C: concentration of the carbon % (in mass), C_0: initial concentration of the carbon in steel % (in mass).

C_{atm} : Potential carbon of the atmosphere %C.

t: time in second(s).

D: coefficient of diffusion of the carbon in the austenite (cm²/s).

x: co-ordinated in sense perpendicular to the surface (cm).

$$\beta' = (0.25 + 0.63(\omega - 0.2)^{0.66})10^{-6} g / cm^2 s\%C .$$

ω varies between 0.2 and 3.2 m/s. In the practice of the gaseous cementation of the steel, the speed of the gaseous flux is usually between 0.3 to 1.5 m/s. The solution of the unidirectional equation differential is:

$$C(x,t) = C_0 Erfc(x / (2\sqrt{Dt}))$$

$$Erfc(x) = 1 - Erf(x) = 1 - 2/(\sqrt{\pi}) \int_0^x \exp(-u^2) du \quad (1)$$

We used several analytic methods to elaborate this paper:
- Gauss's method, which is the general case for the resolution of the equation of heat propagation [8].
- Taylor's method of development, who assures the partial derivatives of the function C(x, t) up to the order (n+1).
- Korn's method of development: When the speed of gas supplying is limited, the quantity of element (dm) coming from the atmosphere toward the surface of the piece is proportional, unlike the concentration of balance C(0, ∞) and the concentration at the moment given C(0, t), at the F surface and at the time dt [7]:

$$dm = \beta' (C(o,\infty) - C(o,t)) . F . dt$$

Where (β': proportionality coefficient, which characterizes the speed of the element surrounding the piece toward the surface). If m is expressed in gram, F in cm², t in second and C in gram by cm², the β' unit is cm/s. If the carbon concentration doesn't exceed the limit of its solubility in the austenite, and the speed of the absorption is not infinite, the change in concentration in the steel is given by the following equation:

$$\theta(x,t) = \frac{C(x,t) - C_{init}}{C_{atm} - C_{init}} Erfc(x/(2\sqrt{Dt})) - Exp(\alpha^2 Dt + \alpha x).Erfc((x/2\sqrt{Dt}) + (\alpha\sqrt{Dt}))$$

Where $\alpha = \beta'/D$ cm^{-1}: relative coefficient of transfer. We can to represent this function under another form:

$$\theta(x,t) = Erfc(B_i/2T_i) - Exp(T_i^2 + B_i).Erfc(B_i/(2T_i) + T_i) \text{ Where}$$
$$B_i = \beta' x/D \text{ and } T_i = \beta'\sqrt{Dt}/D$$

On the surface of the metal, i.e. in x=0, we have [10]:

$$\theta_{surf} = \theta(0,t) = 1 - Exp(T_i^2).Erfc(T_i)$$

The coefficient β' characterize the speed of the interaction of the atmosphere with metal. To facilitate the programming and have values discrete, we use the formula of Korn, who serves to the passage of a continuous sum (integration) to a discreet sum (sum), which we can write:

$$2/(\sqrt{\pi}) \int_0^\mu exp = (-v^2)dv = 1 - (A_1.z + A_2 z^2 + A.z^3 + ...)exp(-\mu^2) = 1 - exp(-\mu^2)\sum_{i=1}^n A_i.z^i. \quad (2)$$

i= 1, 2, 3... n. Where $z = 1/(1 + A_0\mu)$

the constants A_i whose equals to : A_0=0.3275911 ; A_1=0.254829592 ; A_2=-0.28449 ; A_3=1.42141341 ; A_4= -1.453152027; A_5=1.061405429. We limiting the value at n=3, that to make a precision from calculation of the integral to about 10^{-7} [11]. We indicate for $\phi(\mu)$, the

$$Erfc(\mu) = \phi(\mu).exp(-\mu^2)$$

Where $\phi(\mu) = \sum_{i=1}^n A_i/(1 + A_0\mu)^i; \mu = x/(2\sqrt{Dt})$, A_i: have the numeric constants and i=1,2,3.........n

Of another $Erfc(\mu) = 1 - 2/(\sqrt{\pi}) \int_0^\mu exp(-v^2)dv$ are difficult to integrate, it's why one uses an approximate method, that assures a sufficient precision and approach much of exact value of the integral. This approximate method is demonstrated by the Korn mathematician. The detail of this integral is as follows:

expression from the right under the sign sum of the equation (2), then the equation (1) has the following form:

$$Erf(\mu) = 1 - (1 - \phi(\mu)exp(-\mu^2)) = \phi(\mu)exp(-\mu^2) \quad (3)$$

To substitute (3) in (1), we have:

$$\theta(x,t) = \phi(B_i/2T_i).exp(-(B_i/2T_i)^2) - exp(T_i^2 + B_i)\phi((B_i/2T_i) + T_i).exp(-((B_i/2T_i) + T_i)^2) = (\phi(B_i/2T_i) - \phi((B_i/2T_i) + T_i))exp(-(B_i/2T_i)^2) (4)$$

And finally it's the equation that is resolved with computer, on to take the times as a parameter and *x* as a variable. To substitute the value of $\theta(x,t)$ in the following equation:

$$C_{x,t} = \theta_{x,t}(C_{atm} - C_{ini}) + C_{ini} \quad (5)$$

we obtain the value of the concentration on a defined times and in different depths of the cementation layer.

2.2. Acceleration gaseous cementation

The cementation consists in enriching superficially in carbon a metallic piece. The process of the cementation is used extensively in the modern industry because it permitting:
- to get a big resistance against wear.
- to get a better holding to tiredness etc......

In general the process takes place between 900°C and 1100°C, during some hours, the speed of the cementation is located between 0.10 and 0.13 mm/h *ie* to get a layer cemented of depth varying between (1 and 1.5) mm, a variable time was necessary between 10 and 12h.

To solve the problem of the gaseous cementation with change of the potential of the middle it is necessary to solve the second unidirectional Fick's law [12-13]:

$$\partial C / \partial T = D\partial 2C / \partial 2 x \quad (6)$$

or C: concentration of the carbon
t: time of maintain
D: coefficient of diffusion
x: coordinate according to the depth of the layer cemented.

We know that the coefficient of diffusion of the carbon in the austenite during the cementation depends linearly on the concentration of the carbon to a constant temperature, according to formula following:

$$D = (0,04 + 0,08 C) Exp (-Q/RT) \quad (7)$$

And that the flux atomic J distributing in the piece, that is function of the gradient concentration in the layer according to the first Fick's law:

$$J = -D.grad C \quad (8)$$

And since we interested solely in the diffusion according to only one axis, the formula (8) takes the form:

$$J = -D \partial C / \partial x$$

We taking formulas as a basis (7) and (8), and taking some limit conditions, which depend strongly on the type of saturation cycle (combined or unique), one can accelerate the process of the cementation. So in this case it is necessary to solve the linear equation to the partial derived of parabolic type, where D is supposed constant equal to $1,1 \ 10^{-7} \ cm^2/s$, at the temperature of cementation 930°C. The combined cycle and the unique cycle are characterized by the presence, respectively of a constant value during the first stage 1,1%C, and 0,8%C on the surface of the sample, on the other hand during the second stage of the cobined cycle the potential initial is 1,1%C, who change with time. Mathematically these conditions are presented as:

C(0) = 0,8%C for t≥0 (unique cycle)
C(0) = 1,1%C for 0≤t≤t1 (combined cycle).

Where t1 is the time that indicates the end of the first stage of the combined cycle. According to the literature, during the combined cycle, the second stage of the cycle is characterized by the reduction of the potential of carbon, so that the total quantity of carbon that to distribute in metal, during the first stage remains constant, *ie* that the following equation always remained valid:

$$\int_0^\infty C(x,t)dx = Q_0 = cste \qquad for \ t \geq t_1$$

The time of treatment between the potential 1,1%C and 0,8%C is negligible compared with the total time of the process, metal doesn't undergo a decarbonize during the change of the potential, and the following condition will be confirmed:

$$\partial C(x,t) / \partial x = 0 \ in \ x=0 \ and \ for \ t \geq t1$$

The interval 0≤x <∞ indicate that the variable x varies from the surface to an infinite value of the depth. For the numeric realization of an algorithm, as solution of the problem, we use the different limit approximation, for it the domain of integration is supposed fini, *ie* 0≤x≤xmax and conditions to limits of the first and the second genre:
C(∞, t)=C0, ∂C(∞,t) / ∂x=0 become applicable.

The applied data show that the concentration of the carbon, decreases until the initial concentration in steel in

depth. As using a relatively high value of the depth maximal x_{max}, and at x_{max} the concentration will be equal to the one in steel, *ie*:

$$C(x_{max}) = 0,25\%C \ for \ t \geq 0$$

This condition will be verified in the case of the unique and combined cycle. To solve equations of the parabolic type, taking account of the initial conditions, we suppose that the concentration of the carbon is a constant at t=0:

$$C(x, 0) = cste = 0,25\%C$$

Therefore the formulation of the mathematical model will be like pursuit:

2.3. Mathematical model

We are going to solve the second Fick's law in conditions to limits as following:
For unique cycle:
C (x, 0) = 0,25%C t=0
C (0, t) = 0,8%C t≥0
C (x_{max}, t)=0,25%C
For combined cycle:
C (x, 0) = 0,25%C for t=0 and x>0
C (0, t) = 1,1%C for 0≤t≤t1
∂C(x, t) / ∂x = 0 in x=0 for t1≤t≤tf
C(x_{max},t) = 0,25%C

3. Experimental Procedures

These materials were prepared in our laboratory by fusion in a device at a high vacuum (10-5 Torr) using pure materials. After the melting the ingots have underwent plastic deformation by cold rolling before the homogenization treatment in order to accelerate the structure homogenization kinetics. The homogenization temperature and aging time were chosen from the equilibrium diagrams [14].

Specimens in the form of discs of diameter 20 mm and thickness about 1cm were prepared

Table 1 : *Technological parameters T (°C), C_o(%C), C_{atm}(%C) and x_w (cm/s) of some simulations sample.*

	1	2	3	4	5	6	7	8	9	10	11	12	13 1311
T (°C)	900	950	1000	1050	950	950	950	950	950	950	950	950	950
C_o(%C)	0.2	0.2	0.2	0.2	0.2	0.3	0.2	0.2	0.2	0.2	0.2	0.2	0.2
C_{atm}(%C)	0.9	0.9	0.9	0.9	0.9	0.9	0.8	1.0	1.2	0.9	0.9	0.9	0.9
x_w(cm/s)	1.4	1.4	1.4	1.4	1.4	1.4	1.4	1.4	1.4	1.4	1.0	2.0	3.0

from iron (0.10 - 0.15 at. % C; 0.30 - 0.60 at. % Mn; 0.15 - 0.35 at. % Si balance Iron). For microscopic studies specimens were chemically etched with a concentrated solution of Nital at room temperature for 30–90s. Cementation was performed at 1223 K in a gas mixture of H_2, N_2, CO, CH_4 and CO_2. After carburizing, the specimens were moderately cooled.

4. Results and Discussion

We utilise the technological parameters T°C, Co, Catm and xw of the following table, we obtained several results, for plotting the different graphs:

Using the resulting of calculation to plot curves C(x, t), versus x that takes values varying between zero (that correspond to the surface), and 0.15 cm (depth of cementation supposed sufficient in our type of piece, with a step equal to 0.01 and it for different exposure times, that can take values (1h-9h) to deal the effect of various process parameters technological as temperature; the initial concentration of carbon in iron Co %C; the potential carbon Catm %C; and the speed of the flux xw (cm/s). In Fig 1, we present the calculated carbon concentration profiles in iron, carburized at (T=950 °C, C_0=0.2 %C, Catm=0.90 %C, xw=1.4 cm/s) for different times. It is noted that the concentration C(x,t) decreases at a given time, and reaches the initial carbon concentration. It is indicating that C(x,t) increases at a given depth, for various exposure times. Fig 2, show the effect of T °C on xeff, for different exposure times 2, 5 and 8 h, of carburized iron at Co=0.2 %C, Catm=0.90 %C and xw=1.4 cm/s conditions. It will be shown that xeff increase better at high temperature, for the same exposure times. Fig 3, show the plot of xeff vs initial concentration Co for 2, 5 and 8h. It can be clearly seen that xeff increases with

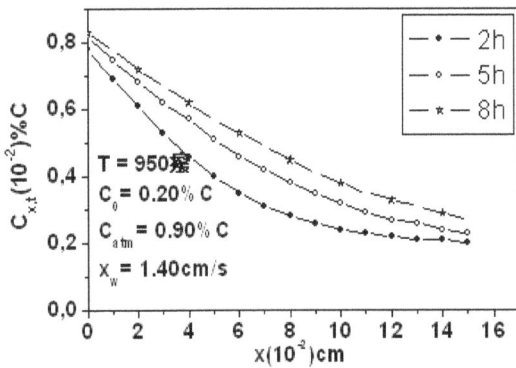

Fig 3. *Effect of initial concentration and ageing time on xeff.*

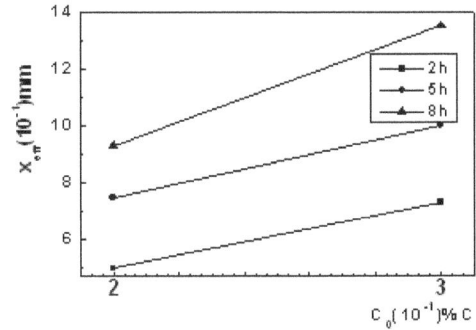

Fig 4. *Effect of atmospheric concentration and ageing on xeff.*

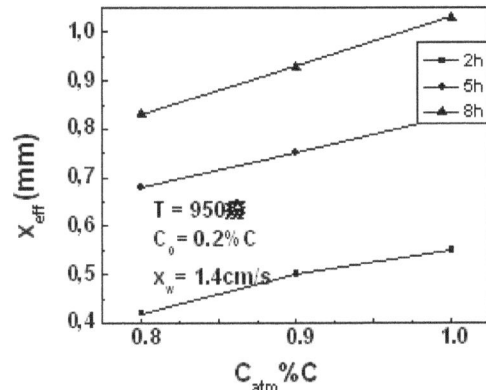

Fig 1. *Calculated carbon concentration profiles in iron ageing 2, 5 and 8h at 950°C.*

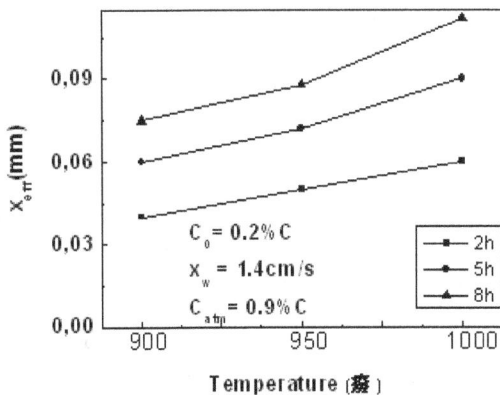

Fig 5. *Effects of speed of the gaseous flux and ageing time on xeff.*

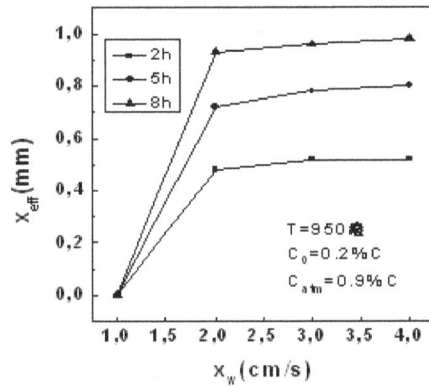

Fig 2. *Calculated effect of temperature and ageing time on xeff.*

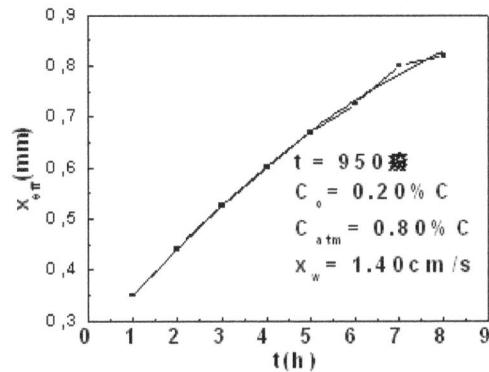

Fig 6. *The penetration depth of the carburized layer as a function of ageing time at 950°C.*

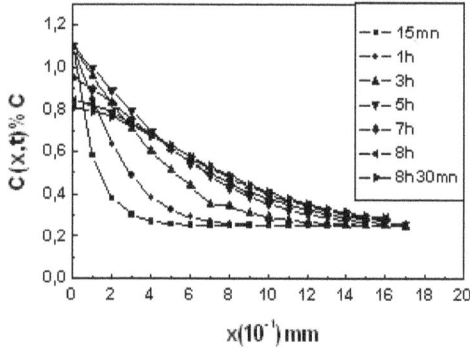

Fig 7. Variation of the carbon concentration according to the depth of the layer cemented, for different ageing time at 950°C.

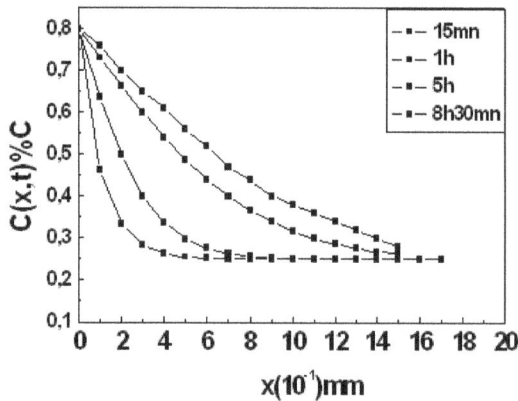

Fig 8. Regime unique: variation of the carbon concentration C(x, t) according to the depth at different time at 950°C.

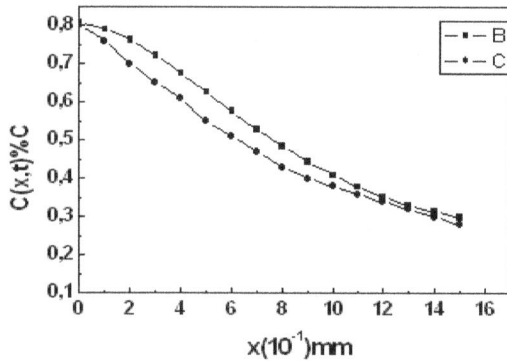

Fig 9. Comparison between curves: diffusion in the unique regime (solid circles) and the combined regime (solid squares).

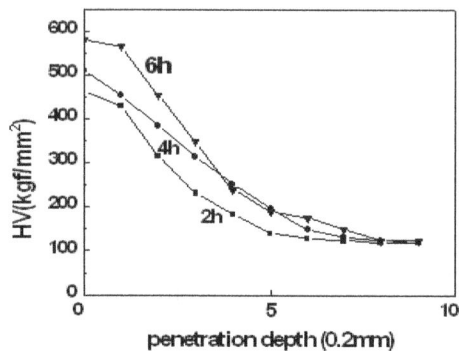

Fig 10. The micro hardness dependence at different ageing time according to the penetration depth.

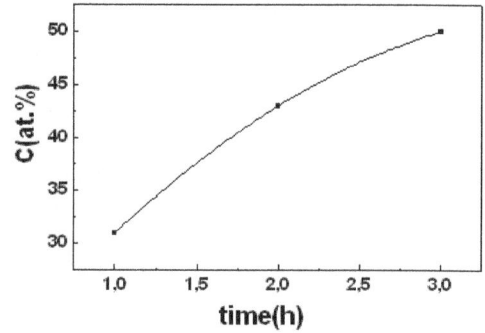

Fig 11. The calculated micro hardness as a function of ageing time at 950°C.

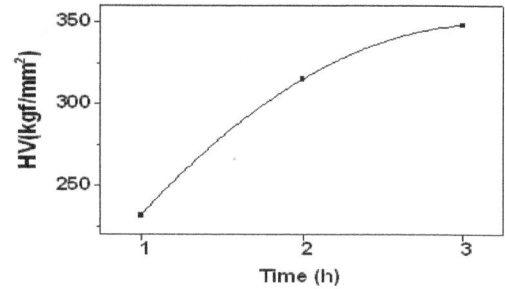

Fig 12. Micro hardness as a function of ageingtime at 950°C.

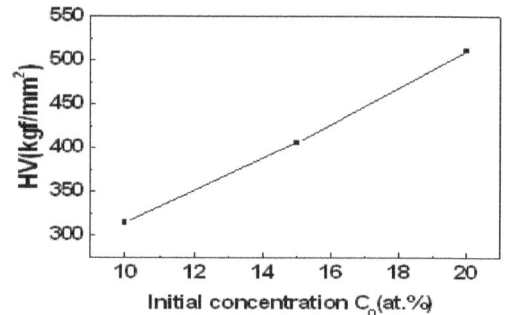

Fig 13. Micro hardness as a function of initial concentration.

Co. It is indicating that xeff increases at a same value of Co, for different exposure times. Fig 4, illustrate the dependence of atmospheric content Catm on the xeff. The simulations were performed at T=950°C, Co=0.2 %C and xw=1.4 cm/s for various times. It is shown that xeff increases linearly with Catm. It is observed that is increases at a same value of Catm for different times. The plots in Fig (3-4), are seen to be linear, indicating that this reaction is first order. It is apparent that the plots xeff *vs* Co (Catm) increase at the same time and for the same Co (Catm) and different exposure time. Fig 5. show the effect of xw on the sufficient depth for various times, at T=950°C, Co=0.2 %C and Catm=0.90 %C. It is apparent that xeff increases quickly for small values of xw, after it increase slowly and reached the constant values for different exposure time. Fig 6. show the effect of carburization time on xeff, same conditions are in Fig 4, but atmospheric carbon content Catm is equal to 0.8 %C, that xeff grows parabolically with time and reaches the constant value of potential surface. Fig 7, represents the variation of the carbon concentration according to the depth of the layer cemented, for different ageing time. We notice that: This concentration (for a stationary depth) increases of one time to another superior. For a concentration fixes, the depth of the layer cemented increases of one time to another superior. Fig 8, represents regime unique: variation of the concentration of the carbon C(x, t) according to the depth of the different ageing time. The concentration decreases in depth since the surface that is equal to 0.8%C, that it reaches the initial steel concentration that is equal to 0.25%C. Fig 9 represents a comparison between curves: diffusion in the unique regime (solid circles) and the combined regime (solid squares). We notices that for a concentration constant, the depth increases with a regime respectively to the other, for the same ageing time. Concentration profiles, as calculated according to the programme, are presented in Fig 1. For a comparison with experiment the micro hardness profiles of the specimens were determined after cementation. The increase in micro hardness with respect to the micro hardness of the non cemented material is given as a function of penetration depth in Fig 10. Fig 11 show the calculated carbon concentration in iron specimens cemented at 950°C, C0 = 10% for 2, 4, and 6h. It grows parabolically with time. In Fig 12. we show the increase of micro hardness *vs.* ageing time. It is clearly seen that an increase of micro hardness corresponds to an increase of carbon concentration, see Fig 11. In Fig 13, we show the increase in micro hardness at T=950°C, t = 4h and thickness 0.6 mm with initial concentration of carbon in iron specimens, that it is sensitively comparable to Fig 3. A sensitive graphic presentation for comparison of the experimental absorption-isotherm data with the model description of the set of equations (5) is provided by Fig (10-13). So a qualitative agreement exists between the calculated results Fig (1-3, 6). and experimental results.

5. Summary

We have elaborate a mathematical model, a program in data processing of the gaseous cementation, that serves to the survey of the concentration in carbon according to the technological parameters, and of the depth of the layer cemented, C%=C%(x, t, Catm, T, Co, xw). While basing on results given by this model, we have study the technological parameters influence on the cemented layer. We deduct some curves presented below that: xeff increases with time (t), with the T°C temperature, with Catm, with Co, and with xw, while taking at every time all constant taken technological parameters. xeff can not increase infinitely because (risk to damage the characteristic mechanical): while increasing the temperature or the time of maintains grains becomes thick (parameter to make fragile). The initial concentration Co is limited (cementation of steels low carbon). We are limited by speed xw that can go until 1.3 cm/s, value taken by the technician like limit.

We know that: the thickness of the layer cemented varies according to the concentration of the carbon according to the equation:

$$C(x, t) = C_0 Erf(x/(2\sqrt{Dt}))$$

The coefficient of the diffusion D in the austenite increases with the content in carbon to a constant temperature according to the equation D=0.04+0.06%C +......

These two phenomena permit to accelerate the gaseous cementation and to get a more preferable carbon distribution (distribution in landing). This method of acceleration is rarely used in the industry, because it is not controlled well.

References

[1]   Chaussin. Hilly: Metallurgy, Volume (1), Metallic Alloys, DUNOD, BET, 1967.

[2]   O. REY, P. JACQOT, Kolsterising: hardening of austenitic stinless steel, *Surface engineering, vol. 18, n°6, pp 412-414, December 2002.*

[3]   P. DYMOND, Kolsterising, Improving Austenitic stainless steel, *ASM Heat Treat 2001, Indianapolis, 15-17 September 2003.*

[4]   Martin. P, J. Alvalez, Fernandez Gonzalez. B. J, Ruis Fernandez. J, Bello Berbegal. J: Cinetica de incorporation de carbone en el acera a partir de las atmosphéras endotermicas, rev: metal. Madrid. 1984, 20.

[5]   T. Turpin, J. DULCY, M. GANTOIS, JFREY, D. HERTZ, Precipitation des carbures au cours de la cementation en phase gazeuse d'aciers inoxydables : approche thermodynamique, cinétique et structurale. *Materiaux* 2002, *Tours*, 21-25 Octobre 2002.

[6]   P. JACQUET, D. ROUSSE, G BERNARD, M LAMBERTIN, A novel technique to monitor carburizing processes, *Materials Chemistry and Physics* 77(2002) 542-551.

[7]   P. JACQUET, D.R. ROUSSE, Mesurements of carbon fluxes during low pressure carburising, *Metallurgy and New Materials Researches*, Vol.X, n°3, 2001,p1-16.

[8]   Tichonov. A, Samarsky. A : Les équations de physique mathématique, MOSCOU, Edition la technique 1951.

[9]   Suzana. Maria. Coelho. Arno. Muller; Metallurgia, ABM, vol 37, N° 282, Mai 1981.

[10] Andreev. U, Potapova. S : Modèle mathématique du processus de la cémentation gazeuse, les fours dans l'entreprise de construction, MOSCOU, TOME (19), 1972.

[11] T. Korn, G. Korn: mathématical Hand Book for scientists and ingineers, 2 ed, Mc Graw-Hill Book Company, (1968).

[12] A. Fick, ann. Der. Physik (1855),94, 59 (in German).

[13] A. Fick, Phil. Mag. (1855), 10, 30. (in English).

[14] B. Massalski (Ed.), ASM, 1990, p. 1471.

3

Design of cylindrical fixed dome bio digester in the condominium houses for cooking purpose at Dibiza Site, East Gojjam, Ethiopia

Molla Asmare

Centre of Competence for Sustainable Energy Engineering, Institute of Technology, Bahir Dar University, Ethiopia

Email address:

mollaasmare98@gmail.com

Abstract: Organic Waste is undesirable matter, which is most frequently generated by human activity that causes environmental pollution. Therefore, domestic biogas production is one of the most promising method of biomass wastes treatment because it provides a source of energy while simultaneously resolving ecological, environmental and agrochemical issues. The provision of bio - energy tackles both energy poverty and the reliance on polluting and Non - Renewable fuels as a result matured biogas production technology has led to the development of a number of biogas appliances for lighting, power generation, and cooking. The most promising among them is the biogas energy in order to meet the energy requirement for cooking application at domestic and community level. In this paper, an attempt has been made to design and develop a cylindrical torpispherical fixed dome bio digester for cooking application in the condominium houses at Debiza site in Debre Markos, East Gojjam in Amhara Region. The size of biogas plant is $53m^3$ and the input materials are different wastes such as kitchens, food waste and the human excreta from a total of 357 people living in four building of 120 residence. The gas production rating of the developed biogas plant is $25.36m^3$/day, which accounts 60.73% of the energy consumption that covers all the energy demand of firewood, charcoal and animal dung cakes that used for baking Injera and bread. The amount of gas obtained averagely, $0.211m^3$/ per household per day for cooking purpose.

Keywords: Design, Fixed Dome, Biogas, Condominium, Cooking

1. Introduction

Sustainable energy production and its use are important features for adequate energy services for satisfying basic human needs, improving social well-beings, achieving economic development and keeping the quality of life of current and future generations without exceeding the carrying capacity of the ecosystems. However, it has been shown that current energy consumption patterns are aggravating various global problems, leading to further unsustainable. On the other hand, energy can also contribute to the solution of problems particularly poverty, the situation of women, population growth, unplanned urbanization, and excessively consumptive life styles. Thus, poverty alleviation in Developing Countries like Ethiopia should involve in energy strategy of universal access to adequate, affordable, and reliable, high-quality, safe, and environmentally benign modern energy services, especially for cooking, lighting, heating, and transport [1].

In Ethiopia 85-94% of the population is dependent on traditional biomass (firewood, charcoal, agricultural residues, and cow dung) to meet their household energy needs. As reported by the Ethiopian Rural Energy Development and Promotion Center (1998), 77% of total final energy consumption consisted of firewood and charcoal, 15.5% consisted of agricultural residues, roughly, 6% was met by modern energy sources such as petroleum, and only 1% of the population-utilized electricity for cooking. Of the total energy demand, approximately 89% was consumed by households, and 4.6% was to industry. While Ethiopian demand for modern energy sources is expected to grow faster than for any other energy sources, biomass fuels will continue to dominate total energy consumption [2].

Biogas is one of a renewable energy that can be used as an energy resource to substitute firewood in public area such as hotels, hospital, universities, and prisons and new developed in common (condominium) house in this paper. Condominium house is one of the public areas where many people live together in the same room or in different section with the same building in more populated cities or towns of our country. A project was established in Debre Markos Town East Gojjam, Amhara Region in a Kebele 04 (Debiza site). At present, the number of population living in condominium house in Debre Markos is around 3532 in1413 condos house in four different site (Debiza, Bole, Road Construction and Hospital direction) [3]. Where, the energy demand is quite high for heating and cooking purpose.

Although the country has abundant energy resources, its potential is not yet well developed due to lack of financial capital and professional experts. As a result, all the demand of fossil fuel and gas is imported from abroad, and its fuel wood resource are exploited unsustainably which led to deforestation and land degradation. The paramount solution is developing renewable energy sources. During the last years, anaerobic fermentation has developed from a comparatively simple technique of biomass conversion with the main purpose of energy production, treatment of organic wastes, improvement of sanitation, and production of high quality fertilizer. Moreover, it has high economical benefits compared to other fuel sources. Because it requires limited financial capital in construction and maintenance, ease of raw materials availability in villages and towns, has less impact on the environment. Therefore, use of biogas energy has to be prioritized in Ethiopia to reduce deforestation, land degradation, and improve the living condition of the society (i.e. health and socio-economic situation of the households, including gender issues). Furthermore, it reduces the contribution of the green house gasses to climate change and Keeping the quality and sanitation of cities and towns by treating the waste using biogas technology.

Biogas is a potential source of energy, particularly where there is an abundant supply of organic waste. It is expected that, the greatest potential of biogas plant have to be realized through community (institution) and condominiums in large capacities. Therefore, the motivation of Biogas technology should be developed in Ethiopia at condominium house in addition to household level. The survey covered four condominium buildings with 120 residences, which accommodate 357 people. In this condominium house there are a plenty source of human excreta, food and kitchen waste that can produce alternative renewable fuel using a fixed dome bio digester.

This paper focuses on the design of cylindrical fixed dome bio-digester for condominium house in Debre Markos at Debiza site for cooking application from different wastes (food, kitchen, and human excreta). In these building, there are excess dry wastes, human execration as well as food waste. These wastes and human execration have highest hydrocarbon composition, which can be converted to flammable organic component to produce biogas. However, these wastes and human execration have environmental negative impact. Hence, biogas technology when properly utilized improves the sanitary and health conditions of the society and helps the people living there in wide aspects ranging from health to socio-economic improvements.

2. Data Collection and Analysis

2.1. Data Collection

The current number of population in the condominium residence is around 357, as the result of such increment, high amount of kitchen, food wastes and human excreta can be collected. The appropriate amount of waste available and its type should be known before put the design of the digester. Hence, collecting data have been made using interview, questionnaire, direct measurement of the waste using balance for consecutive five month, and different literatures. From this survey, different waste in different state is obtained in the common house of Debiza site in Debre Markos. The solid wastes in the condominiums are two types.

Table 2.1. Type of solid wastes available and its description in condominiums

No	Group of was	Description
1	Food waste	Consists of animal ,fruit, vegetable garbage or peel , obtained from preparation or cooking and eating of food including food scrape remaining in each house hold
2	Rubbish waste	Combustible and non combustible waste such as paper, carton ,plastics , wood chips ,broken glass Etc

2.2. Solid Organic Waste

The maximum amount of solid organic waste obtained per month that contains both biodegradable and non-biodegradable from direct measurement using balance recorded as follows for only five month in 2004E.C.

Table 2.2. Recorded sample of dry waste for five month at Debiza condominiums.

No.	Month	Amount of waste collected (kg)	Remark
1	October	4943kg	
2	November	4913kg	
3	December	4923kg	
4	January	5400kg	Christmas
5	February	4953kg	
6	Sum	25132kg	
7	Average	5026.4kg	

2.3. Water Source

The water source of the condominiums is tap water, a great care must be taken during shower taking and cleaning (washing) a cloths because the detergent in the waste has negative impact by increasing the pH value. The detergents are basic soaps that are unfavorable for bacterial activity because the water that flows after showering contains detergent mixes with the toilet water in the safety tank. Therefore, it needs little modification to separate them.

2.4. Amount of Waste Obtained from human being

✓ Total number of toilets =120
✓ Maximum amount of water added to the toilet averagely=2.2L/flash after use, from direct measurement at regular time for one month and taking the average value.
✓ The amount of urine discarded per day per person is 1L or 1kg [4] & the amount of manure discarded per day per person is 0.7kg [4, 6].

Data's are obtained from different literature and direct measurement in order to determine the physical properties

of the waste obtained from condominiums house are illustrated in the following tables.

One cubic meter of biogas=1ib of LPG=0.454kg, 0.43kg of Butane gas, 1.4kg of Charcoal, 38MJ, 12.30kg of cattle dung cakes, 3.5kg of Firewood [Biswas, 1977]

Table 2.3. Basic information on the physical properties of the substances

No.	Type	Quantity
1	Density of food waste	1160kg / m3
2	Density of human waste	1000kg/ m3
3	Water content of food scrape	70%
4	Water content of the human manure	90%
5	Water content of urine	94%
7	Organic content of food waste	85%
8	Mesospheric temperature	25^0c
9	Averagely, solid concentration	8%
10	Retention time	60day
11	Energy content of biogas	$38MJ/m^3$

Table 2.4. Energy consumption of household's in condominium house from interview and EEPCO bill price

No.	Types of fuel	Uses(purpose)	Average Amount per day per household	Amount per Month per household	Total amount of consumed per month per number of residence	Unit price in birr	Cost (in birr)
1	Charcoal	For cooking and heating purpose	0.21kg	6.30kg	100*6.30=630kg	1.68	1061.53
2	Dung cake	For cooking	0.20kg	6kg	6*100=600kg	0.5	300
3	Fire wood	For baking purpose	0.25kg	7.50kg	7.50*100=750kg	1.25	937.5
4	LPG	For cooking stove	0.04kg	1.20kg	1.2*100=120kg	10.45	1254
5	Butane Gas	For cooking stove	0.02 kg	0.60kg	0.6*100=60kg	16.22	973.2
6	Total	-	0.72kg	22.81kg	2218kg	-	5077.83
7	Electricity	For cooking stove only	0.3333kwh	15.76kwh	1576kwh	0.35	551.6

2.5. Energy Cost and Source Related Data

The cost needed to discard the collected waste is 10 birr per month per household or room. The total amount of money spent for this purpose (10*120=1200) per month. Therefore the total amount of cost expense per month is the sum of the two that accounts (1200+5077.83) with a total of 6277.83ETB

2.6. Data Analysis

2.6.1. Waste Obtained from Food and Kitchen
The data obtained from table 2.2 shows, 70% of the

wastes are biodegradable and the remaining one is non-biodegradable which accounts 30%. Therefore The total mass of biodegradable solid wastes (peel of potatoes, onion, can sugar, Oaf etc) per month can be calculated as follows:

❖ Total mass of biodegradable waste =0.7*5026.4kg/month=3518.2kg/month =117.27kg/day
❖ The total mass of non biodegradable waste =0.3x5026.4kg/month=1507.9kg/month=50.26kg/d
❖ Daily volumetric flow rate (Sd) in cubic meter per month

$$Sd = \frac{\text{Total mass of biodegradable wastedkk}}{\text{the density of food waste}} = \frac{3518.2kg/month}{1160kgm3} = \frac{3.03m3/month}{30day/mont} = 0.101m3$$

2.6.2. Amount of Human Waste Based on Literature

❖ The number of population living in condos=357
❖ The amount of manure per day (d) =1kg/day/adult

and 0.4kg / d between 10 to 15 age [5, 6] so considering the average value of the two, it gives = (0.4+1)/2=0.7kg/d. Therefore, the total amount of human waste per day=0.7*357=249.9kg/d.

❖ Amount of urine per person per day=1.L/d=0.001 m^3. Therefore, the total amount of urine waste per day=1L*357=357L/d=0.357m^3/d

So, the daily volume flow rate of human waste (Sd)

$$Sd = \frac{\text{total mass of manure waste}}{\text{the density of human waste}} = \frac{\frac{249.9kg}{d}}{1000kg/m3} = 0.249.9 m3/d$$

Therefore, Total amount of waste obtained (TW) =Waste obtained from food (WF) + Waste obtained from urine (WU) + Waste obtained from human manure (WM)

TW = (0.101 + 0.357 + 0.249)m3/d = 0.7079m3/d

Since the waste is wet, applying one to one dilution ratio to achieve the required solid concentration that accounts 8%, so 0.7079m3/d of water is added. Therefore, Volume of the daily charge (Sd) = 1.4158m^3/d

The volume of the digester (slurry) (VD) is defined as the product of the Volume of the daily charge (Sd) and hydraulic retention time (HRT).

The volume of digester = Total waste per day * retention time/ density of waste in one retention time.

VD=Sd*HRT=1.41m^3/d*60d= 85m^3, therefore, the volume of the tank (VD) =43m^3 of two digester to minimize the risk of damaged.

2.7. Balancing Biogas Production and Energy Demand

Determining the Biogas Production: The quantity, quality and type of biomass available for use in the biogas plant are the basic factor of biogas generation. The biogas incidence can be calculated according to different methods applied in parallel.

❖ Measuring the biomass availability
❖ Determining the biomass supply via pertinent-literature data
❖ Determining biomass supply via user survey

2.7.1. Assumptions Based on Literature

Gas production rate from human waste= 0.078 m^3/kg (1Kg of human excreta produces 0.078m^3 of biogas [4])
Gas production rate from food waste = 0.05 m^3/kg [4]
Retention time =60 days, selected
Ratio of manure and water 1:1dilution
Heat content of methane= 38.13 MJ/m^3

2.7.2. Total Gas Produced

Gas produced from food waste per day =amount of food waste per day*its Gas production rate.

$$=93.94*0.05=5.8635m^3$$

Gas from human waste per day = No. of population * human waste per day * its Gas production

$$=357*0.7*0.078=19.4922m^3$$

Gas produced from Total waste per day = 5.8635m^3 +19.4922m^3 = 25.3557m^3/d

2.7.3. Determining the Energy Demand

The energy demand of any given farm is equal to the sum of all present and future consumption situations for cooking and heating purpose. Determining biogas demand based on present energy consumption, e.g. for ascertaining the cooking-energy demand. This involves either measuring or inquiring the present rate of energy consumption in the form of wood, charcoal, kerosene and bottled gas. Calculating biogas demand via comparable-use data: Such data may consist of:

❖ Empirical values from neighboring systems,(biogas consumption per person and day)
❖ Reference data taken from literature, although this approach involves considerable uncertainty, since cooking energy consumption depends on local cooking and eating habits.

2.7.4. Total Gas Required

The total energy consumption for cooking and heating purpose (Ec) =energy from fire wood +energy from dung cakes + energy from charcoal +energy from LPG+ energy from butane gas+ energy from electricity

$$EC = (750kg + 150kg + 540kg + 120kg + 60kg + 1500kwh)/month$$
$$= (25kg + 20kg + 21kg + 4kg + 2kg + 50kwh)/d$$

One cubic of biogas
=1.4kg of charcoal, implies, 21kg charcoal=1m^3 biogas*21/1.4=15m^3
=0.43kg butane gas, implies, 2kg of butane=1*2/0.43=4.65 m^3
= 0.454kg of LPG implies, 4kg of LPG=1*.4/0.454=8.81m^3
= 3.5kg of fire wood implies25kg of wood=1*25/3.5=7.14m^3
=12.30kg of Cattle dung cake implies 1*20/12.30=1.62m^3
= 38.13MJ, implies= (1*50kwh*3600sec/hr)/38.13MJ=4.72m^3

Therefore, the total energy consumption (EC) = 7.14 + 1.6 + 15 + 8.81 + 4.65 + 4.72) m3/d=41.75m^3/d

The amount of gas produced from organic waste covers 60.73% the energy consumption of the people living in condominiums. If the stove works for a maximum of 3 hours per day, the energy consumption becomes $8.45m^3/hr$ and averagely, 120 house hold obtained $0.0704m^3/h$ gas .The amount of shortage of gas daily equals to total amount of gas consumed daily minus the amount of gas produced daily, EC =41.75-25.36=16.39m^3 /d. Therefore, the total amount of energy demand obtained from firewood, cattle dung cakes and charcoal is nearly substituted by the methane gas produced from different waste material. The remaining energy demand should covered by electricity in order to have free indoor air quality and to contribute in reducing green house gas emission mostly generated from fossil fuel such as carbon dioxide.

3. Design and Sizing of Bio Digester

Among the various types of digesters, in this section of the design, fixed dome torpispherical a continuous feed (displacement) digester is selected for the reason that relatively small amounts of slurry (a mixture of manure and water) are added daily. This enables that gas and fertilizers are produced continuously and predictably. After selecting the type of digester, the retention time, which is a key parameter in determining digester size, is chosen to maximize the percentage of production of biogas with respect to the retention time. Sixty days is chosen as the minimum amount of time for sufficient bacterial action to take place to produce biogas and to destroy many of the toxic pathogens found in human waste.

3.1. Dimension of the Main Parts of the Digester

Dimension of Mixing Pit: The mixing pit of the digester should have a size slightly greater than the daily input and better if no corners.

Assumption: Cylindrical shape is selected based on its advantage as explained on geometrical shape of digester with retention time of 60 day. The diameter (d) and height (h) of mixing pit are equal (d=h)

$V= \pi d^2 h/4$, Where V is the daily substrate flow of the waste after providing 10% safety factor, then the Volume including safety(V)

$V= (10\%*1.41m^3/d)+1.41m^3/d=1.551m^3/d$

In addition, volume can be written in terms of diameter as follows:

$$V=\pi hd^2/4=\pi d^3/4, \text{ solving the diameter (d)}$$

$$d = (4*1.551/3.14)^{1/3}=1.25m$$

Therefore, the mixing height (h) =d=1.25m

Dimension of Compensation Tank: The dimension of the compensation tank (V_{tank}) is around 20% of the volume of the Digester (slurry).

$V_{tank}=0.20*43m^3=8.6m^3$, for efficient spaces assume cubic shape with length of 'a' then

$V_{tank} =a*a*a =a^3$ implies $a=\sqrt[3]{Vtank} = \sqrt[3]{8.6}$ =2.05m

3.1.1. Cross Section of a Digester

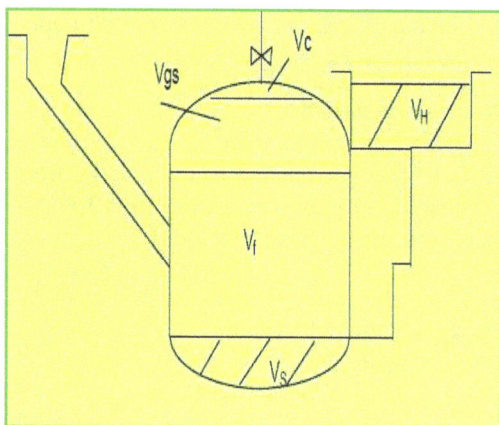

Figure 3.1. *The cross section of digester specifications*

1 Volume of gas collecting chamber at the top layer=Vc
2 Volume of gas storage chamber Vgs, V1=VC+Vgs
3 Volume of fermentation chamber=Vf=V3
4 Volume of sludge layer= V2, Slurry volume V2+V3
5 R1 and R2 is the crown radius of the upper and bottom spherical layer of the digester respectively
6 S1 andS2 art the surface area of the upper and lower dome respectively
7 f1 andf2 are the maximum distance of upper and lower dome

Therefore the total volume of the digester (V) =VC+Vgs+V2+V3=V1+V2+V3, Substituting V1=VC+Vgs

3.1.2. Geometrical Dimensions of the Cylindrical Shaped Biogas Digester

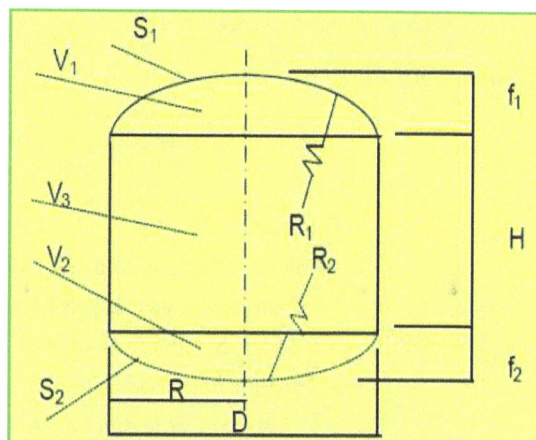

Figure 3.2. *Geometrical dimension of the designed cylindrical fixed dome digester*

For structure stability and efficient performance, fixed dome digester is expressed by the following correlation [7, 8].

For volume	For geometrical dimension
$Vp \leq 5\%V$	$D=1.3078*V^{1/3}, f=D/5$
$V3 \leq 15\%$	$V1=0.08227D^3, R1=0.725D, S1=0.911D2$
$Vgs+Vf=80\%V$	$V3=0.3142D^3$
$Vgs=VH=k(V2+V3)/(1-0.5k)$, here ,$k=0.4m3/m3d$,is gas production rate per m3 digester volume per day	$V2=0.05011D^3, R2=1.065D, f2=D/8.S2=0.83445D^2$, where D is the diameter of digester

3.2. Volume Calculation of Digester Chamber

From the above geometrical dimension correlations: volume of digester (VD) =V3+V2

$$=(0.05011+0.3142)D^3=43m^3$$

implies $D= \sqrt[3]{\dfrac{43m3}{0.36431}}$ =4.9m.Therefore the value of each parameter is calculated by substituting the value of diameter in the above geometrical relation presented in table 3.1 and obtained as V1=9.72m³ ,V2=5.89m³, V3=36.96m³, VC=2.65m³, f1=0.98, f2=0.61m, S1=21.87m², S2=20.03m², R1=3.55m, R2=5.21m

Then the total volume (V)

$$V=V1+V2+V3=9.72m+36.96m+5.89m=52.57m^3=53m^3$$

$$Vgs=VH=k (V2+V3)/ (1-0.5k), =$$
$$\frac{0.4(5.89+36.96)}{1-0.5*0.4} = 10.71m3, V3 =\pi D^2H/4,$$

$$H=4V2/\pi D^2 = \frac{4*36.89}{4.9*3.14}=2.01m, h = h3/+F1+H1= 159cm$$
(water volume which is fixed or standard)

From assumptions:

VC = 0.05 V = 2.65m3,

Vgs = 10.71m3,

VC + Vgs = 2.65m3 +10.71m3 =13.36m3, the value of V1 again calculated as follows.

V1 = [{(VC + Vgs) - {π *D2 H1}/4] = [13.36 - {3.14 x (4.9) 2x H1}/4] which gives H1 = 0.19m=190mm.The value of the height of the above dome up to the end, have fixed h =159cm water volume (1 mm = 10 N/ m2) h = h3 + F1 + H1 = h3+0.98 +0.19m which gives,

$$h3 = 0.42m =42cm$$

Finally, the complete drawing looks like these having all dimensions calculated above or we can take dimensions from standards' having the same plant sizes.

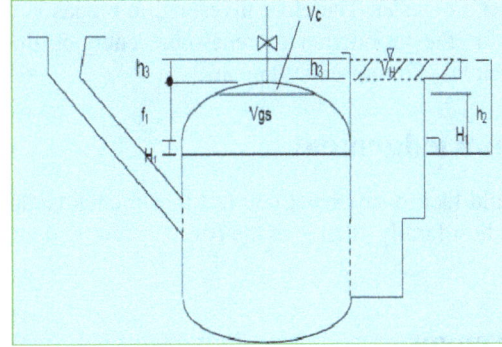

Figure 3.3. *Complete drawing of cylindrical torpispherical fixed dome digester*

Pressure in the Gas Holder or Container: Assume the gas in gasholder obeys ideal gas law i.e. PV=nRT, where n is the mole of gas (n),

$$mole(m) = \frac{mass\ of\ the\ gase(m)}{molecular\ mass\ of\ gase(M)}$$

And

$$mass(m) = \frac{density\ of\ biogas(\rho)}{volume\ of\ gase(V)}$$

Substituting all values it gives,

$$P = \frac{\rho RT}{volume\ of\ gase(V)}$$

The value of M=25.8 (65%CH4 and 35%CO2), universal gas constant(R) =8.314J/mole K, temperature (T) =298K (25⁰C) density of biogas (ρ) =1.15Kg/m³, hence P=1.1bar

Piping: The piping system connects the biogas plant with the gas appliances or users. It should be safe, effectively gas-tight and allow the required gas flow to the appliances during the life span of biogas plant. In this paper, it is recommended to use polyvinyl chloride (PVC) pipe that are economical.

4. Conclusion

The installation of biogas plant leads larger cost saving in addition to environmental and sanitation benefits. The gas produced replaced the total amount of their energy demand that was previous obtained from (charcoal, firewood, and animal dung cake) which accounts 60.73% of their energy needs. The bio-slurry produced after methagonic process also contributes for plantation and garden area present in the condominium house as organic fertilizer. Even thought the initial investment is high to put the plant in to reality, the overall result shows the project is feasible. Therefore taking in to practice the project saves the people living in condominiums from energy cost and from the cost that expense to avoid the collected waste. When the households implement this project, they also play a great role in improving their living standards, environment and creating jobs opportunities for those construct a digester. Therefore investing on biogas is viable option for the production of renewable energies that are important task for our environment.

Acknowledgement

I would like to say thank you to Mr. Simeneh Gedefawu and my best family   that give me constant love and help.

References

[1]   Website: http:// www. Ethiopia HEDON Household Energy Network.htm

[2]   Dr. Getachew Eshete, Dr. Kai Sonder Felix ter Heegde; Report on the feasibility study of a national program for domestic biogas in Ethiopia. SNV Ethiopia, Addis Ababa, 2006

[3]   The Ethiopian News Agency and Debre Markos House development office(2012)

[4]   Jenniy Gregory, semida Silveira, Anthony Derrick, Paul Cowley, Catherine Allinson, Oliver Paish; Intermediate Technology Publications in Association with the Stockholm Environment Institute, Financing Renewable Energy Projects, A guide for development works, 1997.

[5]   Mahin, D.B. (1982). Biogas in Developing Countries. Bio energy System Report to USAID, Washington, DC.

[6]   Fergusen, T., and Mah, R. (2006).Methanogenic bacteria in Anaerobic digestion of biomass.

[7]   Dr. Amrit B. Karki, Prof. Jagan Nath Shrestha, Mr. Sundar Bajgain, (2005). Biogas as Renewable Source of Energy in Nepal theory and Development.

[8]   Design of Biogas Plant, Bio-gas Project, LGED: Preparing this training material all the important information have been collected from the booklets & research materials of Biogas Training Center (BRC) Chendu, Sichuan, Chaina.

4

Cogenerations of energy from sugar factory bagasse

Assefa Alena, Omprakash Sahu[*]

Department of Chemical Engineering, Wollo University, Kombolcha, South Wollo Ethiopia

Email address:
ops0121@gmail.com (O. Sahu)

Abstract: During sugar production, bagasse (waste) is produced which is used as energy resource in the sugar mill. Co-generation power plants using bagasse as the feedstock are attached to several sugar factories in Thailand. These produce steam and electricity for use in the sugar mills and also sell the excess power to the grid. Bagasse, being a by-product of sugar production as well as of biomass origin seems to be a suitable candidate for sustainable energy production. However the case is quite different in Shoa Sugar Factory, which suffers from lack of bagasse during stoppage of mill and as a matter of fact it is forced to cut trees of the surrounding to deliver it to its boilers during stoppage of mill. It is a crystal clear fact that cutting trees without replacement causes the desertification, which is currently the case in Shoa Sugar Factory. It is from this fact that the objectives of the research work emanate. The first part of the study deals with bagasse and its properties, this part of the study focuses on determining the quality and quantity of bagasse that has been used by the factory as a fuel for boilers with respect to the conventionally accepted standards. The outcome of the study indicates the bagasse produced by the factory fulfills all the requirements as a boiler fuel both in quality and quantity wise, during milling time and stoppage of mill without the supply of any additional fuel. The second part of the study focuses on the steam generation and utilization unit. The study conducted in the steam generation unit shows the steam generation unit (boilers) has very low efficiency (on average 56%) when compared to the minimum accepted efficiency of boiler that uses bagasse as a fuel (70 %). The low efficiency is manifested by large quantity of heat losses that should be transferred to steam. The investigation on the steam utilization unit shows it operates without problems. In general the outcome of the study proves that the low efficiency of the boiler resulted in shortage of surplus bagasse. The research output indicates existing surplus bagasse shortage can be solved by improving the efficiency of the steam generation unit. The proposed solutions to the problems are optimization of excess air supply in the combustion chambers, application of bagasse drying system, increasing the capacity of evaporators, efficient operations, maintenance of boilers and its accessories.

Keywords: Bagasse, Energy, Fuel, Heat

1. Introduction

The development of sugar industry in Ethiopia has been remarkable since 1954 when the Dutch Company, HVA established Wonji Sugar Factory with a crushing rate of 1400 tons of cane per day (TCD). Wonji Sugar Factory is the pioneer sugar factory in Ethiopia. Eight years after the establishment of Wonji Sugar Factory, in 1962 Shoa Sugar Factory was established with a crushing rate of 1600 TCD. However, the present crushing capacity of the factory is 1700 TCD. The two Factories are located in the rift valley, Orimia region, in East Central Ethiopia, some 107 Km South East of Addis Ababa at an elevation of 1540 above sea level[1,2]. They are under one management and collectively called Wonji Shoa Sugar Factory (WSSF). Since then

the sugar sector has played a vital role in the country's economy and sugar has become one of the major commodities for local consumption and foreign exchange earnings. The production of sugar from sugar cane is a distinctive process in that it involves no element of synthesis [3]. However, sucrose comes in to the factory and subject to some unit operations [4]. The present system of sugar manufacturing process is essentially a combination of juice extraction from sugar cane by milling or diffusion or a combination of both, clarification of juice, concentration of juice by evaporation to syrup, crystallization of sucrose by vacuum pan boiling, centrifugal separation of sugar and molasses from massacuites, drying and cooling of sugar and sugar grading and packing [5, 6]. All the process needs relatively large amount of energy in different forms such as power for prime movers. This power is needed mainly in the form of electricity for the electric motors and medium to high-pressure steams for the steam engines and steam turbines [7]. The processing of raw cane also requires a large amount of heat provided by low-pressure steam for the processing of juice in to commercial sugar [8, 9]. Sugar cane industry is one of the few industries, which are able to supply its own fuel (in the form of bagasse). The fiber in the cane is generally sufficient to supply all the steam necessary for power production and for manufacture of sugar when utilized as a fuel in the boiler furnaces. With normal fiber content in cane (12-14%) a well balanced factory will be left with a surplus bagasse [10]. Nevertheless, the energy crises of 1974 and 1979 and the uncertainty in the future pricing of fossil fuel has led most sugar producers to think in terms of improved energy balance in their factories. Technological advance in the utilization of by- products of the cane sugar industry have reinforced the policy of maximum efficiency in the use of fuel [11].

The present objective of cane sugar industry is to extract the maximum power from the available fuel by efficiently utilizing the latent heat of steam and reducing the losses to a minimum. The losses are inevitably sustained in the flue gases of steam boiler, gas turbines and in the cooling water of condensers. Significant improvements can be achieved by improving boiling house, and other factory departments leading to maximum process heat economy, bagasse drying to improve its calorific value as fuel, generation of extra electricity from excess bagasse for export to the grid. Shoa sugar factory is the factory that was established to supply its own fuel from the byproduct bagasse and remains with some excess of this bagasse, which will be used during stoppage of mill so that no additional fuel would be used whatever the case may be. At the present condition the reality reveals that the factory is using additional fuel during mill stoppage.

2. Material and Methods

In this research work the detail assessments of the factory has been made by giving due attention on review of existing steam generation plant , present condition and performance of boiler capacity, drum boiler heating surfaces, fittings for boilers attemperator, soot blowers grate, economizer, induced draft(ID) fan , combustion air fan (FD fan), chimney, instrumentation and controls, steam distribution system, steam pressure reducers and de-superheating station, various indicators and recorders installed, lagging and insulation, evaporator station, thermo compressor, juice catcher after last body ,condensate extraction system, material and energy balance based on design parameters. The flow diagram of power plant is shown in Fig. 1. The research work, therefore, emanates from current energy imbalance, shortage of excess bagasse, and wastage of steam. To arrive at the solution the number of works was conducted, among which are; Investigating the obscure operating characteristics, which are specific to steam generation from bagasse as a boiler fuel and utilization of steam by the

factory, identifying the root causes of the problem, examining the extent to which the energy system analysis can be used to know the energy impact of the factory, quantifying and evaluating the performance parameters of the factory with respect to steam generation and utilization, studying the effects of the change in the performance parameters which accounts to prevent the existing problems, recommend appropriate solutions[12].

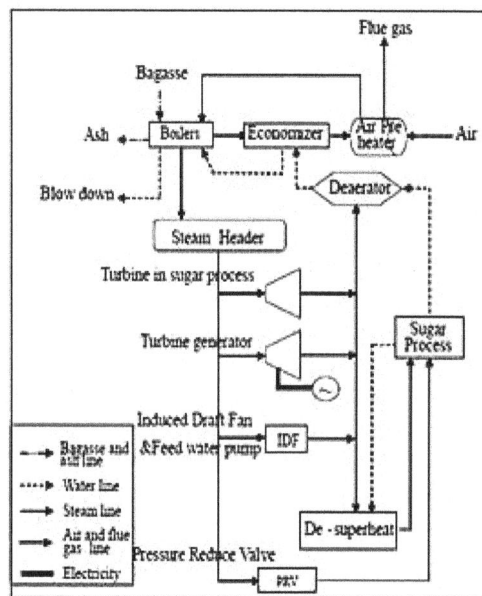

Fig. 1. *Power plant of Sugar Industry.*

2.1. Physical Testing

Materials used for experiment are, Chopping machine, disintegrator, balances, sodium bicarbonate, and mercuric chloride to determine fiber content of the cane. Perforated tray, balances and Oven were to determine the moisture content of bagasse. Rapi-pol extractor, lead acetate, filter paper and balances: it is used to determine the Pol in bagasse. Bomb Calorimeter, type C210, Make: Junk & Kunkel K.G, IKA WERK Stauten Breisgan, measuring cylinder, balances and ignitions wire were used to determine the high heating Value (HHV) or gross calorific value of Bagasse of different Varieties and different moisture content. Combustion (Flue Gas) Analyzer, KM9104 was used to determine the composition of the flue gas leaving the boiler and its temperature [13].

3. Results and Discussion

3.1. Physical Property of Bagasse

The physical property of bagasse after testing is shown in Table no. 1. In which moisture content in bagasse, fiber content in cane, bagasse content in cane and percentage of pol in bagasse was determine.

Table 1. *Physical Properties of Bagasse.*

Variety	Moisture content in bagasse [wt/wt%]	Fiber content in cane [wt/wt%	Bagasse content in cane [wt/wt%]	Pol %in bagasse
N-140	50	14.52	32	3.8
NCO334	50	14.08	30	3.4
B52-298	50	13.67	29	3.2
B41-227	50	14.78	33	3.7
CO-421	50	14.13	31	3.6
Average	50	14.24	31	3.54

3.2. Gross Calorific Value

The heating value of bagasse from cane was shown in Table no.2. In this test determine percentage of moisture, percentage of fiber in cane and gross calorific value was determined from different varieties of cane.

Table 2. *Heating Values of Bagasse Obtained from Major Commercial Cane Varieties.*

Variety	Moisture %	Fiber % Cane	GCV[KJ/Kg]
N-140	50	14.52	8795
NCO334	50	14.08	8430
B52-298	50	13.67	8396
B41-227	50	14.78	8894
CO-421	50	14.13	8469
Average	50	14.24	8597

GCV Gross Calorific Value.

3.3. Calorific Value of Bagasse

The calorific value of bagasse before and after drying was presented in Table no. 3. In this calculation both gross calorific value and net calorific value was determine.

Table 3. *Calorific Values of Bagasse with Different Moisture Content.*

Bagasse	Before drying	After drying	
Moisture (%)	50	35	30
GCV (KJ/Kg)	8597	12544	13602
NCV (KJ/Kg)	7447	10517	11532

GCV Gross Calorific Value, NCV Net Calorific Value.

3.4. Composition of Gasses

The analysis of flue gasses was presented in Table no. 4. The percentage of different gaseous oxygen, carbon monoxide, carbon dioxide, nitrogen and water at boiler temperature was determined.

Table 4. *Results of Flue Gasses Analysis.*

Description	% of flue gases compositions		
	B#1	B#2	B#3
Oxygen	9.4	10.5	11.3
CO	0.89	1.23	1.84
CO_2	10.9	9.8	9.1
N_2	65	76.97	63
H_2O	13.8	12.4	14.3
Temperature	202	218	220

3.5. Bagasse and Steam Balance based on Actual Daily Production

The balances conducted based on the bagasse and steam demand of the factory on daily basis of its present operation condition is intended as one of the factors to address the present fuel problems of the factory. The factory crushes on average 22 hours a day in which this balance is based on. In the determinations of the amount of bagasse and steam produced and consumed, the first step is the determination of loses [14, 15].

1) These loses includes bagasse loses due to vacuum filter, stoppage of mill and other undetermined loses.
2) Heat loses in the processes of steam generation comprises of loses due to Latent heat of vaporization of the water formed by combustion of the hydrogen contained in the bagasse. This is lost in the flue gases with the water vapor if the latent heat is not condensed.
3) Latent heat of vaporization of the water content of bagasse, which in the same way is lost with the flue gases, Losses in ash content and brix of bagasse.
4) Sensible heat losses in the flue gases and fly ash.
5) Losses by radiation
6) Losses by unburned solid,
7) Losses by incomplete combustion of carbon giving CO instead of CO2.

Except loses due to sensible heat in the flue gases and fly ash, radiation, unburned solid, and incomplete combustion of carbon giving CO instead of CO2, the rest loses are included in the determinations of the N.C.V of bagasse. The rest loses will therefore determined here under, which presented in Table no. 5.

Table 5. Bagasse and steam balance based on actual daily production.

4. Descriptions	Unit	Quantity		
1. Cursing Rate	T/h	77		
	T/d	1700		
2. Fiber in can	%	14.24		
3. Average Bagasse in cane	%	31		
4.1. Bagasse Analysis				
(i) Pol in bagasse	%	3.54		
(ii) Moisture in bagasse	%	50		
(iii) Fiber in bagasse	%	46.46		
4.2. Bagasse loses				
(i) For Vacuum. Filter and other losses in cane	%	2.38		
(ii) For stoppages of cane shortage (8% on production and 85% steam usage)	%	58.59		
4.3. Net Bagasse available % cane	T/n	18.48		
4.4. Calorific value of Bagasse				
(i) G.C.V	KJ/kg	8594		
(ii) N.C.V	KJ/kg	7447		

5.1. Steam Condition	Unit	Boiler 1	Boiler 2	Boiler 3
Pressure	Kg/cm^2(g)	21	21	22
Temperature	0C	310	330	320
Temp. of feed water tank	0C	90	90	90
Temp. of feed water after economizer	0C	187	187	190
The gas exit temp	0C	202	218	220
Ambient Temp	0C	30	30	30
CO$_2$	%	10.9	9.8	9.1
Air fuel Ratio		1.79	1.89	1.98
Heat-losses in flue gases	kJ/kg	1399	1605	1685
Capacity	T/h	15.002.03	15.00	15.00
Steam/Bagasse Ratio			1.99	1.98
Efficiency				
Factor- Alpha (solid unburned)	%	95	95	95
Factor -Beta (radiat CO$_2$ & Conv.)	%	95	95	95
Factor - Neta (In.Comb.)	%	95	95	95
Product : A*B*N	%	85.7	85.7	85.7
Boiler Efficiency				
(i)On G.C.V	%	58	56	55.5
(ii) On. N.C.V	%	69	67	66
Boiler Efficiency				
(i)On G.C.V	%	58	56	55.5
(ii) On. N.C.V	%	69	67	66

5.2 Steam Generation Possible	Unit	Boiler 1	Boiler 2	Boiler 3
Capacity	T/h	15	15	15
Steam pressure	Kg/cm^2	21	21	21
Steam temperature	^{o}C	310	330	3201.98
Steam/Bagasse Ratio		2.03	1.99	

6.Steam Consumptions by prime movers

6.1. Mill TurbinesCapacity	hp	300	450	450
Live steam condition				
Pressure	Kg/cm^2	21	21	21
Temperature	°C	300	300	300
Exhaust steam pressure	Kg/cm^2 t/n	1.25	1.25	1.25
Steam flow at operational load		Total12.32		

6.2. Power turbine		Turbine 1	T turbine 2
Pressure		21	20
Temperature	Kg/cm^2	300	300
Exhaust steam pressure	0C	1.25	1.25
Operation Av. Load	Kg/cm^2	1.4	1.4
Specific steam consumption	MW	1.68	11.68
Steam flow at operational load	Kg/MWh T/n	18	18

6.3 Steam to process		
Exhaust from Mill turbine	T/n	12
Exhaust from Power turbine	T/n	18
Total	T/n	30

6.4. Steam Generation (Boilers)		Boiler1	Boiler2	Boiler3
Pressure	Kg/cm2	20	20	20
Temperature.	°C	310	310	310
Installed capacity	T/h	15	15	15
Current Capacity	T/h	12.37	12.37	12.37
Steam/Bagasse Ratio		2.03	1.99	1.98

6.5. Steam Consumption		T/h	% cane
Inlet steam to mill Turbine		12	15.6
Inlet steam to power Turbine		18	23.4
Live steam make up		4.22	5.5
Sugar Drying, sulfur, remelt		1.5	2.0
Losses		1.4	1.8
D. heating water		00	00
Total steam Demand		37.12	

Exhaust balance		T/h	%cane
			48.2
Exhaust produced		37.22	39
Exhaust Required		30	6.65
Surplus /deficit		5.22	

Summary of Bagasse Balance	Boier1	Boiler2	Boiler3
Steam GenerationRequired		1 2.37	12.37
Steam/Bagasse Ratio	12.37	1.99	1.98
Bagasse Requirements	2.03	6.22	6.25
Bagasse Available	6.09	18.48	
Net Surplus Bagasse (Negative)		-0.08	

The bagasse produced from the major commercial sugar cane varieties have on average 50% moisture, 14.5% fiber on cane. For a bagasse with moisture content of 50 % the accepted standard fiber content in cane ranges from 12 to 16%. The experimental result indicates the fiber content in cane of major commercial sugar cane varieties ranges from 13.67 to 14.78 %. The test result also indicates the quantity of bagasse obtainable from a unit weight of cane ranges from 29 to 33 %wt/wt of cane. This gives on average 31 % weight of bagasse per weight of cane. The standard value is 24 to 30 % wt/wt of cane or approximately a quarter. The fiber content % in cane and the quantity of bagasse pro-

duced by the factory is much better to supply the required quantity of bagasse for the boilers both during milling time and stoppage of mill.

6.5.1. Heating or Calorific Value of Bagasse

The results of experiment done on the higher heating value of bagasse at moisture content of 50 % indicate the higher heating value ranging from 8396 to 8894 KJ/kg which is within the standard range. The calorific value of bagasse has also been determined for varying the moisture contents of bagasse. The purpose is to quantify the impact of the moisture content of bagasse on the amount of heat transferred to steam from bagasse during combustion in the boiler furnaces. The experience so far in the factory reveals moisture content has a significance influence on the performance of a boiler. This effect would be manifested when the moisture content of bagasse increase above 50 %; the boiler needs more bagasse or the boiler losses its pressure stability. The test is, therefore, intended to evaluate the boiler performance under a much wider spectrum of bagasse moisture. The test has been conducted on bagasse moisture content of 30%, 35% and 50%.

The result shows the increase in the calorific value of bagasse with the reduction in its moisture content. It indicates it has necessary to dry bagasse to reduce the moisture content so as to improve its calorific value. This would improve the boiler efficiency. Moreover, the efficiency can be improved further by recovering the sensible heat that has been lost with the flue gas by employing it as the heat source in bagasse drying operation.

6.5.2. Flue Gases Analysis

The flue gas analysis indicates the carbon monoxide compositions of the flue gases on three boilers are 890, 1230 and 1840 ppm, respectively when compared to the maximum allowable 480ppm.The test also indicates the percentage of excess air in the combustions chamber, i.e., 79%, 89.7% and 98.7% and the composition of carbon dioxide in the flue gas is 10.9, 9.8 and 9.1 percent, respectively. However, the percentage of excess air to support complete combustion should lie in the range of 30 - 50%. The analysis can be due to the addition of too excess air into the combustion chamber. This too excess air cools the chamber which in turn needs more energy to heat the air before combustion is started. The excess air due to its high pressure and velocity takes some of the carbon monoxide with it in the flue gases before it is completely converted into carbon dioxide. Moreover, the excess air that leaves with the flue gases takes more sensible heat with it than the conventional losses to the atmosphere. It is, therefore, very essential to optimize the percentage of excess air in the combustion chamber. This can be done by setting the percentage of carbon dioxide that leaves the combustion chamber to 15%, which is the optimum percentage to support complete combustion in the boiler furnace, Hugot ET. Al. (1993). The maximum percentage of CO2 in the flue gases that can be theoretically achieved is 19.8% Hugot ET. Al. (1993). If it is needed to obtain complete combustion, without appreciable formation of carbon monoxide, it is desirable to work with the minimum amount of excess air, which will yield the optimum carbon dioxide percentage in the flue gases. By carrying out the optimization work on boiler furnaces, the optimum percentage of excess air in the combustion chamber is 32 % as a result the optimum percentage of oxygen in the flue gases is 5.1 %.

The heat losses in the flue gases are determined using the ratio of the optimum excess air to the theoretically required air to support complete combustion. The optimum ratio is 1.32. The heat losses with the flue gases from each boiler

are, therefore, 273kcal/kg (1150Kj/kg), 278kcal/kg (1164 KJ/kg) and 285kcal/kg (1193Kj/kg) respectively. The percentage of heat recovered by optimizing the quantity of excess air in the combustion chamber for each of the three boilers are 18%, 27.4 % and 29.1% respectively. After optimization of the excess air in the combustion chamber, the quantities of heat transferred into the three boilers are 1289 kcal/kg (5395 kJ/kg), 1285 kcal / kg(5384 kJ/kg) and 12798 kcal/kg (5359 kJ/kg), respectively.

6.5.3. Proposed Possible Solutions

To overcome the existing surplus bagasse problems of the factory and thereby to avoid the usage of wood as a boiler fuel during stoppage of mill, the combined improving actions were proposed as:

- Optimization of excess air supply in the combustion chambers.
- Application of bagasse drying system
- Increasing the capacity of the evaporators.
- Efficient operations and maintenance of boilers
- Optimization of Excess Air in the Combustion Chamber

By optimizing excess air in the combustion chamber the boiler efficiency increases 4-6 %. The gain in the above efficiency due to optimization of excess air will lead to save 1.56 tons of bagasse per hour after generation of the same quantity of steam as before.

6.5.4. Bagasse Drying Application

The experimental test results and the operational data have clearly indicated that the energy potential of bagasse has been under utilized. The result of the experiment shows currently, the drying of bagasse has taken much attention, as it is a power full tool to save bagasse energy in sugar industry. In the bagasse draying, the gases, which is the gaseous products of combustion leaving the combustion chamber is employed as a heat source to remove the water from bagasse. In doing so large quantity of sensible heat that has been lost with the flue gases are recovered. By applying the bagasse drying system, the gain in system efficiency is calculated from the change of the stack gas and is found to be 10.4 %. The calculation for the system assumes that there are no losses in the dryer. These are estimated to be about 4% losses in the dryer and are made as a correction on the final efficiency. The actual gain in system efficiency would then be, therefore, 6.4% and this leads to the savings of 2.2 tons / hr bagasse after generating the same quantity of steam.

6.5.5. Increasing the Capacity Evaporator

Increasing the capacity of the evaporator and utilizing the surplus exhaust steam that has been lost through blow down leads to increasing the crushing capacity of the mill to 90 tons /hr. The increment in crushing capacity of the mill leads to the production of 3.57 tons of bagasse per hour. This increment in evaporations and crushing rate of the mill don't affect the quantity of live steam produced and consumed.

6.5.6. Efficient Operations and Maintenance of Boilers

In sugar cane boilers proper operations and maintenance (O & M) procedures must be followed to insure safe and efficient operations. It is often assumed that good O & M provides no energy savings because it simply "what should be done." In Shoa sugar Factory due to lack of proper O & M energy consumption can increase dramatically as much as 10 to 20 present. Boilers suffer frequently from failure of the various systems including the pressure parts such as tubes, super heaters, air heaters etc. Thus, by carrying out only proper operation and maintenance the factory saves 10 to 20% of its energy consumption. The maintenance includes keeping physical components in good working order and within design specifications. This includes cleaning heat transfer surfaces, controls tuning, and maintaining insulations. Before boiler tuning system diagnostics should be performed and any deficient equipment brought back to specifications.

- By optimizing the excess air in the combustion chamber, additional 1.56 tons/hr of bagasse would be obtained as a surplus bagasse.
- By employing bagasse dryer the factory will save 2.23 tons/hr of bagasse.
- By increasing the capacity of the evaporators to utilize the existing 5.12 tons/hr of exhaust steam which leads to the increment in crushing capacity of the mill as a result of which surplus 3.57 tons/hr of bagasse is produced.

7. Conclusions

Based on the forgoing out puts of the research, the following inferences are made. All the experimental results, investigation and the analysis made on the physical properties of bagasse reveals that, its heating (calorific) value is quite enough to produce the amount of steam required by the factory both during milling time and stoppage of mill without requirement of any additional fuel such as wood like the current situations in the factory. The flue gases analysis made on three boilers of the factory shows high heat losses with the flue gases due to incomplete combustion of bagasse and excess air in the combustion chamber consequently the reduction in the boiler efficiency in which additional fuel is supplied to produce the required quantity of steam. The balances conducted on daily basis reveals there is even a deficit of bagasse at normal operations. The hourly live steam demand of the factory is 37.12 tons. However, at normal operation condition on average additional 0.08 tons of bagasse is demanded to produce the quantity of live steam required. Based on the balance conducted 35.12 tons/ hr of exhaust steam is produced by mill turbines and power turbines, however the exhaust steam demand of the boiler house is on average 30 tons/hr. This illustrates 5.12 tons/hr of exhaust steam is left as waste energy. Since mill has the capacity to crush up to 90 TCH, however, the capacity of the boiler house do not go beyond 77 TCH. On the other hand, the review of the evaporator

shows there is ways of improving its capacity. Increasing the capacity of the evaporator and there by utilizing the surplus exhaust steam that has been lost through blow down leads to increasing the crushing capacity of the mill. The increment in the crushing capacity of mill to 90 TCH yields the production of 3.57 tons of bagasse per hour. This increment in evaporations and crushing rate of the mill don't demand additional live steam requirements.

To overcome the existing shortage of bagasse problems of the factory and thereby to avoid the usage of wood as a boiler fuel during stoppage of mill, the researcher believes and strongly recommends to materialize the proposed solutions as they will improve the fuel economy of the factory there by alleviate the mentioned problems.

Acknowledgements

Authors acknowledge to department of chemical engineering, KIOT Wollo University and Shoa Sugar Industry for providing facilities.

References

[1] Meade-Chen, Sugar cane Hand book, a manual for sugar cane manufacturer and their Chemists Tenth Edition, 1997.

[2] J.Maurice Peturau, By-Products of the cane sugar industry, an introduction to their industrial Utilization, third completely revised addition, Elsevier, Amsterdam-oxford Newyork-tokyo, 1989.

[3] E.Hugot, Hand Book of cane sugar engineering, third completely revised,edition,ELSEVIER, Amstar Dam-oxford-New York-Tokyo, 1986.

[4] John Howard Payne, Cogeneration in the cane sugar industry, sugar series 12,Elsevier Amsterdam-oxford-New York-Tokyo, 1991

[5] RSTCA, Regional sugar cane Training center for Africa Training Manual, Mauritius, 2001.

[6] N.J.Themelis, P.A. Ulloa, Methane generation in landfills. Renewable Energy, 2007, 32 (7):1243–1257.

[7] Villanueva, H. Wenzel, Paper waste – recycling, incineration or land filling, A review of existing life cycle assessments. Waste Management, 2007, 27:29–46.

[8] R.P.Beeharry, Carbon balance of sugarcane bioenergy systems. Biomass and Bioenergy, 2001, 20 (5):361–370.

[9] W.Chaya, S.H.Gheewala, Life cycle assessment of MSW-to-energy schemes in Thailand. Journal of Cleaner Production, 2007, 15(15):1463–1468.

[10] Eriksson, M.C. Reich, B.Frostell, A.Bjorklund, G. Assefa, J.O. Sundqvist, J.Granath, A.Baky, L.Thyselius, Municipal solid waste management from a systems perspective. Journal of Cleaner Production, 2005, 13 (3):241–252.

[11] M.Hauschild, H. Wenzel, Environmental Assessment of Products: Scientific Background, vol. 2. Chapman and Hall, London, UK, 1998.

[12] D.Janghathaikul, S.H. Gheewala, Environmental assessment of power generation from bagasse, 2005.

[13] Bhattacharyya, C.Subhes, Applied general equilibrium models for energy studies: a survey" Energy Economics, 1996, 18:145–164.

[14] Moreira, R.Jose, J. Goldemberg, The alcohol program" Energy Policy, 1999, 27:229–245.

[15] K.Deepchand, A Note on the Pyrolysis Behaviors of Sugar Cane Fibrous Products. Biological Wastes, 2007, 20:203-208.

5

Ethanolysis of calabash (*Lageneria sinceraria*) seed oil for the production of biodiesel

Muhammad Mukhtar[1], Chika Muhammad[1], Musa Usman Dabai[1], Muhammad Mamuda[2]

[1]Department of Pure and Applied Chemistry, Usmanu Danfodiyo University, P. M. B, 2346, Sokoto, Nigeria
[2]Sokoto Energy Research Centre, Usmanu Danfodiyo University, P. M. B, 2346, Sokoto, Nigeria

Email address:
mmukhtar02@gmail.com (M. Mukhtar)

Abstract: Biodiesel production from plant seed oil and animal fat is not a new technologies, though recently searching for alternative renewable sources of fuel is receiving much attention due to global energy demand and increase in environmental pollution. Currently biodiesel is largely produced from edible oil feedstock which may not be sustainable in the longer term due to its competition with food, thus lead to a search for not edible oil feedstock for the production of green fuel. In view of this, homogeneous transesterification of *Lageneraria sinceraria* seed oil has been carried out using NaOH catalyst at 65°C with ethanol which produced a good biodiesel yield of 78% with HHV of 36.34 (MJ/Kg), 0.02% low total water and sediment level, 0.80g/cm^3 density, 0.82 g/cm^3 specific gravity, 27.20 g/cm^3 API gravity, 0.44 mg NaOH/g Acid number and 144°C Flash point. The ethyl ester biodiesel produced, therefore, promises to be a viable source of energy for future use.

Keywords: Ethyl Ester, Quality Parameters, HHV, *Lageneraria Sinceraria,* Seed Oil

1. Introduction

Depletion of the world's petroleum reserves, increase in environmental pollution and increase in demand of energy due industrialization and population lead to search of alternative renewable sources of fuels [1, 2, 3, 4]. Globally transport sector has the highest rate in green house gases emissions (20% CO_2) in the last ten years[5] and it was projected to increases by 60% by year 2030 due to increase in higher energy use and carbon emission[6]. The world focus now is to produce energy from low carbon sources which is environmentally friendly [7],and thus the production of transportation fuels from renewable sources of energy is one of the most important challenges for the sustainable development [8].

Liquid transport biofuels are predominantly produced from biomass which is renewable, sustainable, biodegradable, carbon neutral and environmentally friendly [9]. Thus biodiesel is a renewable alternative source of energy which is non-toxic, biodegradable, with low carbon and particulate matter emissions compared to petroleum-base diesel and carbon dioxide produced by combustion of biodiesel can be recycled by photosynthesis and has become a popular and environmental friendly alternative fuel [10, 11].

Biodiesel is an alternative fuel which can be produced from different mechanisms such as transesterification process, in which mono alkyl ester compound of long chain fatty acid is produced when alcohol, such as methanol or ethanol, in the present of acid or base catalyst chemically react with triglycerides (vegetable oils or animal fats), which can be used in compression ignition engines for energy generation [9, 12, 13]. Recently homogeneous transesterification with NaOH and KOH catalyst is major technologies involved in production of biodiesel [14]

Developing countries such as Nigeria have a comparative advantage for biofuel production because of greater availability of land, favourable climatic conditions for agriculture and lower labour costs [15], but yet many of the present problems of Nigeria are closely related to the problem of energy distribution. Energy plays a vital role in the economic, social and welfare development of any nation, insufficient supply of energy restricts socio economic growth and adversely affects the quality of life [16]. The rising cost of petroleum products and climate change is a problem facing many developing countries and solution to this problem can be found in the explorations of renewable or alternative energy resources and one of such alternate energy source in transport sector is biodiesel which can prepared from triglycerides [17, 18].

Biodiesel is an important fuel option for rural and urban populace, it is relatively easy to manufacture, renewable and has better lubricating and environmental properties than mineral diesel [19, 20]. Biodiesel production is an important component of a combined strategic approach to diversify the sources of fuels, reduce green house gases emission to the environment, decrease current total dependence on fossil fuels and also serve as a tool for poverty reduction through job creation to the timming populace.

In biodiesel production, it is preferable to use available local raw materials for the country where biodiesel plant will be installed [21], therefore, exploration of local renewable resource of energy such as calabash (*Lageneria sinceraria*) seed oil, produced from north-western Nigeria, is of great important. *Lagenaria siceraria* is a climbing ornamental plant which is largely cultivated in the northern Nigeria, it grown as shrub with its fruit hanging on a flat bed and its fruits usually harvested between 90 to 120 days after planting. Calabash is majorly used in rural settlement as container and storage vessels[22, 23]

The aim of this study is to explore the use of calabash (*Lageneraria sinceraria*) seed oil for the production of Biodiesel and to assess it biodiesel properties in order to compliment the effort of Nigerian government on biofuels research.

2. Materials and Methods

2.1. Sampling

The calabash (*Lageneria sinceraria*) seeds were purchased from the Sokoto Central Market, north-western

Nigeria and authenticated at the Biological Science department, Botany unit, Usmanu Danfodiyo University, Sokoto. The seeds were sorted out, dehulled, ground into powder and the oil was extracted using soxhlet extraction method using n-hexane solvent.

All the reagents used in these studies were of Analar grade and distilled water was used throughout the experiment.

2.2. Experimental Procedure

2.3. Transesterification of the Oil

A 500 cm³ 3-necked round bottom flask (set with thermometer, stirrer and condenser with guard tube to prevent moisture entering into the system) was heated to drive out residual moisture. On cooling, 175g of *Lageneria sinceraria* seed oil (crude grade) was added to the flask. The oil was stirred and heated in a water bath to 65°C at which freshly prepared sodium ethoxide (40cm³ ethanol and 1g NaOH) was added rapidly under stirring condition and the reaction continued for two hours at the same temperature. Two layers were observed clearly on cooling. The top layer was biodiesel and the bottom denser layer was glycerine. The top layer was neutralized by diluted acetic acid and washed with distilled water [24, 25].

The CHNS analyses were performed on the ethyl-ester biodiesel using FLASH 2000 Organic Elemental Analyser: 3.0mg of ethyl-ester biodiesel were weighed onto a tin sheet capsule with microbalance and equivalent amount of vanadium (v) oxide (as a source of oxygen) was added and sealed. BBOT {2, 5 – Bis (5 - tert-butyl-2-benzexazol-2-yl) Thiophene [$C_{26}H_{26}N_2O_2S$]} was used as a reference standard [26].

3. Results and Discussion

3.1. Results

Table 1. Biodiesel Quality Parameters.

Parameters	Unit	Test Method	Calabash Seed oil Biodiesel	Limits
Total H₂O and sediment level	(v) %	D 2709	0.02	0.05 max
Density	g/cm³	D 7371-07	0.80	0.90 max
Specific gravity	g/cm³	D 7371-07	0.82	0.90 max
API gravity	g/cm³	D 7371-07	27.20	30-40
Acid number	mgKOH/g	D664	0.44	0.50 max
Flash Point	°C	D 93	144	130 min

The Quality assessments results was compared with biodiesel, ASTM limits reported by Gerpen et al., [27].

Table 2. FAEE CHNS Analysis

[%] C	H	N	S	O*	HHV[MJ/Kg]
73	10.1	4.4	0.2	12.3	36.4

* indicate that Oxygen value was obtained by difference.

Boie equation HHV^a [MJ/kg] = [351.60C +1162.25H – 110.90O + 62.80N + 104.65S] X 10^{-3}

3.2. Discussion

The Table 1 above shows the properties of the biodiesel produce and its various blends specified by the ASTM, which were examined based on B100 biodiesel.

3.2.1. Water and Sediment Level

The water and sediment level is an important industrial indicator of cleanliness of the fuel because water reacts with biodiesel to form free fatty acid and it may lead to microbial growth in storage tank and can contribute to filter plugging and fuel injection system wear [27,28]. Higher level of water content lead to severe corrosion of fuel system components and reduce the heat of combustion [28]. Water and sediment was found to be 0.02% which agreed with the ASTM standard of 0.05% maximum volume [27]. The result indicates that ethyl biodiesel produced is clean and may not form free fatty acid due to hydrolysis [29] and may reduce the microbial growth during storage [27].

3.2.2. Densities, Specific and API Gravities

Densities, specific and API gravities were found to be 0.80g/cm³, 0.82 g/cm³ and 27.20 g/cm³ respectively and all the values are within the ASTM limits of biodiesel [27]. These values must be within tolerable limits to allow optimum air to fuel ratio for complete combustion [30] because high density biodiesel or its blends can leads to incomplete combustion and particulate matter emission .The results, therefore, indicate that biodiesel produced from NaOH-ethanolysis of Lageneria sinceraria seed oil may undergo complete combustion and produce less particulate matter. The lower density of ethyl biodiesel produced in this research may be due to its degree of unsaturation or chain length [31, 32].

3.2.3. Acid Number

The acid number is an important fuel property; it is a measure of free fatty acids in a given product. [28], Acid number of the biodiesel produced from Lageneria sinceraria seed oil indicates that it contains higher fatty acid composition of 0.44 mgKOH/g but is within the acceptable range (0 -0.50 max) specified by the ASTM [27]. This result indicates that ethyl ester biodiesel produced may not cause severe corrosion in internal combustion engine and fuel system [28].

3.2.4. Flash Point

Flash point is another important fuel property and is the temperature at which fuel will ignite when exposed to flame. Biodiesel usually has high (more than 150°C) flash point than petroleum diesel (55-66°C) [28]. Flash point is not directly related to engine performance but is inversely related to fuel volatility [30], it is controlled to meet safety requirement for fuel handling, storage and transportation[33]. The 144 °C value obtained is in agreement with ASTM biodiesel standard (130 °C min) [34] and result revealed that the biodiesel produced contain low volatile impurities (majorly ethanol) [32] and can be useful in arctic region [27].

3.2.5. Ultimate Analysis

Table 2 gives the ultimate analysis of Lageneria sinceraria seed oil ethyl ester biodiesel, the results show low sulphur content of 0.2% which may cause low sulphur oxides emission during combustion [28], high nitrogen content of 4.4% which may lead to the formation of NOx during fuel combustion [35], and high oxygen content of 12.3% which may reduce the particulate emission during fuel combustion. The heating value is an important parameter in the selection of fuel [28] and estimation fuel consumption, and determines the energy content of fuel. The greater heating value (HHV) of a fuel, the lower the fuel consumption and vice versa[36,37]. The produced ethyl ester biodiesel of Lagenaria siceraria seed oil shows slightly higher energy content of 36.34 MJ/Kg as compared to the 35.56 MJ/Kg of Pongamia pinnata (Karanja) biodiesel [38]. Usually biodiesel calorific value is lower than that of conventional diesel due to its high oxygen content [30]

4. Conclusion

The ethyl ester biodiesel produced from Lageneria sinceraria seed oil has acceptable properties approved by the American Society of Testing Materials (ASTM) and National Biodiesel Board (NBB) with high energy content. The high nitrogen content balanced with high oxygen content may reduce the particulate emission during fuel combustion while high acid value may be reduced by the use of acid pre-treatment such as neutralization before oil transesterification. The use of ethyl biodiesel from Lageneria sinceraria seed oil as a transport fuel in the future will, therefore, drastically reduce the overwhelming energy crisis and carbon dioxide emissions in both developed and developing countries of the world.

References

[1]   A.B.M.S. Hossain and A. Salleh, Biodiesel fuel production from Algae as renewable energy, American Journal of Biochemistry and Biotechnology, 4 (3): 250-254, (2008).

[2]   B. R. Dhar and K. Kirtania, Excess methanol recovery in biodiesel production process using a distillation column: A simulation study, Chemical Engineering Research Bulletin, 13; Pp. 55-60, (2009).

[3]   Y.C. L. Dennis, X. Wu and M.K.H. Leung, a review on biodiesel production using catalyzed transesterification, Applied Energy, 87: Pp. 1083–1095, (2010).

[4]   D.Y.C., Leung, X, Wu and M.K.H., Leung, A review on biodiesel production using catalyzed Transesterification, Applied Energy, 87:Pp. 1083–1095, (2010).

[5]   R. Luque, L. Herrero-Davila, J. M. Campelo, J. H. Clark, J. M. Hidalgo, D. Luna, J. M. Marinasa and A. A. Romero, Biofuels: a technological perspective, Energy & Environmental Science, 1(5); Pp. 513–596, (2008).

[6]   B. Metz, O. R. Davidson, P. R. Bosch, R. Dave and L. A. Meyer, IPCC, Climate change 2007: Mitigation, Contribution of working group 3 to the Fourth Assessment Report of the Intergovernmental Panel on Climate Change,Cambridge University Press, Cambridge, United Kingdom and New York, USA, (2007).

[7]   A.P. S. Chouhanand A.K., Sarma, Modern heterogeneous catalysts for biodiesel production:A comprehensive review, Renewable and Sustainable Energy Reviews, 15: Pp. 4378 – 4399, (2011).

[8]   W. D. Huang and Y. H. P. Zhang, Analysis of biofuels production from sugar based on three criteria: Thermodynamics, bioenergetics, and product separation, *Energy& Environmental Science*,4:Pp.784-792, (2011).

[9]   N.N.A.N. Yusuf, S.K. Kamarudinand, Z. Yaakub, Overview on the current trends in biodiesel production, *Energy Conversion and Management*,52 , Pp. 2741–2751, (2011).

[10] Y. Zhang, M.A. Du, D.D. McLean and M. Kates, Review Paper; Biodiesel production from waste cooking oil: Process design and technological assessment, *Bioresource Technology*, 89 ,1–16, , (2003).

[11] R. Wang, W. W, Zhou, M. A. Hann, Y. P. Zhang, P. S. Bhadury, Y. Wang, B. A. Song and S. Yang, Biodiesel preparation, optimization, and fuel properties from non-edible feedstock, *Daturastramonium* L. Fuel, 91, Pp.182–186, (2012).

[12] O. S. Valente, V. M. D. Pasa, C. R. P. Belchior and J. R. Sodré, Physical–chemical properties of waste cooking oil biodiesel and castor oil biodiesel blends, *Fuel*, 90, Pp. 1700–1702, (2011).

[13] G. Hincapié, F. Mondragón and D., López, Conventional and in situ transesterification of castor seed oil for biodiesel production, *Fuel* 90, Pp. 1618–1623, (2011).

[14] H.V. Lee , J.C. Juan a Y.H. Taufiq-Yap,(2015), Preparation and application of binary acid–base CaO–La$_2$O$_3$ catalyst for biodiesel production, Renewable Energy, 74:Pp. 124–132(2015)

[15] M. Balat, Potential alternatives to edible oils for biodiesel production – A review of current work,, *Energy Conversion and Management*, 52, Pp. 1479–1492, (2011).

[16] B. Garbaand A.M. Bashir, Managing energy in Nigeria: Study on energy consumption pattern in selected Rural Areas in Sokoto State. *Nigeria Journal of Renewable Energy*, 10. (1&2):97-107, (2002).

[17] B. Garba and U. P. Ojukwu, Biodegradation of water Hyacinth as an alternative source of fuel: A review. *Journal of Renewable Energy*, 6 (1&2): Pp.12-15, (1999).

[18] N. K. Sahoo, A. K. Satyawati and S.N. Naik, Interaction of *Jatropha curcas* plantation with ecosystem: proceedings of international conference on energy and   Enviroment: Pp. 19-21, (2009).

[19] T. Whittington, Biodiesel Production and use by farmers: is it worth considering? Department of agriculture and food, government of western Australia, available at: http://www.bebioenergy.com/documents/Onfarmbiodieselprod.pdf accessed on 10/03/2013, (2006).

[20] B. R. Moser, and S. F. Vaughn, Efficacy of fatty acid profile as a tool for screening feedstocks for biodiesel production, B*iomass and Bioenergy*, 3 7, Pp.31-41, (2 0 1 2).

[21] A. Anastasov, Biodiesel-Basic characteristics, technology and perspectives, Biotecnol&Biotechnol Anniversary scientific conference, available online at www.diagnosisp.com/dp/journals/view_pdf.php?...id..., *Accessed on* 14/03/2013, (2009).

[22] O. Olaofe, H.N Ogungbenle, B.E Akhadelor, A.O Idris, O.V , O.T Omotehinse and O.A Ogunbodede, Physico chemical and fatty acids composition of oils from some legume seeds, International Journal of Biological, Pharmacy and Allied Science, 1(3): 355-363,(2012)

[23] N. A., Sani , L. G., Hassan, S. M., Dangoggo , M. J., Ladan, I. Ali-baba and K.J. Umar, Effect of Fermentation on the Nutritional and Antinutritional Composition of Lagenaria Siceraria Seeds, Journal of Applied Chemistry,   5(2): Pp. 01-06 , (2013).

[24] S. Puhan, N. Vedaraman, V.B, Boppana, G. Ram, Sankarnarayanan and Jeychendran, Mahua Oil (MadhucaIndica Seed Oil) methyl ester as biodiesel: preparation and emission characteristics. Biomass and Bioenergy, 28:87 – 93,(2005).

[25] S. S. Rahayu and A. Mindaryani, Methanolysis of Coconut Oil: the kinetic of heterogeneous reaction, Proceedings of the world congress on engineering and c omputer science 2009, Vol I, WCECS 2009,   October 20-22, 2009, San Francisco,USA, available online at;http://www.iaeng.org/publication/WCECS2009/WCECS2009_pp134-138.pdf, *Accessed on* 14/03/2013, (2009).

[26] Mukhtar, M. Dangoggo, S.M. and Ross,A.B. (2012),Low Temperature/Pressure Hydrothermal Microwave as a Potential Alternative Method of Processing Microalgae, Proceeding of the 35th Chemical Society of Nigeria, Annual and International Conference, 1, Pp. 510-515.

[27] J.V. Gerpen, B. Chanks, R. Pruszo, D. Clements and G. Knoth, Biodiesel Analytical Methods Subcontractor Report, National Renewable Energy Laboratory. August, 2002-January, 2004, NREL/SR-510-36240, available at; http://www.bentlybiofuels.com/pdfs/NREL_BD_Analytical.pdf accessed on 16/01/2012, (2004).

[28] A. E Atabani, A. S. Siltonga, H. C Ong, T.M.I. Mahila and H.H. Masjuki, Non-edible vegetable oils: a critical evaluation of oil extraction, fatty acid compositions, biodiesel production, characteristics, engine performance and emissions production, *renewable and sustainable energy review*, 18; Pp. 211—245, (2013)

[29] S.K, Hoekmana, A, Broch. C., Robbins, E., Ceniceros, M., Natarajan, Review of biodiesel composition, properties and specifications, *Renewable and sustainable Energy Reviews*, 16. P. 143-169, (2012)

[30] L.F. Ramirez-Verduzeo, J.E. Rodriguez-Rodriguez, A.R. Jaramillo-Jacob, Predicting cetene number, kinematic viscosity, density and higher heating value of biodiesel from its fatty acid methyl ester composition, *Fuel,* 91, P. 102-111, (2012)

[31] A. Javidialesaadi and S. Raeissi, Biodiesel production from higher free fatty acid content oils: experimental investigation of the pretreatment step, APCBEE procedia,5; Pp. 474-478, (2013)

[32] P. Sexena, S. Jawale and M. Joshipura, A review on prediction of properties of biodiesel and blends of biodiesel, *Procedia Engineering*, 51; Pp. 395-402, (2013)

[33] K. Sivaramakrishnan and P. Ravikumar, Determination of Cetane number of biodiesel and its influence on physical properties, ARPN Journal of Engineering and Applied Sciences,   7(2):Pp. 205-211,(2012)

[34] I.M. Atadashi, M.K. Aroua, A.R. Abdul Aziz and N.M.N. Sulaiman, (2011),Refining technologies for the purification of crude biodiesel, Applied Energy, 88 : 4239–4251

[35] Ross, A.B. and Biller, P. (2011), Potential yields and properties of oil from the hydrothermal liquefaction of microalgae with different biochemical content, Bioresource Technology, 102: Pp. 215–225.

[36] Knothe, G. (2008). "Designer" Biodiesel: Optimizing fatty ester composition to improve fuel properties. Ener. Fuels 22: 1358-1364.

[37] K. Sivaramakrishnan and P. Ravikumar, Determination of higher heating value of biodiesels, International Journal of Engineering Science and Technology, 3 (11):Pp. 7981-7987,(2011)

[38] S. A Karmee and A. Chadha, preparation of biodiesel from crude oil of Pongamia Pinnata, *Bioresource technology*, 96(13); Pp. 1425- 1429, (2005).

6

Investigations of mathematical models in solar collectors

Nuru Safarov, Gurban Axmedov, Sedreddin Axmedov

Department of Electronics, Telecommunications and Radio Engineering, Khazar University, Baku, Azerbaijan

Email address:

nsafarov@khazar.org (N. Safarov), exmedovqurban@rambler.ru (G. Axmedov)

Abstract: There has been prepared the calculations methods of heat productivity of sun collectors which enables to determine the expedient of maintenance of polymers in sun collectors in the temperature mode given for different region and climate conditions. Show that, thermo endurance - T_0 maybe using as characteristic of thermo endurance of optic materials. If heating flow, destruction temperature and internal surface temperature is measured during test, it is possible to determine value T_0 and other necessity characteristics. As a result of the taking test was lead to comparison evaluation of considered materials. Working range of heating flow and up level heating embark have been determined.

Keywords: Solar Collectors, Polymer, Heat Transfer, Dynamic Model

1. Introduction

Majority of authors researching polymer collectors demand the comparative analysis and quantity estimation among the construction versions according the heat productivity in different climate conditions. There has been necessity of wide usage of report dependency as the most important period of technical preparation of solar system on the basis of sun collectors with new construction which are created [1,2].

In the investigations held up to the present time, the calculation methods of heat-productivity in sun collectors had been prepared in the calculation machines. At that time, the program maintenance had been written in Pascal language. But nowadays with the integration of high-speed computers to the science there has been appeared the necessity of correction to such kind of calculations and program maintenance. So, the program maintenance worked out by us is considered for modern new-generated computers. This program maintenance enables to create both, non-heat capacity and single -element_mathematical methods. According to the statistical data of direct and scattered sun-rays, the program maintenance also enables to get the distribution of sun-ray pressure getting onto the surface of sun collector for per hour. Besides, it helps to get the heat-productivity for the non-capacity and single-element models and also determines the total heat loss factor, the average temperature of heat transit, the ratio of convective heat-transfer factor and the temperature of transparent cover [3,4].

On the basis of numbers for per hour, the average heat-productivity of ratio of useful work and efficiency of sun panels are determined.

Except the modeling of work in sun collectors in a real climate conditions and the basis of its construction parameters, this prepared method enables the numeric investigation of efficiency of heat-capacity to its heat-productivity. All these estimations are held in the cooling and heating regime.

Obtaining knowledge about solidity and heating endurance of materials is tremendous important during material selection for preparation optical details which work under intense heating.

Lens and filters of projectors, heating equipment's of optic covered are considered as optical details. There isn't this information for many new materials. The reason of this situation is not only unique view to methodology of defining category of thermo endurance but also specifying of the materials. Comparatively low thermo endurance and solidity, weaker resistance to extraction rather than pressing include this specification. By this reason, traditionally extraction methods (for example: for various temperature) complicate process identification of characteristics for embedding optic materials [5,6]. In this situation direct test thermo endurance of optic materials become more important.

As mentioned above, specification of optic materials during their testing highly require heating base and method of their transferring. As specific factor base stable equal heating pollute additive and contactless with foreign substances, ordinary deformation and other similar requirements are demanded. In this view, it presents huge interest testing of one side heating of free placing flatness layers. As a source of heating, concentrating radiation and also solar radiation are utilized.

2. Data Collection and Analysis

Let's look at such a sun collector made of polymer in which heat-absorbing panel is consisted of two boards which are joined by sides themselves with heat-transferring channels. Such kind of construction has been considered as a basis by some investigators.

Non-heat capacity models can be explained like below [5]:

$$\frac{Q_u}{A} = F'[I(\tau\alpha) - U_L(T_m - T_a)] = 2Gc_p(T_m - T_i)/A \quad (1)$$

Here is Q_u – valuable heating, Wt
$(\tau\alpha)$ – optical ratio of useful work
(Gc_p) – water equivalent of circulated liquid, joule/kg
A – surface space of the collector, m^2
U_L – the total heat lost factor, Wt/(m^{2}xK)
T_m, T_i, T_a – the average temperature of heat transfer of solar collector, the temperature in the entrance of solar collector, and the temperature of the air in the environment
F' – the efficiency of solar collector
For the construction above one can be written:

$$F' = \frac{1/U_L}{1/U_L + 1/\alpha_{ds} + l/\lambda_p} \quad (2)$$

Here is a α_{ds}- heat-transfer ratio of inner space of panel to liquid, Wt/(m^{2}xK)
λ_p- heat transfer of panel material, Wt/mK

Heat-lost is composed of radiation on surface, radiation and convective based parts and heat-transfer lost in the back side part.

$$U_L = (q_L^t + q_L^b)/(T_m - T_a) \quad (3)$$

Heat-lost in the back side will be like this:

$$q_L^b = \frac{\lambda_t}{l_t} \quad (4)$$

Here λ_t- is heat-transfer of heat-isolations, Wt/(mxK) and l_t- is are thickness of heat-isolations, m.
The heat-lost of the outer surface $- q_L^t$ are determined by solving the heat balance of the system equations of absorbing surface and a transparent shells:

$$\left(\begin{array}{c} q_L^t - \left[\frac{\sigma(T_m^4 - T_g^4)}{1/\varepsilon_p + 1/\varepsilon_g - 1}\right] - \tau_t\varepsilon_p\sigma(T_m^4 - T_a^4) - \alpha(T_m - T_g) = 0 \\ q_L^t - \sigma\varepsilon_g(T_g^4 - T_a^4) - \alpha_h(T_g - T_a) = 0 \end{array} \right) \quad (5)$$

Here is σ – Stefan-Boltzmann constant, Wt/(m^{2}xK^4)
T_g – the temperature of cover, K
τ_t – is a heat-releasing ability in infra-red diapason
ε_p, ε_g – is a radiation ability of panel and cover

α, α_h – heat convective coefficient between the panel and coating, between the coating and the environment panel, Wt/(m^{2}xK)
α and α_h - ratios, are calculated according the calculation dependence $U_L = (q_L^t + q_L^b)/(T_m - T_a)$.
Taking into account the temperature spreading along its length, the equation which explains the work of solar collector composes the basis of single element model.

$$c_a\frac{dT_m}{dt} = F'[I(\tau\alpha) - U_L(T_m - T_a)] - 2Gc_p(T_m - T_i)/A \quad (6)$$

c_a – it's an imaginary heat-capacity characterizing the whole collector, kJ/(kgxK).
c_a – quantity is defined with the correction of $U_L = (q_L^t + q_L^b)/(T_m - T_a)$ of the total heat lost ratio to the proportion of lost ratio out of the transferring cover to the environment.

In the numerical investigations of solar collectors there have been accepted two operating working modes. First mode coincides to 310 K (normal hot-water supply in the water reservoir), but the second mode to 330 K (hot-water supply in the mode of life). The reporting time was July. In that case there had been worked up Baku-Absheron climate and the initial data were taken since [7]. The thickness of panel walls was 0,001m, the thickness of an air lay was 0,04m, and the thickness of back-heat isolation was 0,03m.
Other quantities had been like this: $\tau\alpha$ – 0,85; τ_t – 0,2; λ_p – 0,2; λ_t – 0,04Wt/(mxK); ε_p – 0,95; ε_g – 0,88; α, α_h – 1500Wt/m^{2}xK; G – 0,083kg/s; c_a – 34400 kJ/(kgxK).
In this job, methods and some results of thermo endurance research are commented during heating optic materials in solar equipment's. Length of the equipment is based on parabolic concentrator which consists of 40 cm length and 30 cm focal distance organic glass. Thermocouple sample is located in focal zone of concentrator. Changing of heating parameters is implemented through breaking of sample focuses. This obtains with help of replacement of thermocouples sample by axis of mirror.
Sample which is provided with thermocouple and entire registration located on focal surface of concentrator. Replacement of heating parameters is obtained by breaking of focuses of sun ray which reflection on sample, because moving sample through mirror. Following concentrator of sun is implemented with help of azimuthally-zenithal photoelectric system which consist optic photo head, automat block and electric engine. Duration of experiment is registered via stopwatch.
Maximum capability of density of radiation on this equipment is 14KV/m^2. It is sufficient requirement for surplus practical cases. As sample for testing were using epoxy adhesive and optic glue materials. Elements are bordered by metal frames and provided complete and temperature sensors. Moreover, they registered separate moment and internal surface temperature of sample. Free settled sample surface happens with stretch-deformation

during non-straight heating which characterize as whole bending. Its internal surface has extensibility tensions which undesirable issue [7].

Spreading of energy on focal surface and through concentrator axis is learned with help of colorimeter which cooled with water, imitated absolutely black matter.

Dependence of raying density from normal solar radiation for diaphragm with 200mm diameter on focal surface is presented in figure 1. Changing of radiation density through concentrator axis is exhibited in figure 2. Theoretically, the concentrator's focal shadow is guessed to be approximately 10 mm but it is 50mm in fact because it geometrically shape is non-precise. That is why maximal radiation density has had 200kkal/m^2 which is sufficient meet requirements of some practical cases. Using of focus breaching is seemed as correct method in regulation heating parameters because flow of heating less changed through concentrator axis.

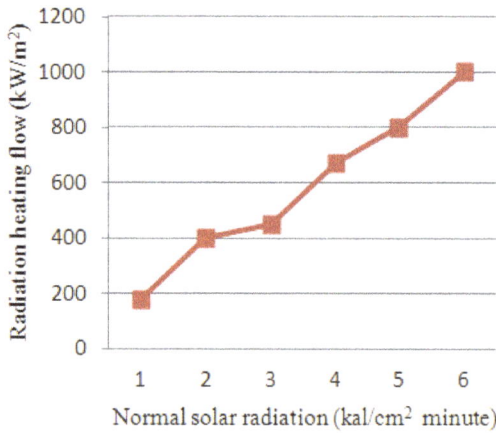

Figure 1. *Dependency of normal solar ray radiation from density of radiation on focal plane.*

Spreading of temperature on focal surface and its parallel surfaces per concrete case is implemented with help of thermo couple. Because used colorimeter type diaphragm is defined radiation density due to size. This measure is indicated that exceeding of temperature was not greater than 10% in whole researched diapason breaching of focusing in 40 mm shadow which is absolutely reasonable.

Figure 2. *Changing density of radiation energy through concentrator axis (q$_0$, ray – heating flow on focal plane)*

Circle samples have been tested which have been made by optic materials with 3-8 mm thickness and 30-50mm diameter. Layers made free settlements in metal frame. It provides with thermo couples and special completing registration which fix temperature of unheated inside layer and destruction moment of sample. Outside surface of sample was covered with swallow layer by 15 mkm thickness due to obstacle reflection radiation energy of sample. Obviously, during the period balanced heating of free settlement circle layer surface there happens tension-deformation case which is clear bend. This period happen pulling tensions in internal layer which cause serious danger.

It is presented that obvious accepting thermo-physic characteristic of material possible to define solid level from pulling due to heating flow of layer surface, temperature of internal surface and destruction period.

Indeed, thermo-flexibility theory of thin circle layer causes that given tensions, deformation of layer equal [8,9]:

$$\sigma = -\frac{\alpha E(t_U - t_i)}{1-\nu}, \varepsilon = \alpha t_U \qquad (7)$$

$$t_U = \frac{1}{h}\int_0^h t(x)dx, \qquad (8)$$

Here h – depth of layer

t(x) spread of temperature due to depth

t_i - temperature of unheated internal surface

E, α, ν – Yung module, linear enhancement coefficient and Poisson's ratio of layer material respectively.

It is not harsh to get expression for average integral temperature due to depth from thermal conductivity equation:

$$t_U = \frac{q\tau}{hc\rho}, q = \frac{1}{\tau}\int_0^\tau q(\tau)d\tau \qquad (9)$$

Here q-heating flow direction to internal surface; c, ρ-heating capacity and density of material respectively; τ-time.

If during some period tension or deformation reach to own limited value then get from (7) and (8).

$$\frac{\sigma_0(1-\nu)}{E} = T_{kr} - \alpha t_i, \varepsilon_0 = T_{kr} \qquad (10)$$

Here σ$_0$, ε$_0$ - solidity border to pulling top level deformation to pulling respectively. Temperature of the inner surface:

$$T_0 = \frac{\alpha q_0 \tau_0}{hc\rho} \qquad (11)$$

From first approach of (10) resulted that

$$(T_0 - t_i) / \alpha = \sigma_0 (1 - \nu) / E\alpha \qquad (12)$$

So, T_0 may be using as characteristic of thermo endurance of optic materials. Resulted from (10) that T_0 is function of temperature. However, (11) and (12) following result: If heating flow, destruction temperature and internal surface temperature is measured during test, it is possible to determine value T_0 and other necessity characteristics.

As a result of the taking test was lead to comparison evaluation of considered materials. Working range of heating flow and up level heating embark have been determined.

Test result of enumerated before materials test result presented in table. Test has been accepted as average value in the result of checking 2-4 times same sample in the same heating flow.

Table 1. Test results of samples.

Material		Epoxy adhesive		Optic glue	
g, Vt/sm²	Depth, mm	τ, sec	t_i, ⁰C	τ, sec	t_i, ⁰C
10	3	-	-	37	130
	5	-	-	42	145
20	3	23	285 130	54	145
	4	14	135	-	-
	5	-	-	15	60
25	8	9	70	-	-
30	3	6	120	-	-
	4	7	95	-	-
	6	4	50	-	-
	8	-	-	7	40

3. Results and Discussion

There had been given results of numeric modeling in figure 3. While according to the stationary -fixed-line model, the heat-productivity is higher than the dynamic one until the break point characterizing the transmitting of heating towards cooling the situation is changed back after the breaking point - the heat productivity in dynamic model becomes higher than stationary one.

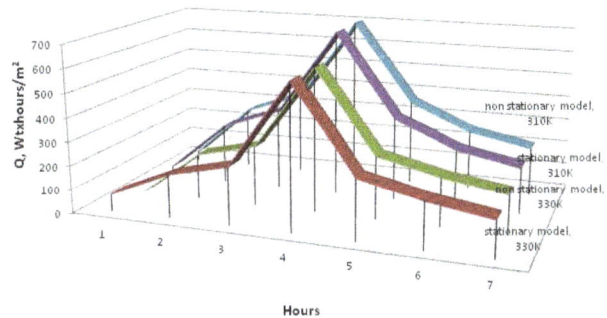

Figure 3. The results of numeric modeling for the solar collectors.

According the stationary and non-stationary -fixed-up model, there had been introduced the density of the total solar radiation onto the surface of the sun collector, heating and cooling intervals of the solar collector and heating productivity in the time-table below.

Table 2. The characteristic parameters for collectors in various modes.

Region	Mode T=310K			Mode T=330K			I, Wtxhours/m²
	Time, hours	Qn, Wtxhours/m²	Qnst, Wtxhours/m²	Time, hours	Qn, Wtxhours/m²	Qnst, Wtxhours/m²	
Baku-Absheron	7-13	2900	3020	8-13	1960	2075	
	14-18	1840	1735	14-17	1100	1034	6700
		4740	4755		3060	3109	

It can be seen from results of tests that optic ceramic is the best endurance to heating. It is important to note that curves which is showed in figure 1 allow to choose materials for develop of some thin wall details which work under one side intense heating (layers, semi spheres and cylinders). If T_0 which is calculated during concrete period for known heating flow during heating of internal detail's surface is placed on or above appropriate curve or then this material are not useful.

Some results of sample testing in 20W/cm² heating flow which is mentioned below:

- epoxy adhesive, depth is 3mm – 130 ⁰C (test period τ = 23sec)
- optic glue, depth 3mm – 145 ⁰C (test period τ = 54sec)

Consequently, T_0 would be used as thermo endurance characteristics of optic materials. If heating flow, splitting period and temperature of in internal surface is measured, and then T_0 indicator would be determined too. As a result of researches compared measures of thermo endurance for tested materials are found. Working diapason of heating has been determined.

4. Conclusion

Analyzing the received results, one can come to a conclusion that the working time-useful heat energy productivity is 9 to 11 hours. Ratio of useful work of collector for 24 hours is approximately 45%.

So, this prepared method enables to compare the mathematical model of static and non-element models of the solar collector. This method also determines the availability of supplying the solar collectors in the given temperature modes for different climate zones and regions.

References

[1]   Mecit A.M., Miller F.J., Whitmore A. Optical analysis and thermal modeling of a window for a small particle solar receiver. Energy Procedia 00 (2013) 000-000

[2]   Mammadov F. Study of Selective Surface of Solar Heat Receiver. International Journal of Energy Engineering, 2012; 2(4): 138-144

[3]   Steinfeld A., Meier A. Solar Fuels and Materials. Encyclopedia of Energy, Volume 5. 2004, 623-637

[4]   Hossain M.S., Saidur R., Fayaz H., Rahim N.A., Islam M.R., Ahamed J.U., Rahman M.M.. Review on solar water heater collector and thermal energy performance of circulating pipe. Renewable and Sustainable Energy Reviews 15 (2011) 3801–3812

[5]   Smirnov S.B., Moyseenko V.V. Numerical study of mathematical models of the collector on the example of the polymer structure. Geliotechnika, 1993, №1, p. 40-43 (in russian)

[6]   Popel O.S., Frid S.E., Sheqlov V.N., Suleymanov M.Dj., Kolomoiech Yu.Q., Prokopchenko I.V. Comparative analysis of construction of solar collectors foreign and domestic production. New technical solutions // Thermal Engineering. no.3. 2006. (in russian)

[7]   Strebkov D.S., Teveryanovich E.V., Tyukov I.I., Irodionov A.E., Yartsev N.V. Solar concentrator technologies for power supply to building. Applied solar energy, vol.38, no. 3, pp. 61-65, 2002.

[8]   Dvernyakov V.S., Zaxarov P.A., Lazarev A.I. Pasichniy V.V. , Smirnaya E.P., Franchevich I.N. Use of radiant heating in solar plants for experimental study of thermal stability of optical materials. Geliotechnika, 1975, no.1, p. 32-37 (in russian)

[9]   Bayramov A., Safarov N., Akhmedov G., Shukurova V.. Thermophotovoltaic converters on the basis AVBVI for high concentration solar applicantions. International scientific journal for alternative energy and ecology. vol.5, no. 37, pp. 85-87, 2006.

7

Survey of airflow around multiple buildings

Ahmed A. Rizk[1,2], Gregor P. Henze[3]

[1]Architectural Engineering Department, Faculty of Engineering, University of Tanta, Egypt
[2]Visiting Professor of Architectural Engineering, University of Nebraska – Lincoln, South 67th Street, Peter Kiewit Institute, Omaha, NE
[3]University of Colorado at Boulder, CEAE Department • 428 UCB, Boulder, Colorado 80309-0428 U.S.A

Email address:

rizk2003@yahoo.com(A. A. Rizk), Gregor.Henze@Colorado.EDU(G. P. Henze)

Abstract: This survey paper offers a review of past and present studies related to air flow around multiple building configurations to achieve energy savings and thermal comfort in hot climate regions in the presence of increased urbanization. The purpose of this review paper is to provide guidelines based on previous studies for the successful design of group housing in a hot and arid climate such as Egypt to improve air flow around multiple rows of buildings. This study presents several types of courtyard designs inside houses that provide direct air flow at windward sides inside compactly planned buildings. Next, air flow is described around one individual building and two buildings that have a passage between them. Furthermore, air flow around buildings is discussed to include several rows of buildings that range from rectangular linear to square shapes with flat and jack roofs, where the main goal is to achieve appropriate wind velocity at inlet surfaces, especially at the second and the third rows of buildings, and to avoid turbulence zones caused by wind around building. Finally, the effect of topography and urban mass on global wind velocity at the city scale is discussed.

Keywords: Hot Climate, Courtyards, and Air Flow due to Wind Pressure around Buildings, Thermal Comfort with Natural Ventilation

1. Introduction

This paper focuses on the previous study that related with wind flow around the different arrangements of buildings. Linear buildings are the main subject of this survey. Configuration of linear buildings forms between buildings around inner-courtyards and rows of multiple buildings. Unfavorable zones due to flow pattern performance are formed as a result of wind faced buildings. This survey study presents the previous solutions that can solve these zones. Aerodynamics shapes of buildings, chess –board arrangements of building rows, and jack roofs can decrease the effects of these zones. Many other solutions as open ground floor to decrease the unfavorable effect of stagnation zones in the wind-ward side of the buildings are not discussed in this review article. Also, the effect of the wind velocity due to Beaufourt scale is not discussed in this paper. Strong wind is completely different than weak wind in air flow around buildings.

1.1. Wind Flow Problem

The article can describe the main problem of air flow around multiple buildings due to wind direction. Multiple buildings have a major problem related to air flow. First row of multiple rows of buildings that faced wind has maximum amount of wind. Next rows without carefully studying the fundamental of wind flow manners with buildings may have minimum amount of wind. Unfavorable wind movement due to wind direction is at inlet openings. A major problem is caused by the first row. Vortex and recirculation movement are a major problem of the next rows of multiple buildings. The methodology to solve the problem is presenting the climatic problem and how can the natural ventilation solve this climatic problem physically. Also, the relationship between air flow and natural ventilation is discussing in this article. The previous studies that deal with buildings around inner-courtyards, row of two buildings, multiple rows of buildings , and jack roofs are analyzed to solve the air flow problem that can affected directly in natural ventilation inside the buildings. Seriously, air flow around building is the main basic of natural ventilation that can achieve thermal comfort inside the buildings in the hot climate.

1.2. Climatic Problem Description

High temperatures and intensive solar radiation in hot arid climates require compact planning of houses with minimum external surface area, small windows that face the wind direction, with most windows facing internal courtyards rather the exterior of the group and finally small courtyards, according to a U.S. military handbook [1]. To temper the effects of these harsh environmental conditions, building design often maximizes the use of compact planning, which reduces and delays the effect of external heat gains [2] due to mutual shading. On the other hand, this compact planning of buildings leads to a significant reduction of available wind velocity for natural ventilation. In addition, urbanization has increased drastically and it is expected that there will be twenty-two mega cities by 2015. Urban areas are warmer than surrounding rural areas due to the urban heat island effect where bare soils and vegetated surface are replaced by concrete structures and pavements which absorb and store more heat from the sun than the original surfaces. Urban heat island effect is further enhanced by exhaust heat associated with human activities such as air conditioning and manufacturing [3].

1.3. Physical Problem Solution

Indoor air velocity in a hot climate acts to cool the occupants by two ways. First, it cools the occupant directly by increasing the convective and evaporative heat transfer from the body surface. Second, it cools the occupant indirectly by removing heat stored in the building [4].

A study on natural ventilation in Thailand [5] revealed that the permissible zone temperature of thermal comfort can be extended to 29°C, 30°C, 31°C for indoor air velocities of 0.2 m/s, 0.4 m/s, 1 m/s instead of air temperatures that range from 21°C to 26°C as suggested by ASHRAE Standard 55-2004. The same study further observed that thermal comfort can be achieved inside houses in a Bangkok suburb during 20% of the year by providing houses with indoor air velocity of only 0.4 m/s. Although this is not a large number, it is worth pursuing, because natural ventilation has very low operating cost in the presence of acceptable indoor air velocities.

Another study on the effect of indoor air velocity on thermal sensation [6] observed that air movement reduced discomfort from heat above 31°C and can achieve a predicted mean vote of PMV=+1 on the sensation scale and for temperatures above 40°C thermal sensations between +2 to +3 can be achieved for velocities over 0.25 m/s.

But in the case of excessive temperatures (≥31°C), air velocities cannot achieve optimum thermal comfort for two reasons: The first reason is related to the occurrence of draft discomfort for high air velocities above 1.5 m/s. The second reason is related to the sensation scale, this scale indicates that completely thermal comfort cannot be achieved at high temperature over 35°C even with higher indoor air velocity as proposed by ASHRAE Standard 55 [7] . Fig. 1 shows the relationship between increasing air

velocity and decreasing sensation scale at the high temperature and also shows the limitation of indoor air velocity (peak point) 2 m/s than portable wind velocity 4 m/s.

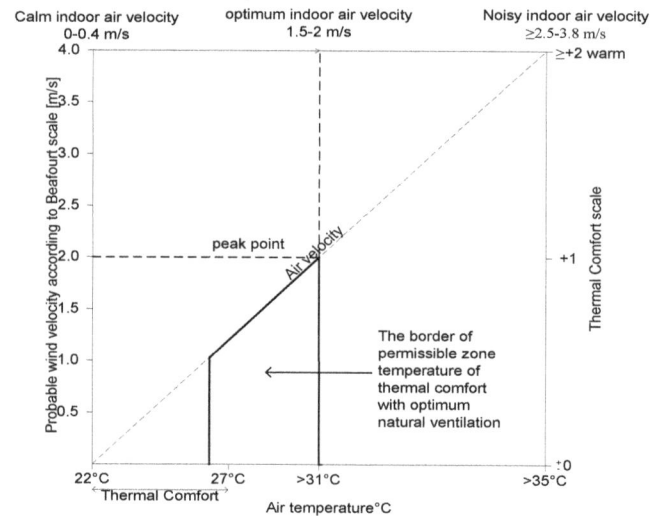

Figure 1: *Relationship between thermal comfort scale and air velocity from [6].*

A U.S. utility company has indicated that each 1°F increase in summer cooling temperature set point will save 3–4 % of air conditioning and has shown also that mechanical air-conditioning of residences costs 11 times more than natural ventilation due to wind pressure [8]. Fig. 2 shows the approximate relationship between both HVAC and natural ventilation with cooling energy cost. The cost increases if using HVAC at high temperature and decrease if using natural ventilation at the same temperature.

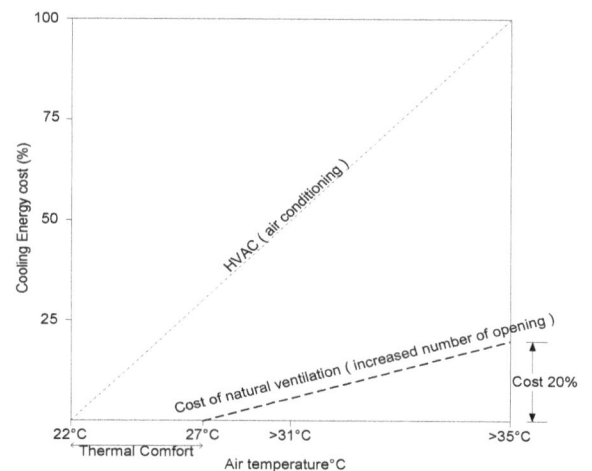

Figure 2: *Relationship between both HVAC and natural air movement with cooling energy cost from [7] and [8].*

1.4. The Relationship between Air Flow around Buildings and Cross Ventilation from ASHREA handbook Fundamentals, [22].

Ventilation depends on the different pressure between intake surface that is at windward side of building and exhaust surface that is at leeward side of building. Recirculation and vortex or stream pattern flows can affect non-preferably on intake and exhaust surfaces of building. Beside air flow patterns around building, climatic conditions, such as wind velocity and direction, and building dimensions can affect also on cross ventilation or infiltration rate inside building.

Equation no -1 shows that the relationship between air flow around buildings and natural ventilation.

$$P_s = C_p P_v$$

P_s = Different pressure between the pressure in the two surfaces of building.

C_p = Local wind pressure coefficient that is according to patterns of air flow around buildings, terrain category,

climatic condition, and building dimensions.

P_v = Wind velocity pressure that is according to The height of building according to the ground level, air density, and gravitational constant.

2. Courtyards

The first approach to improving airflow due to wind pressure is to increase the inlet surface areas on the windward side. The first surface is at the external border of the building that faces wind directly. The second surface is at the internal border of the building that lays on the courtyard and faces the wind indirectly.

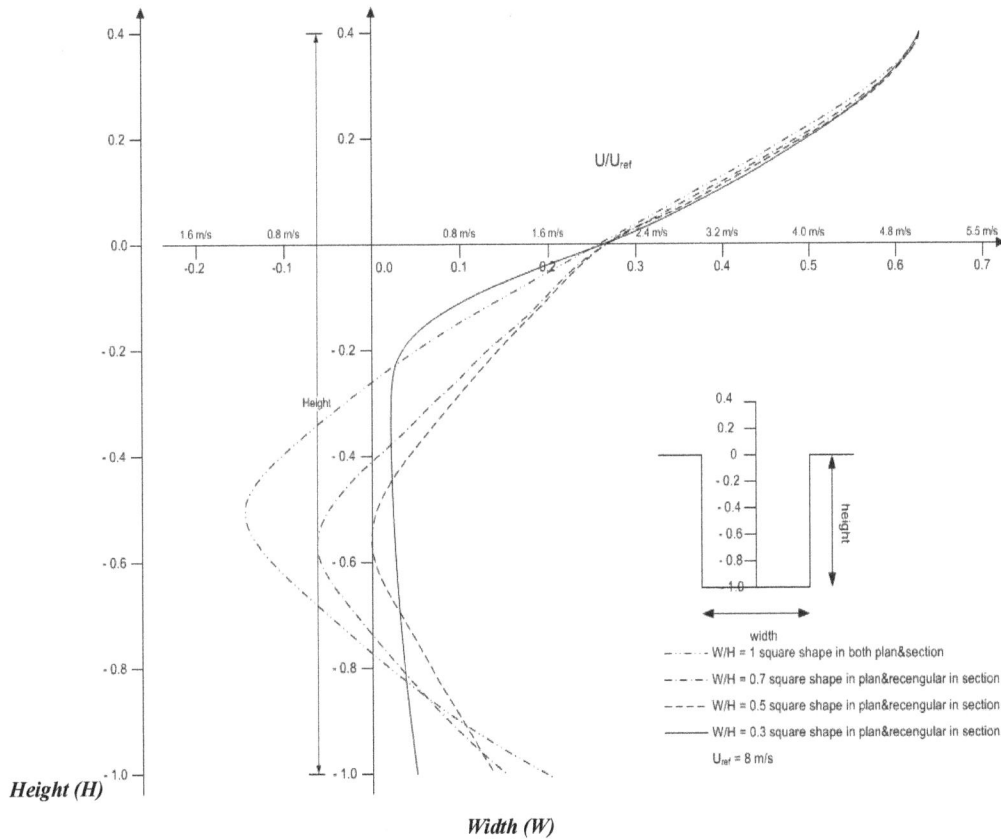

Figure 3: Comparison of U/U ref for four ratios of width/height 1, 0.7, 0.5, and 0.3 for a courtyard from [10].

Naroyan [9] presents an Indian case study in Jaipur City in which multiple elements were integrated (wind tower, courtyard, and chimney) to create natural ventilation inside houses at high ambient temperatures of over 40°C. The integrated courtyard can achieves a fraction of 20% (1.2-2 m/s) of the wind velocity (6-8 m/s) and can reduce the outside air temperature[1] by 20% of the Celsius scale value [9].

A study of an old city center in Havana, Cuba investigates the effect of four ratios of courtyard width to

height: 0.3, 0.5, 0.7, and 1, which achieve air velocity fractions of +4%, zero air velocity, -5%, and -15% of the wind velocity (U_{ref}) at the midpoint of both width and height of courtyard[2]. Fig. 3 shows the four ratios of width to height according to a reference velocity of U_{ref}=8 m/s at over ten meters height [10].

A Chinese study of a courtyard in group of buildings in Beijing suggests a new geometric dimension of courtyard characterized by a small width that faces the wind direction and long length that parallels with wind direction. The

---

[1] This high value of reducing external air temperature is due to building thermal mass which reduces and delays the effect of external heat gains beside natural ventilation.

[2] Although, it is not clear that the represented negative wind velocity due to changed direction or recirculation inside courtyard at the middle height of building. In any case this is not a favorable condition of using courtyard in high wind velocity.

courtyard behaves like an air tunnel to increase low wind velocities in Beijing. The courtyard is used in six-story buildings (20 meters). The six story buildings are in the south and tall buildings (40 meters) are in the north because of the orientation of the prevailing N-S wind in the summer in Beijing. It is observed that in the northern wind case the air velocity ranges from 0.1-0.4 m/s, which is too low to support natural ventilation of the apartments. This reduction of air velocity occurs because the six-story buildings are in a recirculation zone, created by the tall building to the north. When wind comes from the south, the air velocity around the six-story buildings was higher but still low, between 0.3-0.8 m/s because of the small distance between one building to another at the south that deflects wind before it reaches the next one [11]. Fig. 4 shows the layout of two buildings that includes the longer courtyard to increase wind velocity and thus to ventilate the large area as much as possible for these units of six-story buildings.

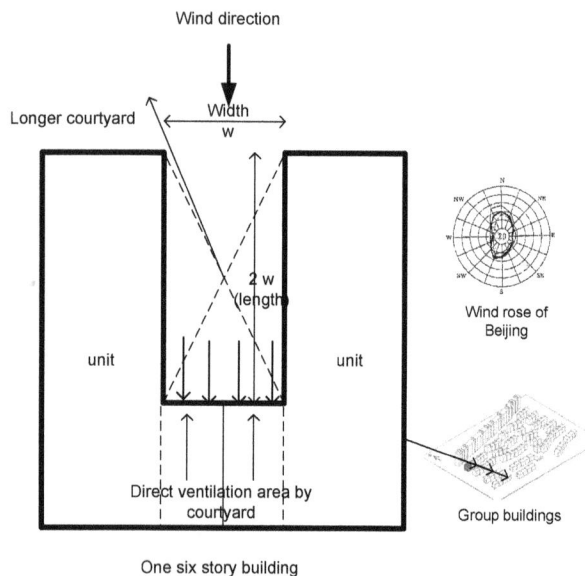

Figure 4: Longer courtyard of two units of six-story building from [11].

3. Air Flow around Multiple Building Configurations

3.1. Air Flow around Individual Rectangular Buildings

To improve air flow around multiple buildings, the different zones around building that occur by wind velocity or wind pressure need to be analyzed. The main turbulence zone that occurs behind the building on the leeward side of the building is called wind vortex or wake. The length of this zone depends on the dimensions of building. The secondary turbulence zones that occur in front of building are at the ground level, roof, and two sides [12].

There are four main pressure zones around a building: For the first zone in front of the building, the maximum positive pressure is at the upper level of the building at approximately 70% of the building height, the negative

pressure is at approximately 30% of the building height from the ground level; between the positive and negative pressures, there is a point of zero velocity and pressure. For the second zone in the back of the building, the maximum negative pressure is at both lower level of building and near the back of the building; the negative pressure decreases and the positive pressure begins to increase with increasing distances from the ground. For the third zone at the roofline of the building, the maximum negative pressure is towards the wind direction, with decreasing negative pressure along the wind direction. The final and fourth zone is at the two corners or sides of the building, standing vortices occur near these corners [13][3]. Figure 5 shows the different pressure zones around the building in a section view.

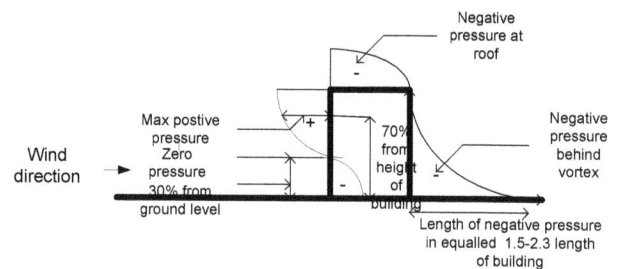

Figure 5: Pressure zones around building from [13]

Table (1) shows the patterns of air flow around building according to ASHREA handbook fundamentals, [22], the pressure zones consists of two main zones. The first zone is at in front of building. The second zone is at behind the building.

Table (1) The pressure zones that caused by wind

Pressure zones	Zones according to building	Patterns	Positions according to ground level
Pressure zones according to wind and building.	In front of building	Upwind vortex	At the lower level.
		Stagnation zones	At the middle level.
		Stream lines	At the roof of the building.
		Unaffected zones	Over the roof of the building.
	Behind building	Recirculation zones	At the lower and middle level.
		Contaminated zones	At the higher level.

Further studies concentrate on aerodynamic shapes such as vessels and wheels shapes. These suggested shapes can decrease all turbulence zones from 50% to 70% because of a reduction of the area that directly faces the wind direction through the use of rounded shape [14], [15]. Figure 6 shows the effect of aerodynamics shapes on wind flow.

---

3 At lower level of windward side of building, recirculation may occur.

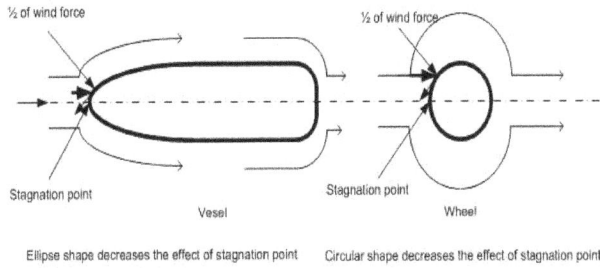

Figure 6: The effect of aerodynamics shapes from [14] and [15].

3.2. Air Flow around Two Buildings

Using the passage between two parallel buildings, similar to a Venturi nozzle, is a solution to improve wind flow on the windward side of buildings, yet two separate standing vortices in front and behind of buildings can be observed, the back vortices are three times as long as the long side that faces the wind[4]; these vortices increase when the passage between buildings decreases and two separate corner streams can be observed also. This result is concluded from two papers, the first one is by Baskarans & Kashef [12] and the second one is by Blocken & Carmeliet [13]. Fig. 7 shows the turbulence zones that occur at different distances of passage.

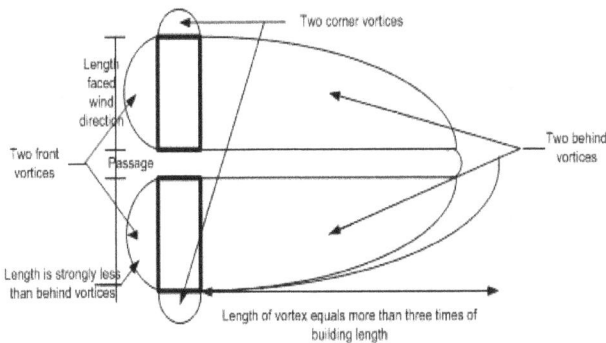

Figure 7 The effect of narrow passage on increasing turbulence zones from [12].

3.3. Air Flow around Multiple Building Rows

The first solution to improve wind flow at windward side of building rows is to provide a narrow and long passage between building rows. It is observed that air velocity increases in the passages between buildings and can achieve 60-150% of the wind velocity. Air velocity at the middle distance of passages between buildings is higher than the other position of inlet surfaces at wind ward side. It is also observed that the air velocity is higher at the first row compared to the second row and similarly for the second and third rows. Fig. 8 shows the effect of narrow passage on increasing wind velocity at different rows [12].

The second solution to improve wind flow at inlet surfaces, i.e., the windward side of a building starting from second row up to seven row is to increase the distance in both

directions x and y. Complex flow pattern may occur with multiple recirculation wakes within the gap regions and channel effect near the inlet plane of the block system [16].

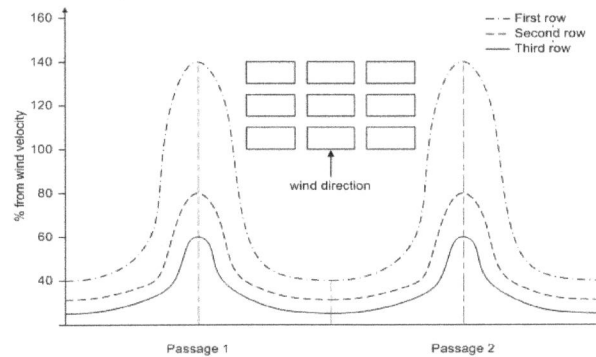

Figure 8: Wind velocity at both inlet surfaces and passages for three rows from [12].

The third solution to improve air flow at inlet surface in the far building row is the chess board arrangement. To improve wind flow in the last row, the solution increases the distances between buildings to achieve two times the length of the building in direction y and one time the length of the building in direction x at a rectangular shape. In a square shape, the solution keeps the distance between buildings to achieve two times of the length of the building in the cavity in direction y but decreases passage in the direction x to less than half of the length of the building. Both square and rectangular shapes arranged in a chess board pattern increase wind velocity. It can be observed that both shapes achieve a suitable wind velocity starting from second row, but rectangular buildings have achieved better natural ventilation than square buildings, although square buildings achieve higher wind velocity than rectangular because of the narrow passage between square buildings [5]. Fig. 9 shows the increasing distance between buildings in the chess board system to arrange buildings that can improve the wind flow at the second and subsequent rows.

The fourth solution concentrates on the effect of the jack roof inside the courtyard between two units in different wind directions. Therefore, this experiment presents two types of roofs, the first one is a jack roof and the second one is a flat roof. This experiment tests the effect of the two types of roofs inside a courtyard. It can be observed that the positive pressure increases more strongly in jack roof arrangement than the flat roof arrangement[56] as shown in Fig. 10 [4].

---

[4] The length of behind vortex depends on the height and the width beside the length of building.

[5] Pressure coefficient depends on many factors such as: 1) pressure of front or back surfaces according to wind direction that this pressure depends on the wind velocity and the surface area that faced wind (constant); 2) wind direction (θ); 3) Height of courtyard (Hc); 4) Courtyard spacing (Sc). Because the coefficient of pressure depends on variety of factors, it has been represented in tabular form for use by designers.

[6] The main aim of increasing positive pressure is to improve air flow at inlet sides of buildings that lay on inner courtyard, because natural ventilation depends on positive at inlet sides of buildings and negative pressure at outlet sides of buildings. In low wind velocity and hot climate prefer increasing air flow inside buildings that depends on the difference between positive and negative pressure.

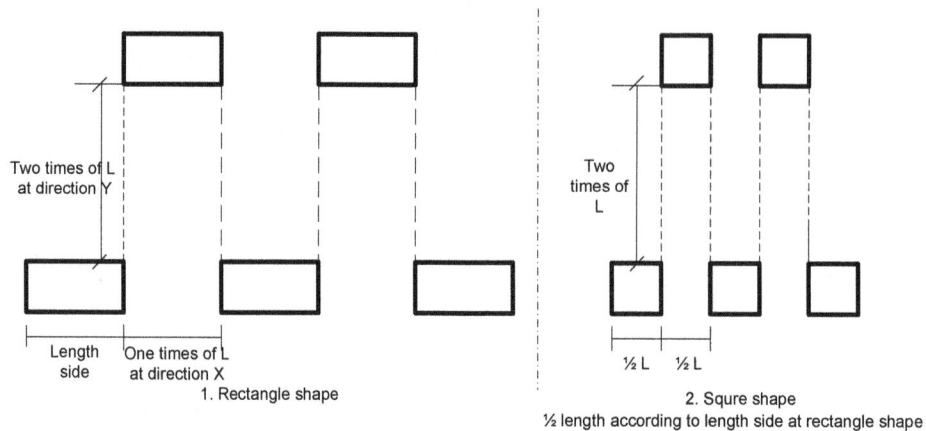

Figure 9: Groups of buildings shows increasing distance, narrow passage, and chess board arrangement from [5].

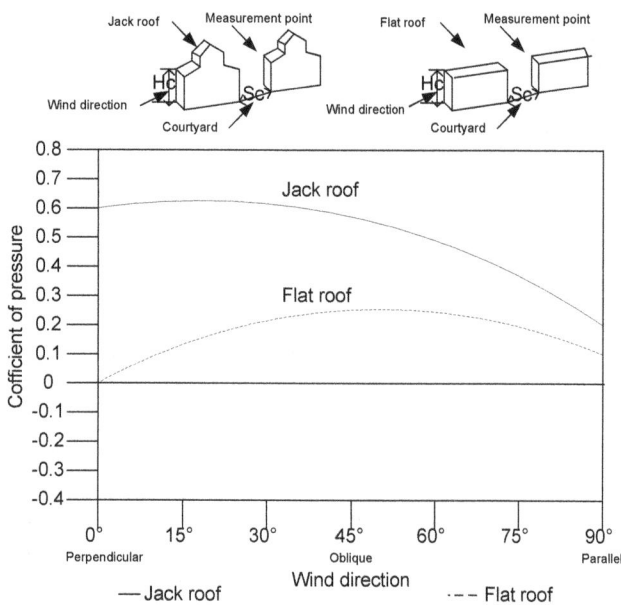

Figure 10: Comparison between jack and flat roof on coefficient of positive pressure [4].

4. Airflow at Urban Area Planning Scale

To improve airflow due to wind pressure at inlet surfaces, i.e. windward sides of buildings and create sufficient positive pressure at these surfaces, the initial prevailing wind with its velocity needs to be maintained and turbulence zones need to be avoided before blowing to the urban area. The main effects on wind are topography and man-made structures, i.e., buildings.

4.1. Topographic Effects

At times, mountains have unfavorable effects on the prevailing wind especially when mountains faced the wind direction, in spite of the fact that the wind velocity increases two times on the top of mountains. It can be observed that the reduction of wind velocity behind

mountains is up to 50% of the initial value. This reduction was observed in the urban area of Seoul according to the Korean experiments [17]. Mountains also cause turbulence zones, i.e., vortices in the zones behind the mountains relative to the wind direction. The large vortices by mountains were shown to affect the urban area in South Korea [18].

Another effect of mountains is that they cause stagnation, i.e., complete elimination of wind velocity and dispersion of wind outside the urban area. These complete reductions of wind velocity were observed at urban area in Salt Lake City [7] [19]. Fig. 11 shows the unfavorable effect of mountains on wind flow at the urban scale.

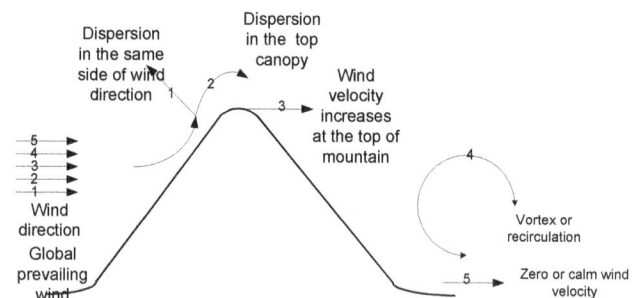

Figure 11: The effect of mountains on global prevailing wind from [17], [18], and [19].

4.2. Building Effects

The first row of buildings affects the second row located behind it relative to the wind direction. It causes recirculation in front of the second row. This vortex occurs because the wind converged downstream at the windward side of building and diverged upstream at the roof of building, this diversion occurs especially with flat roofs. This vortex may occur even when increasing the distance

---

[7] The unfavorable effects of mountains depend on many factors such as wind direction and site of urban mass. But the mountains sometimes can create better conditions for natural ventilation, for example, if the layout of urban mass lies between two mountains. The mountains can be favorable for natural ventilation because they can create air passages (channel) between them.

between multiple rows to achieve three times of the height of building. Fig. 12 shows the effect of first row of urban mass to cause recirculation at the cavities between second and third rows [3].

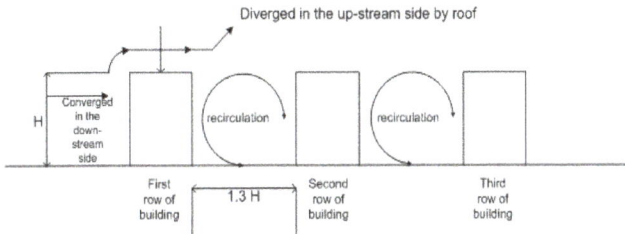

Figure 12: Effect of roofs on the cavities between multiple rows from [3].

Another effect of buildings is the effect of urban mass. To decrease unfavorable effect of building rows and urban masses, one solution is to stagger the building rows or urban masses. Fig. 13 shows the unfavorable and favorable of urban masses effect.

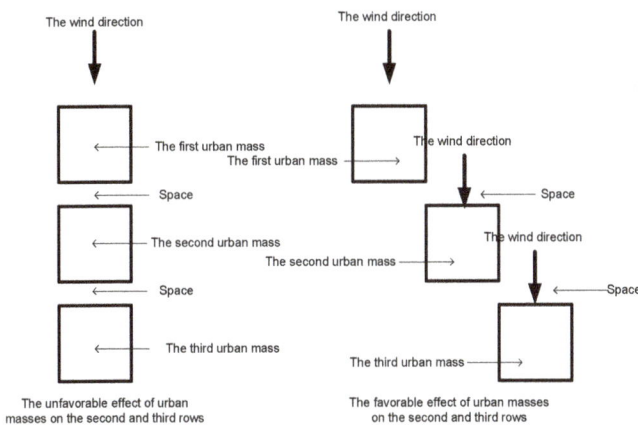

Figure 13: Effect of the first urban mass on the second one

The effect of urban masses can decrease the prevailing wind to less than one quarter of the global wind at the height of 20 meters from the ground level [20]. Fig.14 illustrates the comparison between urban wind and global wind.

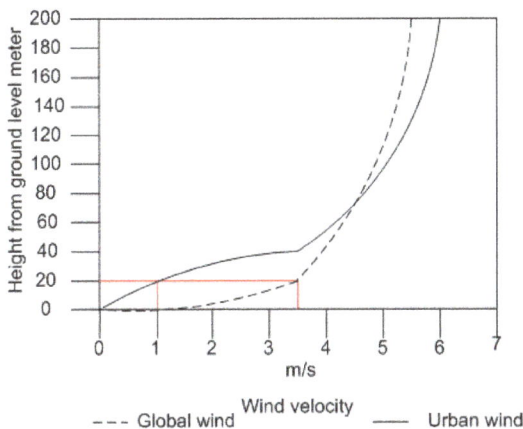

Figure 14: Comparison between global and urban wind from [20].

5. Conclusions and Summary

Thermal Comfort & Wind Velocity

Natural ventilation due to wind induced airflow can achieve both thermal comfort and reduction of cooling energy cost over the year with only 0.4 m/s in slightly hot districts. This reduction depends on both outdoor air temperature and wind velocity in hot and arid areas. There is a limitation of using wind velocity that is over 2 m/s. If more velocity is needed, it should be provided above the height of a person (2 m/s) from the floor or integrated with other natural passive cooling strategies such as high thermal mass and evaporative cooling or integrated with mechanical cooling. When applying these methods, if the outdoor temperature is over 40°C, a value of +2 and +3 on the thermal sensation scale (warm sensation) can be achieved with suitable air velocity.

Air flow Inside Courtyards

The integrated closed courtyard with wind catcher and chimney can achieve only 20% of the freeflow wind velocity along the wide dimension of the courtyard. Another study of closed courtyard observes negative airflow velocities inside the courtyard (recirculation movement) especially in the presence of high wind velocity. The last study of courtyards characterized by a small width and long length may achieve higher velocities than the first solution, this can be a future study to solve hot climate in locations such as Farafra Oasis in Egypt, and this solution is suitable to high outdoor temperature with low wind velocity.

Air Flow around Multiple Buildings

Wind flow causes a vortex behind the building. This vortex depends on the dimensions of building especially the area that faces the wind. The length of this vortex ranges from one and half times to twice of the length.

Wind flow causes four different zones of pressure. The positive pressure is in front of the building and the negative pressure is at the roof, two corners, and behind, when negative pressure increases. The turbulence zone occurs because of the stagnation point that is located at the center of the building at the upper level; this point achieves maximum positive pressure. Therefore, aerodynamic shapes such as vessels and wheels can decrease both of positive pressure especially at the stagnation point in front of this shape and the maximum negative pressure around the other zones of this shape. A future study will concentrate on aerodynamic shapes in districts with high outdoor temperature and high wind velocity such as Farafra Oasis, Egypt. Natural ventilation depends on the difference between the positive pressure in the zone in front of the building and the negative pressure in the other three zones around the building especially behind the building. But airflow depends on both decreasing the turbulence zone that occurs in front of the next row of the building especially with high wind velocity and increasing wind velocity that introduces to these next rows .

The first attempt to achieve high performance both in terms of airflow around buildings and natural ventilation inside building, uses a narrow passage that face the wind between two buildings to increase the Venturi effect; yet the passage also increases the length of the back vortex to more than three times of length of the building. This attempt is suitable to districts that have low wind speed like Aswan.

Table (2) Summary of results

Item	The effects	The evaluation according to natural ventilation
1-Climatic elements.	High wind velocity is preferable for hot climate	To achieve high air velocity at wind ward side of building
1-1- Wind velocity at wind ward side of building.	According to thermal comfort	Tools : wind velocity and direction, passages , cavities , and arrangement of buildings
1-1-1- 0.2 m/s.	It can achieve thermal comfort up to 29 °C.	Narrow passage and chess board arrangement are required to achieve Venturi effect.
1-1-2- 0.4 m/s.	It can achieve thermal comfort up to 30 °C.	Preferable if accurate air flow around building can be achieved.
1-1-3- ≥1.0 m/s.	It can achieve thermal comfort up to 31 °C.	Moderate passage is required.
1-1-4-≥1.5 m/s.	It can achieve draft discomfort.	Avoid Venturi effect.
1-2- Wind direction at wind ward side of building.	Perpendicular wind direction at wind ward side is preferable for hot climate.	It can control in wind velocity at windward side.
1-2-1-Perpendicular with wind direction.	It can achieve maximum air velocity at windward side (50-70 % from wind velocity)	Preferable in extreme hot climate.
1-2-2- Slope with wind direction	It can achieve moderate air velocity at windward side(30-40% from wind velocity).	Preferable in moderate climate.
1-2-3-Parallel with wind direction	It achieves low air velocity at wind ward side(10 % from wind velocity).	Non-preferable in hot climate.
1-3-Outdoor air temperature	It is related to air velocity	Outdoor temperature and wind velocity can determine the required air velocity at windward side
1-3-1-≥31°C	-Air velocity can achieve predict mean vote PMV=+1 with maximum air velocity that equals 1.4 m/s.	It can achieve nearly thermal comfort.
1-3-2-≥40°C	-Air velocity can achieve predict mean vote PMV=+3 with maximum air velocity that equals 1.4 m/s.	It can achieve partly thermal comfort.
2-Building arrangements	According to wind velocity	It can increase or decrease wind velocity.
2-1-Narrow passage	It can achieve over 100% from wind velocity according to Venturi effect.	Preferable in low wind velocity case 0.4 m/s.
2-2- Chess board arrangements	It can achieve around 40% from wind velocity at the second and third rows.	Preferable in hot climate.
2-3- Jack roofs	It can achieve around 40% from wind velocity at the second and third rows.	Preferable in high wind velocity.
3-Courtyards		
3-1-Narrow courtyards	It causes negative pressure at courtyards.	Courtyards as negative pressure (outlet side) at leeward side are preferable in low wind velocity.
3-2- Wide courtyards	It causes positive pressure at courtyards.	Courtyards as positive pressure (inlet side) at leeward side are preferable in high wind velocity.
4- Urban wind (terrain category)	Obstacles that faced wind can decrease the wind velocity up to 50% or more	-Preferable in high wind velocity 3m/s. -Non- Preferable in low wind velocity that is less than 1.4 m/s

The second attempt shows the low wind velocity at the second and the third rows of linear buildings with narrow passage between them and suitable distance (cavity) between them. The wind velocity at the cavity achieves only 10-20% of wind velocity in the next rows and achieves 40 % of wind velocity at the first row, while it achieves more than 60% at the passages. This attempt is suitable to low wind velocity with strong solar radiation[8].

The third attempt arranges rectangular or square buildings in a chess board arrangement and suggests twice the length of the cavity according to wind flow with the passage being one unit length. It can achieve up to 40 % of

---

[8] The narrow passage can cause both high wind velocities because of Venturi effect and low solar radiation that is preferable at hot climate such as most districts of Egypt.

wind velocity at different rows. This is suitable in any climate conditions especially that demands compact planning to avoid strong solar radiation in very hot climates.

The last attempt uses linear building with jack roof. It can achieve 0.2-0.8 according to wind direction of positive pressure at the cavity between rows. This attempt is suitable for high wind velocity with multiple linear building rows.

Air Flow around Urban Area – Regional Planning- at City Scale

Unfavorable conditions:-Mountains- that faced wind – decrease global prevailing wind up to 50% and sometimes cause large vortices, although, the wind velocity increases at the top of mountains.

-Buildings cause recirculation in front of the next rows of buildings and decreases wind velocity up to 50% from real value because of the vortex that happens in the top of building- flat roof. The urban mass affects each other in away similar to what happens in the building rows.

Favorable conditions: According to topographic effects, the wind velocity can increase when the buildings lie between two rows of mountains. According to urban masses effects, the wind velocity can keep its velocity when the buildings are arranged in a staggered way to wind direction.

6. Recommendations for Future studies

This paper shows improved natural ventilation around multiple rows of rectangular buildings in hot climate at Thailand, India, Cuba, South Korea, and Salt Lake City similar to Aswan and Farafra in southern Egypt.

The paper also shows how to select the locations of future buildings according to topographic and building effects. It also shows how to arrange the future buildings, for example, in chess board arrangement to benefit the Venturi (channel or passage) effect to increase wind velocity. It suggests the suitable distance between future buildings, the long length is to decrease wind recirculation and the smaller or narrow passages is to increase wind velocity.

The paper also shows the benefit of aerodynamic shapes to decrease the unfavorable effect caused by stagnation pressure. Courtyards that have narrow width facing wind direction can be used to increase wind velocity and long length parallel to wind to collect wind, [21].

Fig. 15 presents the innovative study of solutions that applied in compact buildings at hot climate. Compact buildings are designed to avoid the unfavorable solar radiation. Chess- board arrangements of multible rows of buildings will be studied in future study at hot climate such as Aswan and Farafra oasis, Egypt.

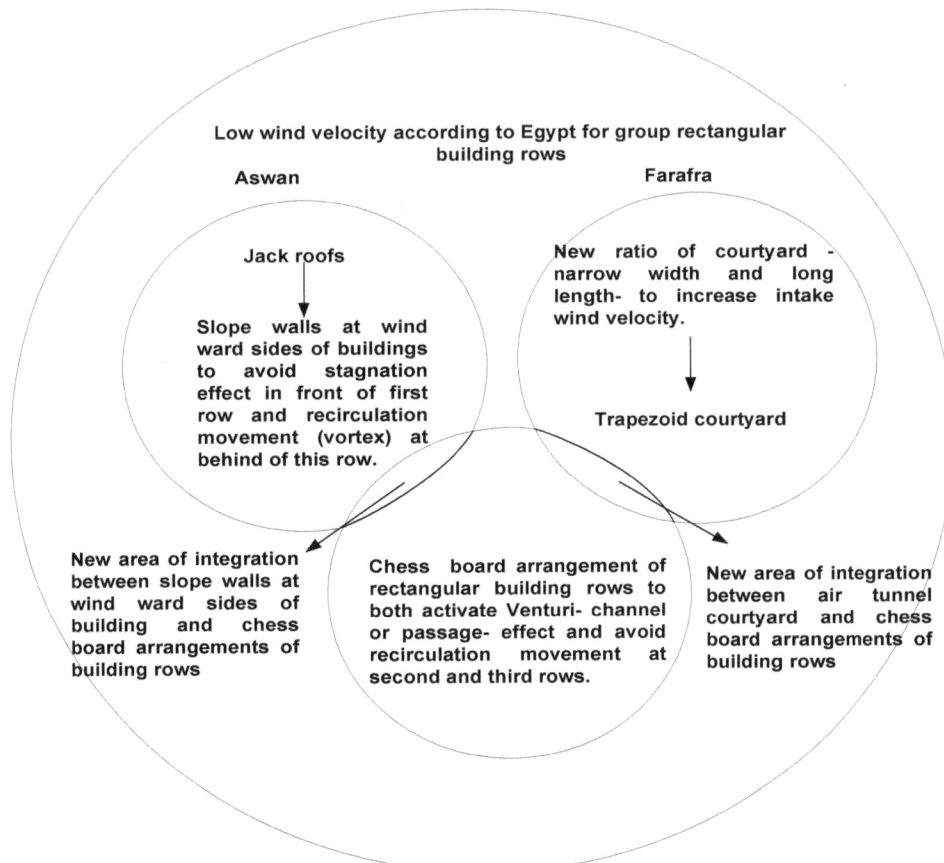

Figure (15) *new interest areas of future studies to improve wind flow around buildings for hot climate cities with low wind velocities at Egypt.*

References

[1] Military Handbook, 2004, Cooling Building by Natural Ventilation, Department of Defense, Approved for Public Release, USACE Publication Depot, pp 7.

[2] Brown, G.P., 2000, Sun, Wind, and Light; Architectural Design Strategies, 2nd Edition, John Wiley & Son Publishers.

[3] Yamada, T. , 2005 , Numerical Simulation of Air Flows in and around A City in A Coastal Region , American Meteorological society , Annual Conf. , San Diego , California .

[4] Bauman, F., Ernest, D. , and Arens , E. A. , 1988 , The Effects of Surrounding Buildings on Wind Pressure Distributions and Natural Ventilation in Long Building rows , Center for The Built Environment, University of California , Berkeley Publishers , Scholarship Repository .

[5] Tantasavasdi, C., Srebric, J., and Chen, Q., 2001, Natural Ventilation Design for Houses in Thailand, Energy and buildings, Vol. 33, Issue 8, pp 815 – 824.

[6] Heidari, S., 2005, Effect of Air Movement in Buildings, Passive and low Energy Cooling for the Built Environment International Conf.

[7] Brager , G.S. , and de Dear, R., 2001 , Climate , Comfort , & Natural Ventilation : A New Adaptive Comfort for ASHRAE Standard 55 , , Center for The Built Environment , University of California , Berkeley Publishers , Scholarship repository .

[8] Aynsley, R., 2002, Energy with Indoor Air movement, International Journal of Ventilation, Vol. 1, pp 33 – 38.

[9] Narayan, T., A Passive Courtyard Home in Jaipur, India: Design Analysis Thermal Comfort in A Hot Desert Climate, Arizona State University Publishers.

[10] Tablada, A., Blocken, B., Carmeliet, J., De Troyer, F., and Verschure, H., Geometry of Building's Courtyards to Favor Natural Ventilation: Comparison between Wind Tunnel Experiment and Numerical Simulation, leaven, Belgium.

[11] Chen, Q., 2006, Sustainable Urban Housing in China, Springer Publishers, Netherlands, pp 116-123.

[12] Baskaran, A., and Kashef, A., 1996, Investigation of Air Flow around Buildings Using Computational Fluid Dynamics Techniques, Engineering Structures Vol. 18, No 11.

[13] Blocken , B. , and Carmeliet , 2004 , Pedestrian Wind Environment around Buildings : Literature Review and Practical Examples , Journal of Thermal Envelope and Building Science 28 (2) : 107- 159 .

[14] Popinet, S. , Smith, M., and Stevens, G. , 2003 , Experimental and Numerical Study of The Turbulence Characteristics of Air Flow around A research Vessel , Journal of Atmospheric and Oceanic Technology .

[15] Xia, J., Hussaini, M. Y., and Leung, D. , 2003 , Numerical Simulations of Wind Field in Street Canyons with and without Moving Vehicles , 16th ASCE Engineering Mechanics Conference.

[16] Lee, R., Chan, S. T., Leone, J. M., Stevens, D. E., 1999, Air Flow and Dispersion around Multiple Buildings, 7th International Conf. on Air Pollution, San Francisco, CA.

[17] Oh , S.N. , A Bio – climatic Assessment on Urban Area at Seoul Based on Observations and Numerical Simulations , Applied Meteorology Research Laboratory Publishers , Korea .

[18] Chung, Y.S. , and Kim, H. S., 2008, Mountain – Generated Vortex Streets Over The Korea South Sea, International Journal of Remote Sensing, Vol. 29, No. 3, pp 867 – 877.

[19] Brown, M., Marty, L. , Calhoun, R., Smith, S., Reisner, J., Lee, B., Chin, S. , and De Croix, D. , 2001 , Multi-Scale Modeling of Air Flow in Salt Lake City and The Surrounding Region , ASCE Structures Congress Conf. , Washington , Dc .

[20] Leach, M .J. Chan, S. T., and Lundquist, J. K., 2005, High-Resolution C F D Simulation of Air Flow and Tracer Dispersion in New York City, Sixth Symposium on The Urban Environment, Atlanta, GA.

[21] http://www.eere.Energy.Gov/buildings/energyplus/weatherd ata/1_africa_wmo_region_1/E .

Effects of atmospheric variables on the performances of parabolic trough concentrating collector

Sadik Umar[1], Umar Kangiwa Muhammad[1], Muhammad Mahmoud Garba[2], Hassan N. Yahya[2]

[1]Department of Physics, Kebbi State University of Science and Technology Aliero, Aliero, Nigeria
[2]Sokoto Energy Research Center, Usmanu Danfodiyo University Sokoto

Email address:
umarmagajisadik@yahoo.com (S. Umar)

Abstract: In recent years solar energy has been strongly promoted as a viable energy source. one of the most simplest and direct application of this energy is the conversion of solar radiation in to heat energy using a devices called sollar collectors [3].In order to enhance this conversion efficiency, important parameters that influence the system performances need to be evaluated. To achieve the state objectives, this paper investigates the effects of some meteorological variables (Wind Speed and Ambient Temperature) on the performance of parabolic trough collector. The results obtained show that wind speed is inversely proportional to the direct solar radiation and it has also shown that a maximum daily average wind speed of 1.39m/s was observed, when ambient temperature reach it lowest value. Furthermore, efficiency is highly correlated with wind speed and negatively correlated with ambient temperature.

Keywords: Parabolic Trough Collector, Direct Solar Radiation, Efficiency, Wind speed, Ambient Temperature, Concentrating Collector

1. Introduction

Energy is one of the most important factors in social and economic development of a country. The amount of energy consumption per capita of a country is the measure of the nation's economic development and it has become an important parameter for sustainable development through the world [7].Energy demands across the world have been on a rapid rise recently due to population growth and industrialization. It is obvious that the conventional energy sources (Fossil Fuels) are under increasing demand and there is a need to examine alternative sources to augment these energy sources [8]. The sun represents one such alternative energy sources, which is presently underutilized particularly in sub Saharan Africa, where there is abundant sun radiation. Solar energy is a renewable and sustainable energy source. It has a promising potential to meet our present and future energy needs. In their course to develop solar thermal technology as an alternative and most prospective, sustainable source of energy for steam production for both industrial and domestic application. "Reference[4], in his effort to generate steam using solar

energy ,parabolic dish collector was constructed using wood, aluminum sheet and plane mirror and a temperature 200°C was obtained".

"Reference [5] investigates the effect of reflective materials on the performance of parabolic trough concentrator by using three different reflective materials, Aluminum sheet, car solar reflector and Aluminum foil reflector, where it was found that Aluminum sheet reflector has the highest performance followed by car solar reflector and then Aluminum foil". There also, to enhance the performance of solar parabolic trough concentrator, numerous research groups and individual around the world have contributed to the performance improvement of the solar parabolic trough concentrating collector, by evaluating the influence of some important parameters on the performance of the system. The present work investigates the effect of some atmospheric parameters (Ambient temperature and Wind speed) on the thermal performance of parabolic trough solar collector.

2. System

2.1. System Description

Figure 1 shows the schematic diagram of a solar parabolic trough concentrator. The overall dimensions are shown in table 1. The system comprised of a parabolic trough shape and receiver tube. The parabolic trough shape was made up of ply wood and soft wood joined together and a mirror of 3mm thickness was cut in to rectangular shape of 8cm x 94cm. The rectangular shaped mirror was fixed with shinny side facing up on to parabolic trough shape made from wood, which in turns formed a parabolic reflective surface. The parabolic trough takes the advantage of all the parallel rays from the sun that incident on it and converge it at the focal point [2]. The black painted copper tube of 28.5mm thickness was placed at a focal point to serve as receiver tube (absorber).

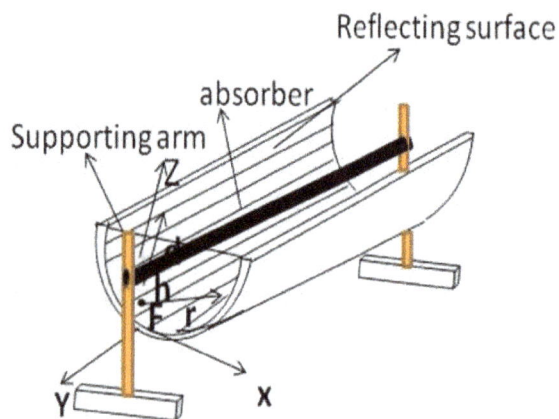

Figure 1. Schematic diagram of a solar parabolic trough concentrator

2.2. Parabolic Trough Collector Working Principle

Figure 2 Shows a typical parabolic trough collector (PTC), which is basically comprised of a parabolic trough shaped concentrator that are capable of reflecting direct solar radiation on to a receiver tube located in the focal line of parabola [8]. Since the collector aperture area is bigger than the outer surface of the receiver tube, the direct solar radiation (beam radiation) is concentrated.

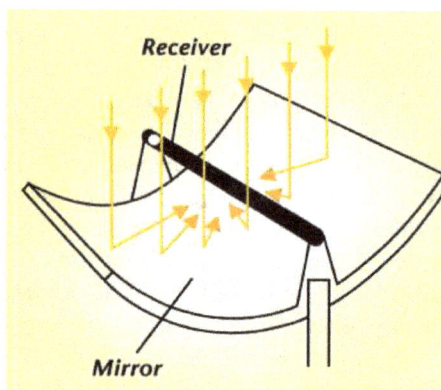

Figure 2. A typical parabolic trough concentrator [1]

The concentrated radiation reaching the reciever tube heats the fluid that circulates through it, thus transforming the solar radiation in to thermal energy in the form of sensible of the fluid [9]

Table 1. Parabolic trough collector system specifications

Parameters	Values
Focal distance F	0.15m
Depth of the parabolic trough	0.33m
Length of collector	1.82m
Absorber Area	2.85cm^2
Aperture Area	1.64m^2

3. Experimental Setup

The Parabolic trough Concentrator was set for performance evaluation test at Sokoto energy research Center testing area, under climatic conditions of Dundaye Village, Sokoto state, Nigeria (Latitude 13.1° North and Longitude 5.2°).The water tank was elevated at 0.5m above the ground level in such a way that water can flow downstream to the receiver tube. The test was done at constant flow rate(m_w) of 0.0177kg/min. The parameters recorded during the test were: water inlet temperature (T_i), outlet temperature (T_f), intensity of diffused solar radiation (I_d), global (I_g) and direct (I_b), ambient Temperature (T_{amb}), and wind speed at parallel to the system using digital Anemometer. The thermocouple data logger was also used for temperatures measurement of four different points, ambient, absorber, inlet temperature of water and out let temperature of water simultaneously. The ambient temperature was measured by placing the thermocouple terminal at the surrounding. The absorber temperature was also read and recorded by using a temperature resistive material to gum the thermocouple terminal on the absorber surface, the inlet and outlet water temperature was measured and recorded for Seven hours (from 9:00am to 4:00) at 30 minutes interval.

The parabolic trough collector utilizes only direct component of solar radiation, therefore in order to measure direct solar radiation, two pyranometers were used one to measure global and another one measured diffuse and then direct was obtained by subtracting diffuse from global.

Therefore Collector Field efficiency was computed using equation (1) as [6].

$$\eta_c = \frac{(T_f - T_i)m_w C_p}{AI_b} \qquad (1)$$

Where A = Area of the collector (m^2)
C_p = the specific heat capacity of fluid (KJ/Kg/°C)
M_w = Collector field flow rate (Kg/s)
T_f, T_i = Collector field temperatures at outlet and inlet (°C)
I_b = direct solar radiation (W/m^2)
η_{field} = Collector field efficiency
The flow rate was converted from Kg/min to Kg/s.

4. Results and Discussion

The significant of carrying out this experimental work was to find the effect of some weather parameters on the performance of parabolic trough collector. In order to achieve this target, the experimental test was carried out for six consecutive days (from 20.10.11 to 25.10.11) for seven hours daily at 30 minutes interval and the average was taken for each day. The daily average data was used in equation (1) to calculate the collector field efficiency for each day and the results is shown in figure 3-5:

Figure3 represents the average daily variation of ambient temperature, wind Speed and efficiency for six days at 30 minute interval. According to results in figure 3 the higher average daily wind speed of 1.39m/s was obtained on the sixth day, when ambient temperature is 42.8°C and efficiency 5.59% was attained. The lowest wind speed of 0.65m/s was observed on the second day at 40.5°C Ambient temperature and efficiency of 7.0% was observed, efficiency had it peak value of 8.36% on the third day when Ambient temperature has its lowest value of 37.2°C and wind speed of 1.30m/s was recorded. The highest efficiency was obtained when the ambient temperature has the least value. This shows that ambient temperature is inversely proportional to the efficiency. Although both the wind speed and ambient temperature fluctuates due to changes in the day but in most of the time an increased in wind speed decreases ambient temperature.

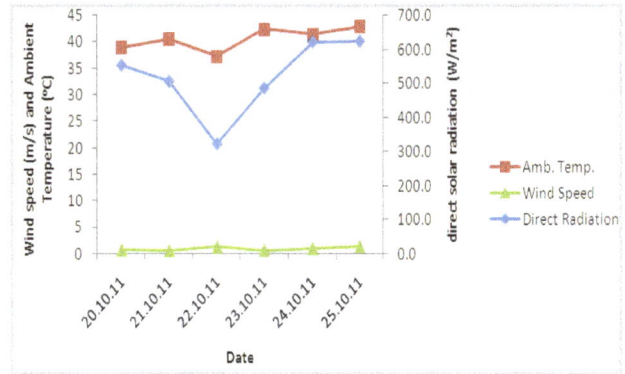

Figure 3. *Average daily variation of ambient temperature, Wind Speed and Efficiency.*

Figure 4 is a graph of daily average wind speed, ambient temperature and direct radiation. The result shows that ambient temperature increases with an increase in direct solar radiation and also decreases with the decrease in radiation, while wind speed increase with the decrease in ambient temperature. This indicates that higher wind speed is observed in the morning or evening period when solar radiation and ambient temperatures are low. Therefore, it has been clearly justified that there exists relationship between direct solar radiation, ambient temperature and wind speed.

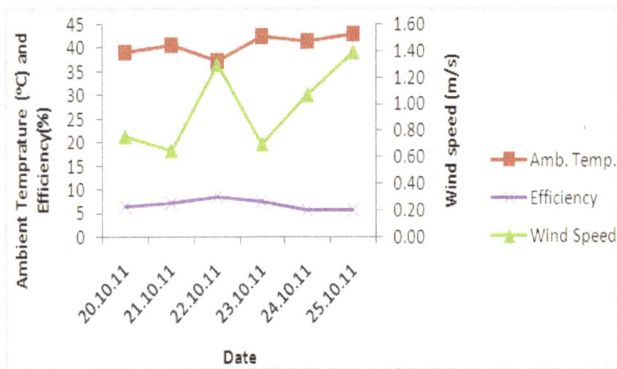

Figure 4. *Average daily variation of Ambient temperature, Wind Speed and direct radiation.*

Figure 5 Shows variations of daily average direct solar radiation, ambient temperature, wind speed and efficiency. In this figure direct solar radiation and ambient temperature are highly correlated, while at the same time efficiency and wind speed are correlated, therefore wind speed shows a negative effect on direct solar radiation and also ambient temperature shows a negative effects on efficiency but positive effect on solar radiation , wind speed also shows a positive effect on efficiency.

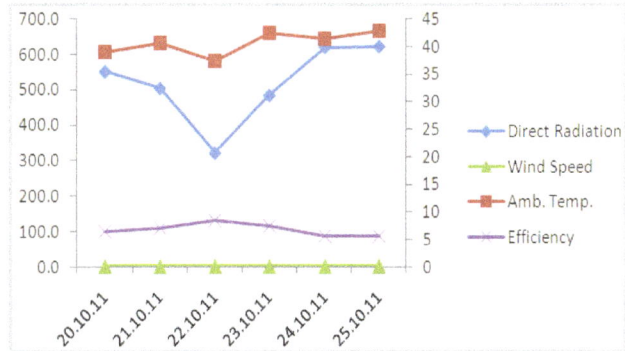

Figure 5. *Daily Average Variation of Ambient temperature, direct solar radiation, wind speed and Efficiency.*

5. Conclusion

Parabolic Trough Collector was tested under the climatic condition of Aliero Town, Kebbi State, Nigeria to see the effect of meteorological parameters on the performance of the system. From the experimental results obtained it has shown that lowest daily average ambient temperature of 37.2°C was recorded at 1.30m/s daily average wind speed while efficiency has its peak value of 8.36%. It is therefore, indicates that efficiency and wind speed has positive correlation and negatively correlated with ambient temperature. Furthermore, wind speed is negatively correlated with direct solar radiation. However it clearly indicates that ambient temperature and wind speed affects the thermal performance of the parabolic trough collector.

References

[1]   J. Dascomb. (2009), "Low cost concentrating solar collector for steam generation". Unpublished Thesis Submitted for the ward of Doctor of Philosophy, Department of Mechanical Engineering, Florida State University. Retrieved on 6th may, 2011. From http://esc.fsc.edu/documents/Dascomb JThesis.pdf

[2]   J.F.M.Escobar, S. Vazquez, Y. Montiel, F. Granados-Agustin, Cruz-Martinez & E. Rodriquez-Rivere,(2011) "Building a parabolic solar concentrator prototype". Journal of Physics, Conference series 274 (2011). Retrieved on 5th August, 2011. From iopscience.iop.org/1742-6596/274/012104. Pdf.

[3]   G. Iordanou, (2009) "Flat Plate Solar Collectors for Water Heating with Improved Heat Transfer for Application in Climatic Conditions of the Mediterranean Region". PhD Thesis, Durham University. Available at durham E-theses online: http://etheses.dur.ac.uk/174/.Retrieved On 17[th]December,2010. (unpublished)

[4]   F. Joshua, (2009). "Design, construction and testing of parabolic solar steam generator". Leonardo Electronic Journal of practices and Technologies. Retrieved on 16[th]December, 2010 from http://lejptacademicdirect.org/A14/115_133.pdf

[5]   M. Kawira, R. Kinyua, & J.N. Kamau, (2010) ".Fabrication and characterization of a prototype parabolic trough solar concentrators for steam production", Retrieved on 18[th]june, 2011.From http://elearning.jkuat.ac.ke/journals/ojs/index.php/sr.

[6]   A. Thomas, (1991). "Operation and performance of the solar steam generation system installed at government silk factory, mysore". Energy conversion management, Vol. 33 (3), Pp191-196.

[7]   Varun and S. K Singal, (2007), "Review of augmentation of energy needs using Renewable energy resources in india". Renewable and Sustainable Energy, 11:1607-15.

[8]   E.Zarza,(2005), "Medium Temperature Solar Concentrators". Solar energy Conversion and Photo energy system. Vol.1. Encyclopedia of support system (EOLSS).

[9]   E. Zarza, L. Valenzuela, J. Leon, K. Hennecke, M.Eck, D.H.Weyers, M. Eickhoff (2002), "Direct steam generation in parabolic troughs". Final result and conclusions of the DISS project. In. Steinfeld, A (eds) Book of proceedings of 11th Solarpaces international symposium on concentrated solar power and Chemical energy Technologies, held in Zurich (Suitzerland) September 4th-6th,2002. Paul Scherer institute, Villigen (Suiza), 2002, Pp 21-27 ISBN 3-9521409-3-7

9

Utilizations of food waste as an anaerobic digester feedstock

Krishna Kumar, Omprakash Sahu[*]

Department of Chemical Engineering, KIOT, Wollo University, Ethiopia

Email address:
ops0121@gmail.com(O. Sahu)

Abstract: For many years, anaerobic digestion has been utilized in order to treat the odorous, pathogenic, and dissolved-oxygen-reducing-characteristics of both anthropogenic and livestock effluent waste streams. In addition to this beneficial biological treatment, such digestion provides both methane gas and digestate which serve as a valuable fuel and fertilizer, respectively. However, food waste also has the potential to serve as a useful feedstock for anaerobic digestion due to its high volatile (combustible) solids content and propensity for rapid biodegradation. There are a number of parameters of concern when using food waste for such digestion which increases the operation complexity of digester systems, but if such devices are properly monitored and adjusted, food waste has the potential to serve as a sole feedstock or as part of a dual manure-food waste input; both cases provide an improvement in gas generation production.

Keywords: Bacteria, Decompositions, Decay, Methane

1. Introduction

In order to fully understand the advantages and disadvantages in the different feedstock used for anaerobic digestion (AD), it is necessary to understand the biology and variables inherent in organic matter decomposition and decay. Towards this end, the different bacterial regimes involved in AD will be discussed, as well as how the particular characteristics of food waste may enhance or disrupt microbial function. The unique considerations necessary for the use of food waste as a feedstock will be elaborated on. Next, both the financial aspects of food waste digestion, as well as the environmental benefits resulting from the diversion and biodegradation of food remains from the municipal waste stream will receive attention. Also, the applicability of using food waste as a feedstock in developing countries will be analyzed from a technical expertise and resource perspective, using Haiti as an example. Finally, a few examples of anaerobic food waste digestion systems will be reviewed as well as some of the challenges encountered during operations.

2. Anaerobic Microbiology

There are four stages in the breakdown of organic matter on the path to methane production; each stage is the result of a different bacterial subpopulation. These stages include hydrolysis, fermentation (or acidogenesis), acetogenesis and eventual methanogensis[1]. Hydrolysis involves the conversion of complex molecules and compounds – carbohydrates, lipids and proteins – found in organic matter into simple sugars, long chain fatty acids and amino acids, respectively. Acidogenesis in turn converts these into volatile fatty acids, acetic acid, carbon dioxide and hydrogen gas. Acetogenesis converts the volatile fatty acids into more acetic acid, carbon dioxide and hydrogen gas. Methanogens have the ability to produce methane by using the carbon dioxide and hydrogen gas or the acetic acid produced from both the acetogenic or acidogenic phases. The Figure 1 for an overview of this decomposition process.

Some food wastes have a tendency to raise or lower the pH of digester solution, dependent upon the specific food remnants involved. Care must be taken to ensure that the digester liquid pH remains near neutral or at least between 6.7 and 7.4 so that the methanogens can survive[3]. When a bacterial food supply, or feedstock, is initially introduced into the digester, it is possible that the rate at which hydrolysis proceeds may overload the system with fatty acids thereby depressing the pH to the point of methanogenesis inhibition[4]. The often elevated levels of ammonia within manure feedstock usually serve as a buffer against drastic pH changes. This safeguard is also present in the anaerobic digestion of food waste when proteins are decomposed to form ammonia. This ammonia elevates levels of bicarbonate through the creation of ammonium salt

by the removal of dissolved carbon dioxide from the digestion solution[5].

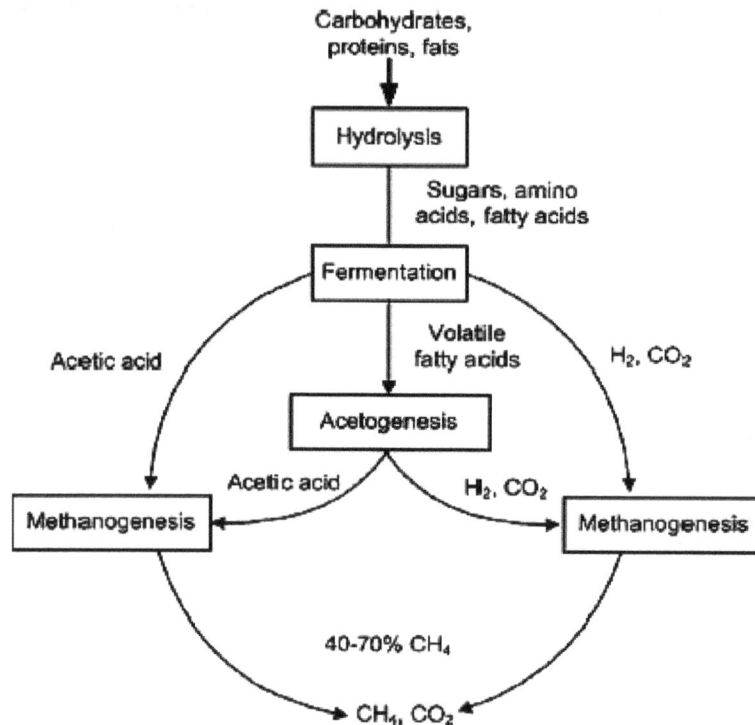

Figure 1. Stages of Anaerobic Digestion[2]

There are additional means of controlling the pH besides relying upon the inherent alkalinity. The possible depression of pH can also be avoided through control of volatile solids/organic loading, since in the first steps of the AD process these will be converted into volatile fatty acids. In a continuously-stirred tank reactor (CSTR) low-solids-content digestion system, it is typically recommended that the feedstock be diluted sufficiently so that the total solids (TS) concentration falls between 6% and 10%[6]. In scenarios where the selected feedstock repeatedly decreases the pH to levels of concern, buffering agents including lime and/or sodium bicarbonate can be added to the digester. Unlike lime however, bicarbonate is advantageous insomuch as it does not have a tendency to form insoluble salts and is very unlikely to increase the pH outside the preferred range of methanogens. Also, it will not create a vacuum in the area over the digestion fluid, or headspace, due to the reaction with carbon dioxide[5].

3. Unique Considerations to Using Food Waste Feedstock

In assessing the feasibility of using food waste as a means of feeding anaerobes, it is important to note the specific characteristics which set it apart from manures and sludge. These include the increased likelihood of the presence of contaminants such as detergents and cleaners,

nonbiodegradable components such as plastics and bones, and nonhomogenous-sized organic matter which will necessitate a form of size reduction. Each of these will require a form of pretreatment before food waste introduction into the digestion chamber(s). Furthermore, both human and animal sludge have one particular characteristic which gives them a unique advantage over food waste feedstock: a bountiful supply of anaerobes[7]. This is important because the initial loading of the digester with methanogens means a greatly reduced lag time between the previous acid generating stages and the methanogenic acid reducing stage. As a result, it is much less likely that the pH of the digester liquid will ever dip below the range preferred by the obligate anaerobes. In order to overcome these potential setbacks while using food waste for AD, all kinds of equipment have been developed for both size reduction and contaminant removal. These can include devices such as trommel screens, hydrocyclones, pulpers/grinders, metal separators and flotation tanks. These devices are reliant upon the different densities and/or metal properties of the various contaminants and are usually able to remove the major undesirable portions of the food waste stream including plastics and grit. Figure 2 provides an example layout of a food waste AD in Korea showing the various pretreatment processes; this facility will be discussed in more detail later.

Perhaps one of the most desirable qualities of these

digester systems is their ability to create useful gas energy with net-zero carbon emissions. But a reasonable question to ask when considering all these food-waste-specific pretreatment devices is whether or not there is actually a net energy gain. There is already energy input involved with both food and wastewater-solids-based AD systems due to the fact they are typically heated in order to maintain a constant operational temperature throughout the year. Digesters typically operate in either the mesophilic (20 °C –

45 °C) or thermophilic range (45 °C - 70 °C), where the latter is generally preferred for its higher pathogen destruction and faster rates of gas production and solids degradation[9]. Whether or not an outside source of energy or power is used will depend on the particular plant arrangement. For the heating and power requirements could (and often do) draw a parasitic load on the combined heat and power (CHP) operations usually found at AD sites.

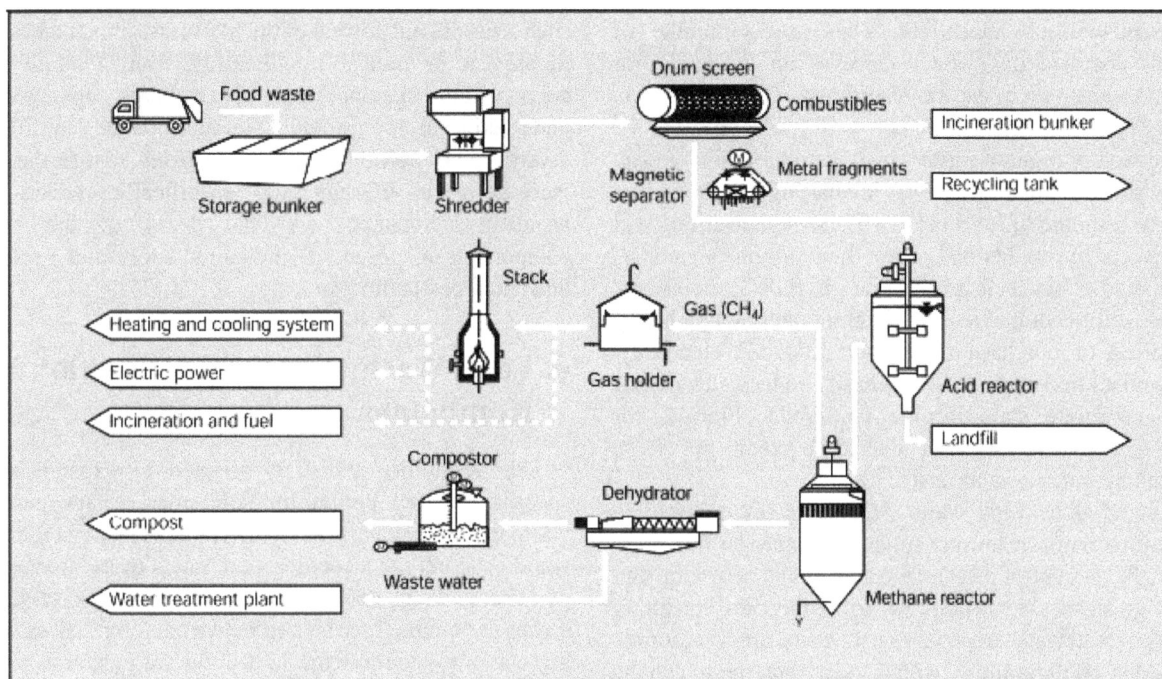

Figure 2. Operation Diagram of Food Waste Digester in Kombolcha[8]

4. Financial Considerations

In many cases, there are other recycling opportunities which will compete with food waste AD. These include using food wastes as either an animal feed or soil amendment through composting. As a result, from an economic perspective, the AD of food waste may not be the most financially sound option for its ultimate disposal. For example, Disney World through a company called Neutral Feed has been converting a significant portion of its food waste into animal feed, and the resulting product has even been approved by the United States Department of Agriculture for human consumption[10]. Due to the complexity of markets associated with the end-products of AD, it is difficult to develop a solitary analysis which would be applicable to any individually proposed system. However, there are a number of factors of concern which should be analyzed for any planned AD construction which include facility space requirements, water demand, wastewater discharge quality and quantity, the quality and quantity of the digestate residual, local biogas markets, and electrical use and electricity production[11]. One such study was performed by a consulting firm, R.W. Beck, to determine

whether or not an Iowa AD facility would be financially justifiable for incorporation into a solid waste agency's refuse management framework[9]. The methodology used in this study was to analyze a number of already existing facilities in Europe, since in 2004, the time of the study, no commercially operated facilities in the U.S. were using the organic fraction of municipal solid waste (OFMSW) for anaerobic digestion. R.W. Beck's sensitivity analysis showed that the present worth of the investment was most sensitive to the growth of the waste stream (feedstock) over a 20 year project horizon. However, some other important parameters analyzed included how much revenue would be gained from thermal and electric sales, as well as how much the facility could charge (while remaining competitive with other waste disposal options) for tipping fees.

R.W. Beck further provided an energy balance table which accounted for two energy inputs and outputs. Energy inputs included electricity and thermal energy. Energy outputs included methane and soil conditioner. Accepting that soil conditioner can be appropriately measured in terms of energy content, the table shows that digester produces roughly ten times the energy it consumes on an annual basis. This table appears to be a result of averaging the values from all the AD facilities surveyed across Europe. However, if this is the case, it is important to note that none of the facilities studied solely processed food waste, but rather typically processed OFMSW which, quite possibly, requires a more intensive pretreatment process due to a more variability within the waste stream. As a result, the electric cost estimated in this analysis is likely an overestimate.

The financial success of any food waste digestion enterprise will be highly dependent on the ability to acquire acceptable rates of revenue from carbon credits, sales of renewable energy back to local power utilities or private purchasers, and tax incentives. There are a number of different organizations which serve as an exchange for carbon credits, such as the California Climate Action Registry (CCAR). The CCAR currently lists a protocol for the digestion of organic waste which would apply to an AD facility for food waste which was attempting to earn carbon credits as a source of revenue[12]. These organizations will typically certify credits and allow their sale to companies who wish to offset their greenhouse gas (GHG) emissions. Different utility companies will often set rates at which they are willing to purchase renewably produced electricity, usually on a kilowatt-hour (kWh) basis. Some states offer a $/kWh corporate state income tax credit. Florida, for example, currently offers a one cent credit per kWh produced by renewable means[13].

For stand-alone food waste AD plants (as opposed to co-digestion with wastewater solids at wastewater treatment plants), the amount of discharge water should be minimized to the greatest extent possible to reduce the resulting discharge wastewater treatment cost. There are a multitude of digester configurations which exist, but these can be broken into high (>25% TS), medium (10%-25% TS), or low solids models (<10% TS)[14]. It is clear that the waste stream most suitable for use in a food waste AD is one which will be highly controlled at the source. Different restaurant and food service industries have different degrees of processing and separation before waste removal occurs. Without this source separation/pretreatment, upfront processing will make up a large part of the operational costs incurred in AD of food waste.

5. Potential Environmental Benefits

Food waste is the second largest component of the U.S. municipal solid waste (MSW) stream, comprising about 18% of the total amount sent to MSW landfills. This computes to about 30 million tons of waste annually[15]. As such, the diversion of this organic component of MSW can serve to greatly reduce landfill loading rates and has a high potential of greatly expanding their service life. As food waste decomposes in landfills, it is typically degraded into both carbon dioxide and methane gas and emitted into the atmosphere. While some of these landfills have gas collection systems to control these emissions, MSW landfills, according to the Environmental Protection Agency (EPA), are not required to install these gas collection and control devices until the capacity of the landfill exceeds both

2.5 megagrams and 2.5 million cubic meters of space (as listed in EPA Title 40 Code of Federal Regulations Part 60 Subpart WWW). By the time (and if) these thresholds are ever reached, it is highly likely significant amounts of methane, a GHG about 21 times more potent than carbon dioxide, has already been released into the atmosphere[16]. As a result, the ability to capture and collect this methane produced from food waste before the waste is sent off to landfills provides a significant means of reducing our overall GHG emissions. Furthermore, since food waste comprises such a significant portion of the MSW stream, it makes sense to prevent its transport to landfills from a purely space conservation standpoint. It is a large and complex undertaking to appropriately site and operate landfills, and diverting any portion of the refuse being sent to them will increase their lifespan and specifically reduce those problems associated with the decay of the organic components disposed of, including odors and vector (or unwanted pest) attraction.

6. Food Waste Anaerobic Digestion in Kombolcha

This feasibility study is intended to expand upon a specific disposal option for solid marketplace generated waste in Kombolcha. Due to the Haitians' heavy reliance on wood charcoal for a cooking fuel, much of the western half of Hispaniola is deforested. Biodigestion can serve as a means to both produce an alternative cooking fuel, as well as serve as a way to reuse the 75% of the waste stream which is estimated to comprise of food cast-offs[17]. While the digestion of food waste can produce approximately four times the methane yield as pig slurry can[6], there are a number of important advantages involved with using pigs in the digestion process in Haiti. Perhaps the most important of these is the lack of expertise and training in AD operation. Most reference literature which discusses AD makes it very clear that once the digestion process becomes unstable, it is very difficult to ensure that the entire system does not go "sour." In other words, even with expensive monitoring equipment which can provide up-to-date readings on such variables as pH, alkalinity and gas production rates, it is almost impossible to stabilize an anaerobic reactor which is starting to become unbalanced. In order to operate and maintain a steady and balanced food waste AD, all food residuals used as a feedstock would need to be well-homogenized, contaminant-free, and lacking such things as bones and shells. These preparations for AD would either be energy or labor intensive. Since Haiti does not have an extensive electric infrastructure[18], it would seem unwise to propose a system dependent on electric pumps or motors. Also, due to the unlikelihood of a solid waste management operation providing enough revenue to pay a sizeable workforce, a system which is reliant upon the smallest number of workers possible would provide the best chances of a financially successful operation. Keeping these obstacles in mind, remembering the bacterial buffering capacity which manures provide, and considering that Haitians are familiar with pig husbandry already, the best proposal for an anaerobic digestion facility in Haiti would involve using swine to consume the organic portion of the marketplace waste stream and provide a homogenous, contaminant free feedstock in the form of pig manure.

6.1. Case Studies

(A) In Oakland, CA, there is currently a wastewater treatment operation run by the East Bay Municipal Utility District (EBMUD), which codigests food waste with their wastewater solids in anaerobic digesters. The EPA provided them with a grant to study the possibility of using pure pulped food waste as a feedstock for AD, and in March 2008 they released a report that provided very encouraging results from their bench-scale studies. These results included that for an equal amount of digester volume, food waste has the ability to produce up to three times the methane volume that wastewater solids can produce[19]. This is partially a result of higher energy content per mass, but also is due to the fact that food waste can be added to the digester at a higher loading rate (per volume) than wastewater solids. EBMUD further discovered that the solids deposition rate for food waste was less than that encountered for the digestion of wastewater solids. This fact could give AD facilities which used food residuals a cost advantage over those which would digest manures since the tipping fee for digester solids is often considerable. However, the opposite may be true if the digester sediment was thermally treated (say in thermophilic conditions) to the point where it could be bagged and sold as a soil amendment. The only other problem temporarily encountered by EBMUD was digester instability due to inversion heater failures. These failures produced digester instability and testing operations had to be restarted as a result. The published report highlighted the importance of keeping a stable operation temperature.

(B) Another example of successful food waste digestion is at the digestion facility in Anyang City, South Korea. This digester system can handle 15 metric tons (mt) per day of food waste. While the waste is processed at an operational cost of U.S. $60/mt, it is almost impossible to dispose of it at a landfill for the cost of $25/mt due to limited space. Originally, the plant ran into difficulties with their conveyor system due to bones and metal fragments, but these problems were addressed and everything is currently operational[8]. This system uses a two-tank design; one is used for acid digestion and the other is used for methane production and capture. One of the interesting aspects of this design is that some of the alkaline liquid from the methanogenic tank is circulated into the acidogenic reactor in order to ensure the pH of the acidogenic tank does not get too low. The acidogenic tank is also used as a means of further feedstock purification. As it is mildly agitated through stirring, floating plastics and other pieces of scum are skimmed of the top while ground up grit and other dense materials are removed from the bottom of the tank.

7. Conclusion

Food waste can serve as high methane-producing feedstock for AD if it is properly pretreated and the digester is carefully operated. However, before construction of such a system, it is important to characterize the potential waste stream and determine the cost and amount of energy needed for proper pretreatment and operations. Also, a financial study of the different tax incentives, utility power buyback rates, local and federal government grants, carbon credit prices, and biosolid treatment regulations and markets should be performed. Different systems will likely need to be determined and developed on an individual waste stream basis. It is recommended that a pilot-scale facility be tested to determine the digestion process which works best prior to full-scale construction, especially considering that there currently is no commercially-operated food waste AD in the United States. For Ethiopia and developing countries, due to lack of financial, educational, and material resources, it is probably best to focus on a more stable feedstock such as pig or some other animal manure for possible AD units. People in developing countries are typically part of a more agrarian society where animal caretaking and consistent manure collection are familiar practices. It is no great step to collect this manure and dilute it by a constant amount before input into a digester. However, having to constantly adjust a food dilution rate because of an inconsistent food waste stream can prove problematic even to people who have experience with anaerobic systems. Regardless, individual opportunities need careful assessment and all feedstock options should be considered both individually and comingled to find the best system.

References

[1] Gerardi, Michael. The Microbiology of Anaerobic Digesters. Hoboken, N.J.: Wiley-Interscience, 2003. eBook.

[2] Li, Yebo, Stephen Park, and Jiying Zhu. "Yebo Li *, Stephen Y. Park, Jiying Zhu." Solid-state anaerobic digestion for methane production from organic waste 15. (2011): 821-826. Web. 26 Nov 2010.

[3] Viessman, Warren, Mark Hammer, Elizabeth Perez, and Paul Chadik. Water supply and pollution control. Prentice Hall, 2008. Print.

[4] McCarty, Perry. "Anaerobic Waste Treatment Fundamentals." Public works 95.9-12 (1964): n. pag. Web. 19 Nov 2010.

[5] Georgacakis, Dimitris, M. Sievers, and E.L. Iannotti. "Anaerobic Waste Treatment Fundamentals." Agricultural Wastes 4.9-12 (1982): 427-441. Web. 17 Nov 2010.

[6] Steffen, R., O. Szolar, and R. Braun. "Feedstocks for Anaerobic Digestion." Institute for Agrobiotechnology Tulln (1998): n. pag. Web. Nov 1 2010. <http://www.adnett.org/dl_feedstocks.pdf>.

[7] Baron, Samuel. Medical microbiology. 4th ed. Galveston, TX: Univ of Texas Medical Branch, 1996. eBook.

[8] "Technical Brochure #66 - Food Waste Disposal Using Anaerobic Digestion." CADDET. CADDET Centre for Renewable Energy, 1998. Web. 25 Nov 2010. <http://www.caddet-re.org/assets/no66.pdf>.

[9] "Anaerobic Digestion Feasibility Study." Iowa Department of Natural Resources. R.W. Beck, June 2004. Web. 25 Nov 2010. <http://www.iowadnr.gov/waste/policy/files/bluestem.pdf>.

[10] Jaworski, Carole. "UF Helps Animals Get Leftover Treats From Theme Parks and Restaurants." University of Florida News. University of Florida, 25 Nov. 1997. Web. 19 Nov 2010. <http://news.ufl.edu/1997/11/25/food/>.

[11] Uhlar-Heffner, Gabriella. "Seattle Studies Anaerobic Solution for Source-Separated Food Residuals." Biocycle 44.12 (2003): 39-40, 42. Web. 19 Nov 2010.

[12] Current Organic Waste Digestion Project Protocol." Climate Action Reserve. California Climate Action Registry, 07 Oct. 2009. Web. 25 Nov 2010. <http://www.climateactionreserve.org/how/protocols/adopted/organic-waste-digestion/current/>.

[13] Serve to Preserve." Renewable Energy Tax Incentives. Department of Management Services - State of Florida, 2010. Web. 25 Nov 2010.

[14] Jewella, William, Robert Cummings, and Brian Richards. "Methane fermentation of energy crops: Maximum conversion kinetics and in situ biogas purification." Biomass and Bioenergy 5.3-4 (1993): 261-278. Web. 25 Nov 2010.

[15] Turning Food Waste into Energy at the East Bay Municipal Utility District (EBMUD)." Region 9: Waste Programs. Environmental Protection Agency, 26 Aug. 2009. Web. 26 Nov 2010. <http://www.epa.gov/region9/waste/features/foodtoenergy/food-waste.html>.

[16] Landfill Methane Outreach Program." EPA Home. Environmental Protection Agency, 13 Oct. 2010. Web. Nov 2010. http://www.epa.gov/lmop/basic-info/index.html#a02.

[17] CHF International. Haiti Emergency Solid Waste Collection, Landfill Rehabilitation and Jobs Creation Program (SWM). (Cooperative Agreement # 521-A-00-04-00028-00). Final Report (July 28 2004 - April 30, 2005). Silver Spring, MD: CHF International, August 15, 2005.

[18] Rehabilitation of the Electricity Distribution System in Port-Au-Prince – Supplemental Financing." Document of the Inter-American Developmental Bank. Inter-American Developmental Bank, 2010. Web. 26 Nov 2010. <http://idbdocs.iadb.org/wsdocs/getdocument.aspx?docnum=35294747>.

[19] Gray, Donald, Paul Suto, and Cara Peck. United States. Anaerobic Digestion of Food Waste. East Bay Municipal Utility District, 2008, Web. 19 Nov 2010. <http://epa.gov/region9/organics/ad/EBMUDFinalReport.pdf>.

10

Design and computational fluid dynamic modeling of a municipal solid waste incinerator for Kampala city, Uganda

F. Ayaa[1], P. Mtui[2], N. Banadda[1], J. Van Impe[3]

[1]Department of Agricultural and Bio-Systems Engineering, Makerere University, Kampala, Uganda
[2]Department of Mechanical Engineering University of Dar-es-Salaam, Dar-es-Salaam, Tanzania
[3]Department of Chemical Engineering, Ku Leuven, Leuven, Belgium

Email address:
fayaa@caes.mak.ac.ug (F. Ayaa), plmtui@yahoo.com (P. Mtui), banadda@caes.mak.ac.ug (N. Banadda),
jan.vanimpe@cit.kuleuven.be (J. V. Impe)

Abstract: In Uganda, the government targeted to produce at least 15 MW from Municipal Solid Wastes (MSW) by end of 2012, which was not achieved. It is against this background that this project's twofold objective is to explore the energy potential of MSW in Kampala and design an environmentally friendly waste-to-energy incinerator for electricity generation. The obtained waste characterization results show that the average composition of MSW in Kampala city varied as follows: food and yard waste, 90.64 %; papers, 1.67 %; plastics, 1.77 %; polyethylene, 2.99 %; textiles, 0.59 %; glass, 1.16 %; metals, 0.15 % and others 1.03 %. The proximate analysis of the food and yard waste component indicated volatile matter of 73.29 %; fixed carbon of 4.36 %; moisture of 8.49 % and ash of 13.86 %. Furthermore, the ultimate analysis of the MSW on dry basis yielded Carbon 22.58 %; Hydrogen 3.22 %; Oxygen 14.06 %; Nitrogen 1.56 %; Sulphur 0.24 % and Ash 58.33 %. The Lower Heating Value (LHV) and Higher Heating value (HHV) of the MSW were 9.49 MJ/kg and 10.19 MJ/kg on dry basis respectively. The HHV and LHV of the food and yard waste determined from the bomb calorimeter was 15.11 MJ/kg and 14.68 MJ/kg, respectively. An incinerator was designed to suit the characteristics of the MSW and optimized using ANSYS Computational Fluid Dynamics (FLUENT Version 14, 2011). The total time needed to incinerate the waste was 31 minutes in comparison to 25 minutes for typical incinerators. The optimal capacity of the incinerator is also 460 kg/hr as opposed to the design capacity of 567 kg/hr.

Keywords: Waste Characterization, Computational Fluid Dynamics, Municipal Waste, Incineration

1. Introduction

One of the biggest challenges to solid waste management in developing countries is the lack of training, knowledge and skills, experience sharing and suitable technologies to handle ever growing volumes of municipal solid waste. According to[17], most municipal authorities in developing countries concentrate their limited finances available for MSW to richer areas of the municipalities where citizens with more political power reside. The urban poor are in most cases neglected and hence suffer from most of the life threatening conditions caused by a deficient solid waste management system. It is therefore important that municipal engineers and planners in developing countries develop and implement solid waste management approaches that are effective, viable and sustainable[8]. It suffices to mention that the current market for solid waste management technologies offers either too simple systems that require large land areas that will be less and lessavailable in urbanizing Africa or too complicated and are capital intensive and complex to operate. They also require a lot of energy input when many African countries are suffering a deficit in energy supply. Moreover, these systems do not benefit from the value of solid waste management i.e. nutrients and energy (biogas) recovery. Municipal waste in Uganda is generally composed of wet carbon and nitrogen rich materials that include: organic waste from households, agro-industrial waste (slaughter houses, food industry) and agro waste: manure and straw [10]. The volume of solid waste generated in urban centers of Uganda is increasing as a result of growing urban population, concentration of industries, consumption of residents, and inadequate finance and facilities to manage waste collection and disposal[12]. Approximately 1,500 tonnes of solid waste are generated in Kampala city daily [7]. Currently, municipal solid waste (MSW) is landfilled, openly burnt or haphazardly dumped. There is no control of greenhouse emissions or leachate at the landfill in Kiteezi, making it an enormous environmental risk. The need for suitable and tailored technologies for solid waste management in Uganda cannot be overemphasized.

The Uganda government targeted to produce at least 15 MW from MSW by end of 2012 and 30 MW by 2017[10]. Uganda has for a long time relied solely on hydropower for grid electricity but prolonged droughts have been a key factor to unreliable power supply [2]. Biomass, which supplies 90 % of the country's energy requirements, is threatened by the rapid deforestation rates. MSW is an untapped energy resource that has the ability to bridge the energy gap in Uganda. Incineration is a proven commercial technology and second most popular after landfilling. Experiences from the developed parts of the world indicate that business opportunities can be created along MSW treatment value chain to support livelihoods and clean the environment in communities by principally recovering materials (recycling), recovering energy, bio-converting to fuel and compost, and land filling of the remaining residues. One of the biggest challenges developing countries like Uganda face is the lack of tailor made technologies that are linked with MSW characteristics so that the output can be exploited in a useful manner to benefit communities.

Computational fluid dynamic (CFD) analysis of thermal flow in the combustion chamber of a solid waste incinerator could providecrucial insight into the incinerator's performance [15]. CFD also makes it possible to evaluate velocity, pressure, temperature, and species concentration of fluid flow throughout a solution domain, allowing the design to be optimized prior to the prototype phase [1]. It is against this background that this project seeks to design an environmentally friendly waste to energy incinerator, which can be used for electricity generation for especially households in Kampala city and use CFD modeling to optimize the incinerator design.

2. Materials and Methods

2.1. Study Area

This study was carried out at Kiteezi landfill, located about 12 km from the Kampala city center as shown in Fig. 1. The Kiteezi landfill covers 20 acres and is currently operated by the private company, Otada Construction Company. Truck crawlers are used to spread and scatter the waste in an effort to stimulate decomposition. Sometimes the waste is sprayed with insecticide to kill off flies before it is covered with soil[11]. The landfill has a leachate treatment plant which

uses mechanical aeration to reduce the Chemical oxygen demand of the leachate before it's released to the environment. Kiteezi landfill receives waste from Kampala city and surrounding areas but for this study the focus was on the waste received from Kampala city. Kampala, the capital city of Uganda is located 0°15′N and 32°30′E. It has a total area of 190km^2[9]. The city is experiencing rapid population growth due to immigration and natural increase[11] and it is estimated to have a population of 1.5 million inhabitants[7]. Kampala city has five divisions with Kawempe division being the poorest and central division which includes the Central Business District (CBD) being the wealthiest of them all[4]. Makindye division is mainly a residential area having a mixture of very low income and medium to high income areas in addition to being generally peri-urban in nature [11]. Kampala's other divisions are Rubaga and Nakawa.

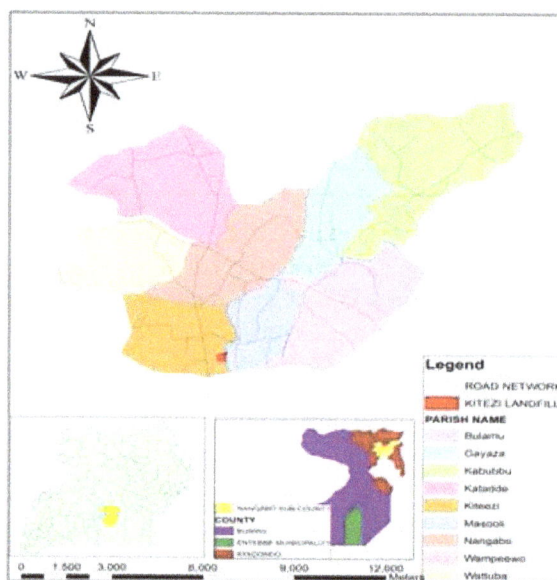

Figure 1. Kiteezi Landfill, Kampala (Uganda).

2.2. Determining the Physical Composition of the Waste

Trucks offloading solid waste at Kiteezi landfill, one from each of the five different divisions of Kampala, were randomly selected on each day of the analysis. The selected truck was then directed to empty its contents in the section of the landfill that was preselected for this study. Waste pickers (also known as scavengers) employed for analysis, assisted in manually sorting the waste into the different fractions given by[17] namely organics, hard plastics, metals, papers, soft plastics (polythene), glass, textiles and leather. All the other materials that were not classified into any of the previously fractions were considered to belong to the faction of others and this included items like medical wastes, inert items like soil and ash etc. The weights of the different fractions were then obtained and recorded in a data sheet by a research assistant before the proportion of the different fractions in the waste was calculated for each division. The organic fraction from each division was then thoroughly mixed up with the aid of spades and fork hoes before being spread out. A 5 by 2 grid was made on the spread out organic waste before 10 samples each weighing one kilogram were randomly picked from each grid. These samples were then mixed together to form one sample. From this sample, a new final sample weighing one kilogram was drawn and then placed into tightly sealed plastic bag before being taken to the laboratory for nutrient and energy content analysis respectively. This procedure was repeated for the organic waste in the other trucks selected from the different divisions of Kampala for thirty consecutive days and thereafter for two consecutive days after every two months. Samples were collected for a period of four months (September to December, 2012) on a weekly basis.

2.3. Proximate Analysis of the MSW

In determining moisture content of the samples, about 5 g of the sample brought in from the field, was weighed in to a dish that had been dried in an oven and weighed. The uncovered dish was then dried in an oven at 105°C for about three hours. The dish was covered and was transferred to desiccators and weighed as soon as the dish was cooled. The heating and weighing procedure was repeated until successive weight did not differ by more than one milligram. Loss in weights was recorded. The sample procedure is repeated for two other sub-samples and the average moisture content of the three sub-samples is taken to be the moisture content of the sample.

2.4. Ultimate Analysis & Heating Value Determination

The ultimate analysis of the MSW was estimated using the heat and material balance software [14]whereas the heating value of the MSW was determined byforming 1 g of the sample into a pellet. The pellet was placed in the sample pan of the bomb calorimeter (GallenKamp autobomb, United Kingdom, CAB001.ABC.C) and the energy content of the sample is determined following the procedure of [6].

2.5. Computational Fluid Dynamic (CFD) Modeling

The incinerator was designed using[16] and the geometry drawn in Solid Edge V 16, meshed and optimized using ANSYS Computational Fluid Dynamics (FLUENT Version 14, 2011). An Eulerian multiphase model was used. The energy conservation equation was solved for both solid and gaseous phases. Heterogeneous and homogeneous reactions were defined in the model to express the detailed combustion kinetics in the furnace. The design was optimized to suit the characteristics of the MSW through several iterations.

3. Results and Discussions

The physical composition of the MSW in Kampala is depicted in Fig 2. The major components of the MSW in decreasing order are agricultural waste, namely, food and yard waste, polyethylene, plastics, papers, glass, rubber, textiles and metals.

The proximate analysis of the MSW shown in Table 1 was used to define the solid waste.

Table 1. Proximate analysis of food and yard waste.

Property	%
Moisture content	8.486
Ash content	13.854
Volatile matter	73.295
Fixed carbon	4.365

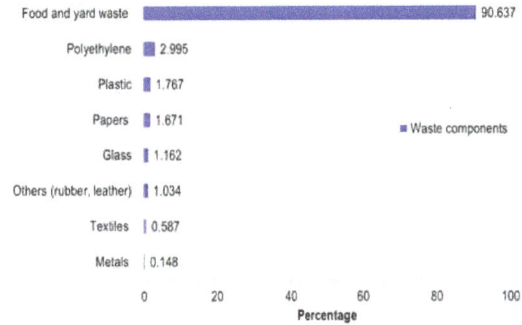

Figure 2. MSW composition of Kampala City (Uganda).

Table 2. Ultimate analysis of MSW (Dry basis)

Component	Weight %
C	22.58
H	3.22
O	14.06
N	1.56
S	0.24
Cl	0.00
F	0.00
Ash	58.33
Total	100.0

Table 2 shows the ultimate analysis of the MSW on dry basis. The ash content of the municipal waste from Kampala is high.The heat content of the food and yard waste and mixed MSW is shown in Table 3. The lower heating value (LHV) of the waste was approximately 9.49 MJ/kg. This energy content is highly suitable for a thermal energy conversion process. The minimum calorific value of waste should be higher than 6 MJ/kg in order to meet the temperature requirement of combustion[3].

Table 3. Energy content of MSW

MSW	Heating value MJ/kg - dry	
	HHV	LHV
Food and yard waste	15.11	14.68
Mixed MSW	10.19	9.49

However, a heating value of about 11 MJ/kg is needed to sustain combustion.An auxiliary fuel is thus required if the waste is to be incinerated.

In this study, the incinerator geometry used for simulation is shown in Fig. 3. The incinerator is symmetrical. MSW is fed from the top right which is the main air inlet. Also, two primary air inlets and a secondary air inlet were incorporated in the design to provide additional air for combustion. The incinerator walls were assumed adiabatic. The grate of the furnace was considered stationary during the simulations. The computational mesh generated from the incinerator geometry is shown in Fig. 4 consisting of polyhedral shapes.

The main model input parameters are summarized in Table 4. The location of the primary and secondary air inlets was optimized iteratively to achieve maximum utilization of air for complete combustion of the MSW. During simulation, Eulerian-Eulerian approach was considered for the gas-liquid phases system.

Table 4. Design criteria and operating conditions for incinerator

Object	Parameter	Unit	Comments
Incinerator	Overall length (m)	10.75	
	Overall height (m)	7.92	
	Overall width (m)	2.01	
	Wall material thickness (mm)	230	Refractory
	Insulation thickness (mm)	38	Clay
MSW feed	Length (m)	1.52	
	Width (m)	2.01	
MSW bed	Bed porosity	0.5	
MSW Feed Material	Feed flow rate (kg/s)	0.1278	(460 kg/ hr)
	Feed material density (kg/m3)	450	
	Feed material temperature (K)	300	
	Thermal conductivity (W/m-K)	0.15	
	Specific heat (kJ/kg-K)	2.5	
Main Air	Flow rate (kg/s)	12.36	
	Temperature (K)	300	Ambient Temp & Pressure
Primary Air #1	Flow rate (kg/s)	7.556	
	Temperature (K)	300	Ambient Temp & Pressure
Primary Air #2	Flow rate (kg/s)	5.107	
	Temperature (K)	300	Ambient Temp & Pressure
Secondary Air	Flow rate (kg/s)	9.188	
	Temperature (K)	300	Ambient Temp & Pressure

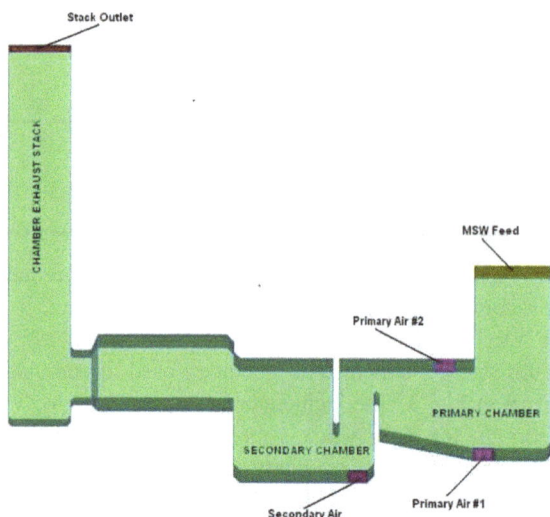

Figure 3. Incinerator geometry.

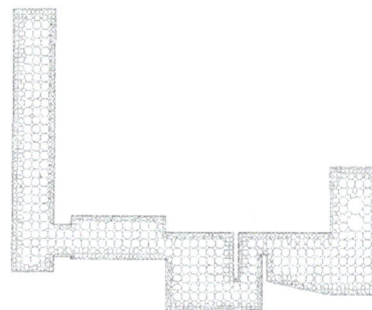

Figure 4. Mesh of incinerator geometry.

In contour plots (Fig 5-14) hereafter, the red and blue color represents the maximum and minimum values, respectively. The temperature profile inside the incinerator is shown in Fig. 5. The average temperature reached in the incinerator is about 2000 K (1727 °C) and the lowest temperature inside the incinerator is 300 K which corresponds with MSW feed temperature. However, from the 3D perspective, there are high temperature zones in the incinerator as shown in Fig. 6. A uniform temperature of the gases is achieved in the exhaust stack unlike in the primary or secondary chambers. The charging hood, as would be expected has a temperature of 300 K because the municipal solid waste is fed in at ambient temperature. There is a burner located inside the primary chamber (Fig. 7). Ignition is initiated at a temperature of 480 K and is increased until self-sustaining combustion of the MSW is achieved.

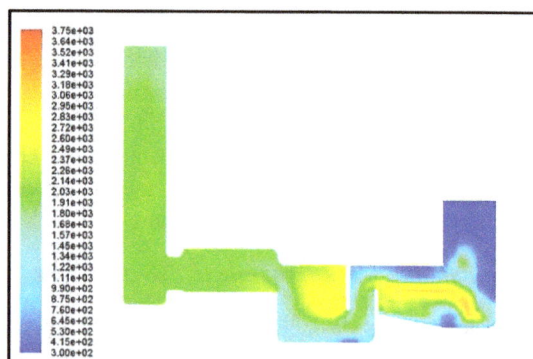

Figure 5. Static temperature of MSW-gas (K) across vertical mid-plane of the incinerator.

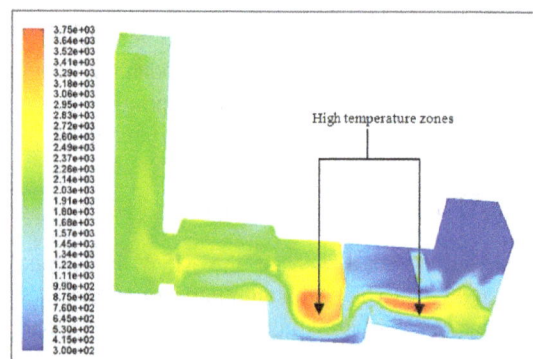

Figure 6. Temperature of MSW in a 3D view.

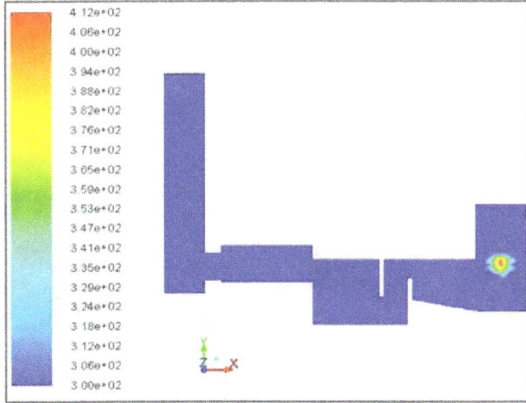

Figure 7. *Position and temperature (K) of burner.*

Fig 8 depicts MSW as enters the incinerator from top right at a concentration of about 0.867 (mass fraction). As expected, the concentration quickly drops down as the MSW devolatilizes as it reaches the secondary chamber where the temperature is high (See Fig. 5 and 6). The volatile species include tar, water, char and non-condensable gases.

Figure 9 depicts moisture concentration in the incinerator. As expected, the moisture is quickly evaporated as the MSW enters the secondary chamber where the temperature is high.Since MSW contains large percentage of volatiles, it produces large quantities of tar when pyrolysed as shown (Fig. 10). The tar falls onto the incinerator bed and quick is cracked to char and non-condensable gases (see Fig. 11). It is seen that large quantities of char is formed in the primary chamber. The char is deposited at the bed where it undergoes heterogeneous reactions (Table 5). Some of the char is combusted to provide heat in order to sustain the incineration process. It is observed that almost all of the char is consumed in the primary chamber.

Fig. 12 depicts very small quantities of carbon monoxide escaping from the exhaust stack. An insignificant amount of CO at exhaust implies all the carboneous material of MSW has been fully incinerated. Fig. 13 shows the concentration (mass fraction) of oxygen in the incinerator. Large quantities of unconsumed oxygen corresponds to the intake of MSW at top right. Also, as expected, higher O_2 concentrations are visibly seen at the locations of primary and secondary inlets. However, it is clearly seen that the oxygen is nearly completed at the exhaust stack, indicating that the MSW is fully incinerated with about 0.5% oxygen at stack exist.

Fig. 14 is a contour plot depicting the concentration of water vapor in the incinerator. It is seen that large quantities of water vapor emanate from the exhaust stack as product of combustion as well as from the raw moist MSW.

Incinerators that have been constructed and tested and have been found to be satisfactory have an operating cycle of 7 h of charging at 25 minute intervals, and a burn down through the next 17 h [13]. The total flow time obtained during the simulations was however 31 minutes, therefore municipal solid waste from Kampala requires a longer combustion period compared to typical solid waste. This is because the waste has a high percentage of organic waste as

previously stated and high moisture content. The incinerator capacity obtained during simulations is 460 kg/hr. unlike the design capacity of 567 kg/hr. This is critical during loading, to ensure complete combustion of the waste.

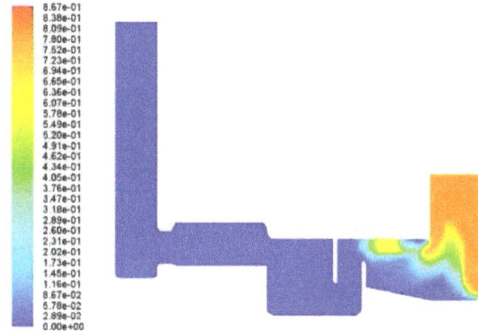

Figure 8. *Mass fraction of Municipal solid waste (Dry Ash free).*

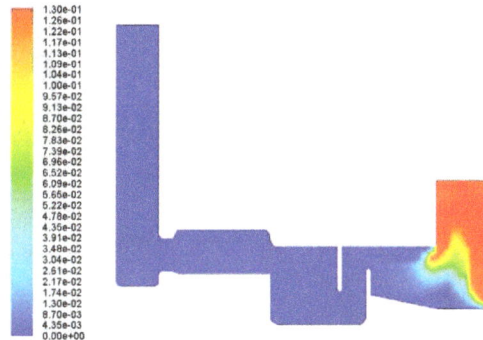

Figure 9. *Mass fraction of moisture.*

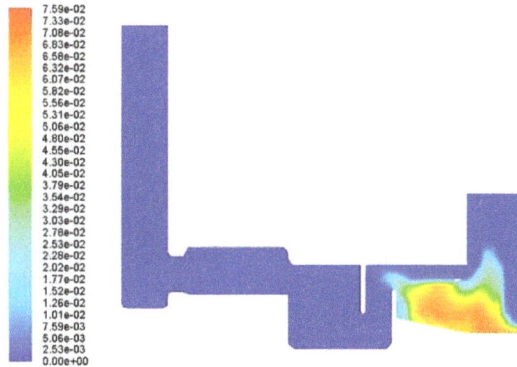

Figure 10. *Mass fraction of tar.*

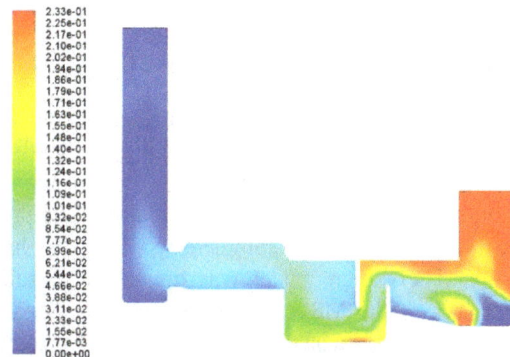

Figure 11. *Mass fraction of Char*

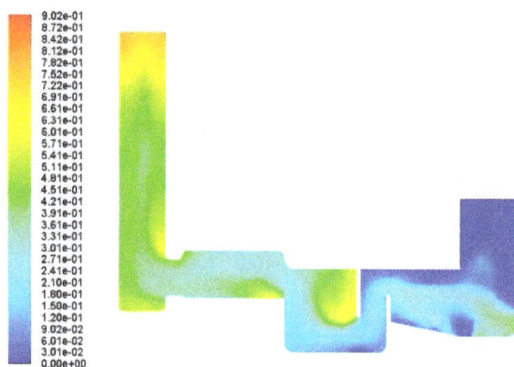

Figure 12. Mass fraction of carbon monoxide.

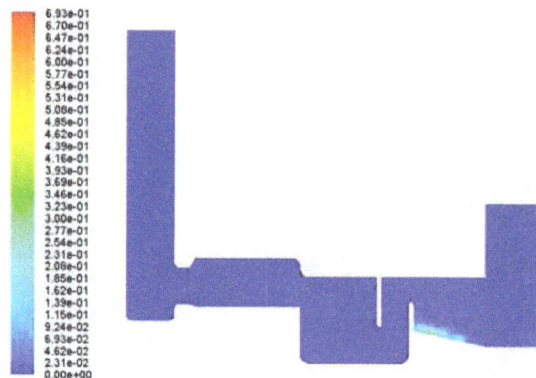

Figure 14. Mass fraction of water vapor.

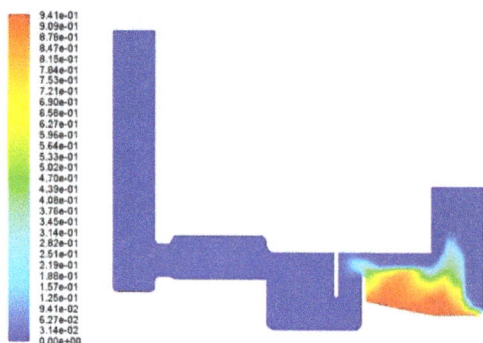

Figure 13. Mass fraction of Oxygen

Table 5. Heterogeneous and homogeneous chemical reactions.

	Process	Chemical Reaction	Pre-exponential factor (s-1)	Activation Energy(J/kmol)
1	Drying	$H2O\ (l) \rightarrow H2O\ (g)$	2.56E+02	8.79E+04
2	Pyrolysis	$C_6H_{10}O_4 \rightarrow 2.33\ CH_{2.92}O_{0.938} + 3.087\ C + 0.0272\ CH_4 + 0.233 CO + 0.3298 CO_2 + 0.6599\ H_2 + 0.9277\ H_2O\ (g)$	1.0715E+2	7.78E+4
3	Tar cracking	$CH_{2.92}O_{0.938} \rightarrow 0.39411\ CH_4 + 0.572\ CO + 0.0579\ CO_2 + 0.282769\ H_2O\ (g)$	1.0715E+2	6.63E+07
4	Gasification 1- Water-gas	$C + H_2O\ (g) \rightarrow CO + H_2$	1.00E+08	1.26E+08
5	Gasification 2 – Bourdard	$C + CO_2 \rightarrow 2\ CO$	1.00E+08	1.125E+08
6	Gasification 3-methanation	$C + 2\ H_2 \rightarrow CH_4$	1.00E+05	1.26E+08
7	Water-gas-shift	$CO + H_2O \rightarrow CO_2 + H_2$	13.89	1.26E+07
8	Oxidation 1	$C + O_2 \rightarrow CO_2$	4.75E+05	2.00E+08
9	Oxidation 2	$CO + 0.5\ O_2 \rightarrow CO_2$	2.239E+05	1.7E+08
10	Oxidation 3	$H_2 + 0.5\ O_2 \rightarrow H_2O\ (g)$	9.87E+07	3.10E+07
11	Oxidation 4	$CH_4 + O_2 \rightarrow CO + 2\ H_2O$	2.11E+10	2.02E+08

4. Conclusion and Recommendation

The major components of the MSW are agricultural waste, namely, food and yard waste. Also in this study, the incinerator design was successfully optimized using computational fluid dynamic (CFD) modeling. The process was iterative to achieve complete combustion in the incinerator. The design model was optimized to suit the characteristics of the MSW through several iterations. The total time needed to incinerate the waste was 31 minutes in comparison to 25 minutes for typical incinerators. The

capacity of the incinerator is 460 kg/hr as opposed to the design capacity of 567 kg/hr. The residence time of the particles inside the chamber was not sufficiently predicted. There is need therefore for further study to optimize the incinerator design with the residence time. It was also assumed that the grate was stationary, but to for proper mixing of the waste, it is recommended that the grate is moving. CFD simulations of a moving grate would give a more accurate solution and should thus be explored. In conclusion, municipal solid waste from Kampala is suitable for incineration and it is recommended for a prototype to be constructed to obtain experimental results. These results can be used to upscale the technology to commercial scale for electricity production.A study to design the auxiliary equipment for the incinerator is also recommended. The socio-economic implications of the incinerator for Kampala were not studied during this research and thus need to be explored in another study.

References

[1]   ALENTEC. Computational fluid dynamics modeling,applications for engineering solutions. [Online] [Cited:January16,2013.]http://www.alentecinc/com/papers/CFD/Statement%20of%20 Qualifications-CFD.pdf.

[2]   Electricity Regulatory Authority. Constraints to investment in Uganda's electricity generation industry. [Online]2008.[Cited:January16,2012.]http://www.era.or.ug/Pdf/constraints to investment-In%20Generation.pdf.

[3]   Green, K. Industrial Ecology and Spaces of Innovation. [ed.] Sally Randles Kenneth Green. Massachusetts : Edward Elgar Publishing, 2006. ISBN-13:978 1 84 542097 0.

[4]   Golooba, M.F. "Devolution and outsourcing of municipal services in Kampala city, Uganda:An early assessment."2003. Vol. 23, pp. 405-18.

[5]   Howard, G., Pedleyb,S., Barrett, M., Nalubega, M., Johal, K. "Risk factors contributing to microbiological contamination of shallow groundwater in Kampala,Uganda." 2003. Vol. 37.

[6]   Jessup, R.S. Precise measurement of heat of combustion with a bomb calorimeter. [Online] 1960. [Cited:November12,2012.]http://digital.library.unt.edu/ark:/67531/metadc13253/m2/1/high_res_d/NBS%20Monograph%207.pdf.

[7]   Kampala Capital City Authority. Services rendered by KCCA. www.kcca.go.ug. [Online] [Cited: February 1, 2013.]http://www.kcca.go.ug/services.php#waste management.

[8]   Kaseva, M.E. and Mbuligwe, S.E. "Appraisal of solid waste collection following private sector involvement in Dar-es-salaam city,Tanzania.". 2005. Vol. 29, pp. 353-366. DOI:10.1016/j.habitatint.2003.12.003.

[9]   Matagi, S.V. "Some issues of environmental concern in Kampala, the capital city of Uganda." 2002. Vol. 77, pp. 121-138.

[10]   Ministry of energy and mineral development. The renewable energy policy for Uganda. s.l. : Ministry of energy and mineral development, 2007.

[11]   Mugagga, F. The public-private sector approach to municipal solid waste management. How does it work in Makindye division, Kampala district,Uganda? Norwegian university of science and technology. Trondheim : s.n., 2006. Doctoral dissertation.

[12]   National Environment Management Authority. "State of environment report for Uganda 2006/07.". Kampala : National Environment Management Authority, 2007.

[13]   Niessen, W.R. Combustion and incineration processes. Applications in Environmental engineering. 4th. s.l. : CRC press group, 2010.

[14]   —. Heat and material balance spreadsheets. Combustion and incineration processes. Applications in Environmental Engineering. [Software]. s.l. : CRC press group, 2010.

[15]   Ryu, C., Shin, D. and Choi, S. "Combined simulation of combustion and gas flow in a grate-type incinerator." Vol. 55., pp. 174-185.

[16]   US Army Corps of Engineers. Engineering and design incinerators mobilization construction. Engineering manual No.1110-3-176. Washington DC. : US Army Corps of Engineers, 1984.

[17]   Zurbrügg, C. "Urban solid waste management in low-income countries of Asia, how to cope with the garbage crisis.". Durban, South Africa. : s.n., 2002.

11

Synthesis and characterization of biodiesel from castor bean as alternative fuel for diesel engine

Molla Asmare[*], Nigus Gabbiye

Centre of Competence for Sustainable Energy Engineering, Institute of Technology, Bahir Dar University, Bahir Dar, Ethiopia

Email address:

mollaasmare98@gmail.com (M. Asmare)

Abstract: This paper deals with the transesterification of Ricinus Communis (RC) oil with methanol to produce biodiesel in the presence of KOH as a catalyst. Moreover, this study analysis the fuel properties of RC biodiesel and diesel fuel blend to use castor oil methyl ester as a possible alternative fuel for diesel engines. Various properties of the RC biodiesel and their blends such as density, kinematic viscosity, iodine value, saponification number, Cetane number, heating value, flash point and acid value were determined. The experimental results were compared well with American Society for Testing and Materials (ASTM D6751) and European biodiesel standards (EN 14214). The experimental design as well as statistical analysis were done and analyzed using design expert 8.0.7.1 version soft ware. The predicted optimum conditions for castor oil biodiesel production were a reaction temperature of 59.89^0c, methanol to oil ratio of 8.10:1 and a catalyst of 1.22 wt% of oil. The methyl ester content under these optimum conditions was 94.5% w/w of oil, and all of the measured properties of the biodiesel met the international standards of EN14214 and ASTM D 6751 with the exception of density and viscosity. Therefore, the viscosity and density of the ester was high and further reduced by blending with diesel fuel up to B45 to satisfy within the ASTM D6751 and EN 14214 limits for biodiesel.

Keywords: Castor Seed, Castor Oil, Biodiesel, Transesterification, Blend, Response Surface Methodology

1. Introduction

The developments of societies have accompanied by an increase in growing energy needs. Their energy requirements have achieved through the combustion of various materials (oil, coal and natural gas) which considered as fossil fuels and therefore non-renewable, which creates environmental problems. These facts have converged in the search for renewable energy sources such as Biofuels: a non-toxic, biodegradable, agricultural source, with a high heating value and oxygen content [1].

Global warming is one of the greatest environmental threats facing our planet caused by increasing in atmospheric Green House Gases (GHG) due to human activities since the start of the industrial era [2]. When fuels were burnt, there are just a few basic types of primary exhaust emissions (oxides of nitrogen (NO_X), Carbon monoxide (CO), hydrocarbons (HC), Carbon dioxide (CO_2) and particulate matter (PM)). In addition to these, primary pollutants reactions in the atmosphere generate secondary pollutants that cause acid rain, photochemical smog and tropospheric ozone depletion. Many of these pollutants have serious implications on human health and the environment. Consequently, many countries have established strict environmental policies and regulations that must meet by all automobile manufacturers.

The search for alternative fuels started when the pollution created by the burning of fossil fuels shows severe environmental problems because biofuels have a significant role in overall reduction of CO_2 emissions [3]. Bioethanol is the most well known biofuels used in gasoline engines. Similarly, manufacturers have worked with biodiesel, as it is the most common alternative fuel for traditional diesel engines.

Among the most promising sources, vegetable oils and animal fats have attracted much attention as a potential resource for the production of biodiesel, which is quite similar to conventional diesel in its main characteristics and can be easily blended with diesel fuel in any proportion with minor or no modifications to the engine as well as fuel system [4]. The production and use of biodiesel have increased significantly in many countries around the world using numerous feedstock sources. Unfortunately, it is in nascent status in many African countries.

Over the past decade, the consumption of transport fuels in Sub-Saharan Africa has increased at a rate of 7% per year in line with increased economic activity [5]. This has had a great economic impact on about thirty-five crude oil importing countries in Africa. However, they have large landmass for farming and abundance resource of edible and inedible oils, some of which grow in the wild. This shows that Sub-Saharan African countries are a region with a high potential feedstock for biodiesel production.

Biodiesel is the general name for fatty acid alkyl esters and the most common alternative fuel for traditional diesel engines. It can be produced by transesterification in which oil or fat is reacted with a monohydric alcohol in the presence of appropriate catalyst. To complete a transesterification reaction Stoichiometrically, a 3:1 molar ratio of alcohol to triglycerides (TGs) is necessary. Practically, the ratio needs to be higher to drive the equilibrium to a maximum ester yield. The process of transesterification is affected by the mode of reaction, molar ratio of alcohol to oil, type of alcohol, nature and amount of catalysts, reaction time and temperature.

Several studies have been carried out using different oils, alcohols (methanol, ethanol, isopropanol and butanol) and catalysts, notably homogeneous ones; sodium hydroxide, potassium hydroxide, sulfuric acid and supercritical fluids or enzymes such as lipases [6]. Consequently, a number of reasons for converting oils and fats into biodiesel have been also discussed. Some of the main problems with oils and fats are high viscosity and low volatility that cause the formation of deposits in engines due to incomplete combustion and incorrect vaporization characteristics [7].

Ricinus communis L (RC), which belongs to the family of euphorbiaceous is one of biodiesel feed stock. The bean contains toxin that makes the oil and cake inedible. It grows very well on marginal land, is drought and pest resistant, and has a yield of about 1413 liter per hectare when cultivated. The beans contain 40 – 60 % oil by weight for high yield breed type. It is one of the highest viscosities among vegetable oils. The oil produced from the seed of the castor plant (*Ricinus communis*) has stimulated some interest as a biofuels. Its oil coloration ranges from a pale yellow to colorless, and has a soft and faint odor with a highly unpleasant taste. The fatty acids in a castor oil contain a hydroxyl functional group, which cause castor oil to be more polar than other vegetable oils. Since the polarity of castor oil is sufficiently high, the oil mixes completely with methanol during the biodiesel reaction. It is made up of TGs; 91-95% Ricinoleic acid, 4-5% Linoleic acid, 2-4% of oleic acid and 1-2% Palmitic and Stearic acids. Ricinoleic acid, a monounsaturated, 18-carbon fatty acid, is unusual in that it has a hydroxyl functional group on the twelfth carbon. As a result, the automotive industry uses castor oil for the production of high performance motor oil and braking fluids [8].

Nowadays, all Ethiopian petroleum products imported either through the port of Djibouti or from Sudan. Besides the cost of fuel, long distance transportation adds to the cost of the fuel getting to Addis Ababa that causes a large burden on Ethiopia's trade balance. One main issue is that around 65% of Ethiopian export earnings are to pay for the import of petroleum products. Despite the availability of huge energy resources, the current level of harnessing this energy is very low. This is due to poor socio-economic situation in the country on the one side, and a low level of awareness about the potential and value of energy by most stakeholders on the other side. Amongst the identified alternative renewable energy sources, biofuels in particular energy crops received attention as a promising and sustainable energy sources, of which, biodiesel has arisen as a potential candidate for a petro diesel substitute that minimize the escalating budgetary pressure for diesel oil.

The development of biodiesel is a recent and at its initial stage in Ethiopia. Ethiopia has ample potential and opportunities for the production and utilization of biodiesel from Jatropha, castor oil, palm oil (edible), and neem instead of petroleum. However, the country did not yet benefited from those biodiesel feedstocks because of educated human power and economical constraints. Among them, Castor bean is a non-edible biodiesel feedstock to substitute the consumption of fossil fuel. In addition, it is widely available and has no any other commercial purpose, has high oil content and yields per hector, grows in marginal land and has a resistance for variable climatic and soil conditions. So far, there is no observable market on the use of biodiesel products. However, within a short period a significant number of foreign, local, and joint companies have invested in the biodiesel industry. Nevertheless, the search for feedstock other than jatropha is still at its ground level. The objective of this work was to synthesize and characterize biodiesel from Castor seed using homogeneous alkali catalyst (KOH) via transesterification reaction and comparing the physico-chemical properties with international biodiesel standards.

2. Materials and Methods

2.1. Chemicals and Apparatus

Chemicals: Castor bean was purchased from local market and processed in to castor oil to be used as a raw material. The crude castor oils were neutralized to reduce the acid value using NaOH pellet and anhydrous Methanol of 99.9% purity; potassium hydroxide and other chemicals were of analytical reagent grade. Those chemicals, which were used during the experiments, were purchased from neway private limited company and used without further purification. Moreover, diesel oil was purchased from total diesel station from local market and used during the experimental studies.

Apparatus: The basic equipments used during the experimentations were oven dry, Glass reactor; temperature controlled hot plate equipped with magnetic stirrer, condenser, Centrifuge, hydrometer, Vibro viscometer, conical flask, different size beakers and flasks, sensitive balance, PH electrode, burette.

2.2. Experimental Set Up and Descriptions

2.2.1. Raw material Preparation

The de-pulped nuts were sun dried on the open floor for 10 days to remove the moisture from the seeds and to ease the detachment of the seeds from the nuts. Then a sample was randomly selected to determine moisture content in the nuts according to AOAC Standard (AOAC, 1980). 100 gm of each sample was oven-dried at 100^0C for 7h in Debre Markos University, Agricultural College, plant and natural science Laboratory. The dry sample was cooled at room temperature in the desiccators and re-weighed to determine the weight loss. The test was replicated three times and the average moisture content of the castor bean was found to be 26%. Then, the dry nuts were bagged in plastic and stored for subsequent use in a moisture free container.

2.2.2. Cooking and Grinding

Mortar and pestle were used to crush the beans into a paste (cake) in order to weaken or rupture the cell walls to release castor fat for extraction. The process of heating breaks down the cells containing the oil and liquefies the oil to improve the extraction process. Hence, the ratio of kernel weight to raw castor bean seed weight (W) was calculated by the following formula:

$$W = \frac{mass\ of\ castor\ kernel}{mass\ of\ raw\ castor\ bean} * 100 \qquad [2.1]$$

2.2.3. Oil Extraction

The grounded fine powders of castor seed kernel was cooked and then dried for 8h at 80^0C in drying oven model 202-1AB. Cooking was done to coagulate protein (which is necessary to permit efficient extraction) and to free the oil for efficient pressing. The Castor bean oil was extracted from the seeds by soaking in hot boiled water until the oil floats and then allowed to settle until the impurities

precipitated. The traditional procedures of castor oil extraction permit the extraction of oil from the kernel. However, it is very inefficient and time consuming. The percentage by mass of crude oil extracted from castor kernels was 19.2%, which is about 34.9% of the oil present in the kernel [8]. The extracted oil was settled for two weeks then decanted and filtered with the help of filter paper to remove all the suspension particles from the extracted oil. After completing separation, the purified oil was stored in closed container at room temperature.

The amount of oil extracted was calculated with the following formula:

$$\%\ of\ castor\ oil = \frac{Mass\ of\ castor\ oil}{Initial\ mass\ of\ castor\ seed} * 100 \qquad [2.2]$$

2.2.3.1. Pretreatment of Crude Castor Oil

The extracted crude castor oil may contain phosphatides (phospholipids), gums and other complex compounds that could promote hydrolysis (increase FFA) of vegetable oil at the time of storage. In addition, during transesterification process, these compounds could interfere. Therefore, they were removed by acid pretreatment (degumming) process.

Figure 2.1. Experimental set up during degumming process for purification of crude castor oil

The acid pretreatment loss was calculated with the following formula:

$$Acid\ pretreatment\ loss = \frac{Weight\ of\ crude\ oil - Weight\ of\ pretreated\ oil}{Weight\ of\ crude\ oil} \qquad [2.3]$$

To neutralizing the acid value and FFA of crude castor oil 0.5N NaOH was added to the degummed castor oil and heated the reaction mixture while stirring until the temperature reached 80°C to break any emulsion that might have formed during neutralization. Moreover, Sodium chloride (10% of the weight of oil) was also added to settle out the soap formed. Then the mixture was

transferred into a separating funnel and allowed to stand for 1hr to remove the soap formed and hot water was added repeatedly to the oil until the soap remaining in the solution was removed. The caustic pretreated oil was then drawn off into a beaker. The final FFA content was determined and the caustic pretreatment loss was then calculated by the following formula:

$$Neutralization\ loss = \frac{weight\ of\ degmmued\ oil - weight\ of\ neutralized\ oil}{weight\ of\ degummed\ castor\ oil} \qquad [2.4]$$

Finally, the neutralized oil was passed through hydration process by the addition of 30% hot distilled boiled water in which the mixture was stirred for 2 minutes and allowed to stand in the separating funnel until two clear phase observed. Thereafter, the aqueous layer was removed at the bottom. The procedure was repeated to ensure the removal

of most gums and soaps. This process continued until the pH of the oil reached almost neutral. Then the oil was dried in oven at 120^0C for 2hrs to remove the water present in it.

2.2.4. Experimental Setup

Batch transesterification reactor system was employed in this work as shown in Figure 2.2. A 500ml capacity three-necked glass reactor equipped with magnetic stirrer that provide the mixing requirement in a temperature controlled hot plate, which was a capable of controlling the temperature with a deviation of 1^0C. In addition, the condenser provides cooling system for the experiment to control the leak of methanol by supplying cooled water in the inlet and the hot water was rejected in the outlet part.

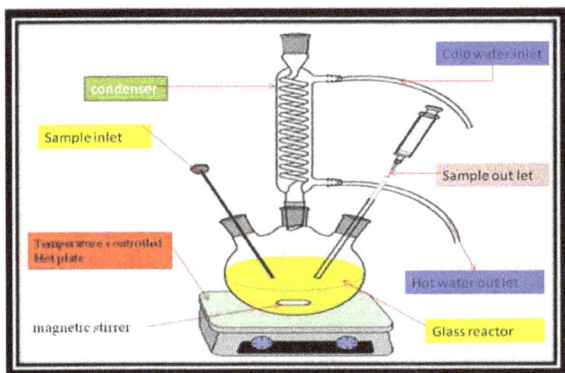

Figure 2.2. *Experimental set-up for biodiesel production through transesterification [15, 16]*

2.3. Characterization of Pretreated Castor Oil and its Biodiesel

The physicochemical properties of pretreated oil have to be determined prior to biodiesel production process. The feedstock status determination helps not only to know the condition of the oil but also helps to make certain decision on whether it requires further treatment or not. The main physico-chemical properties that have to be determined are percentage of FFA content, AV, SN, IV, kinematic viscosity and density. These parameters directly or indirectly affect the quality of the final product the so-called biodiesel.

Moisture content determination of castor seed kernel: Empty dish was weighed with and without cooked, grounded, and dried castor kernel. Then 100gm of cleaned sample was weighed and dried in a digital drying oven model 202-1AB at 80°C for 8hrs and the weight was taken after every 2hrs. After each 2hrs, the sample was removed from the oven and placed in the desiccators for 30 minutes to cool. The procedure was repeated until a constant weight obtained. Finally, the weight was taken and compared with the initially recorded weight. The percentage weight in the kernel was calculated using the formula:

$$Moisture\ content = \frac{Wi - Wf}{Wf} * 100\% \qquad [2.5]$$

Where, Wi = initial weight of sample before drying;
Wf = Weight of sample after drying.
Determination of Specific Gravity (SG): Density bottle (volumetric cylinder) was used to determine the density of the oil. A clean and dry bottle of 50ml capacity was weighed (W_0) and then filled with the oil, stopper inserted and reweighed to give (W_1). The oil was substituted with water after washing and drying the bottle and weighed to

give (W_2). The expression for specific gravity is:

$$SG = \frac{W_1 - Wo}{W_2 - Wo} \qquad [2.6]$$

Similarly, the same procedure was applied to determine the SG of biodiesel using ASTMD 4052.

Determination of Viscosity (μ): Digital Vibro viscometer was used to determine the viscosity of oil and biodiesel. The kinematic viscosity was determined at 40°C followed by ASTM D445-09. The temperature of a water bath was set at 40°C and calibrated. 50 ml of sample was placed into the viscometer and allowed the viscometer and sample to equilibrate to the water bath for 30 minutes. The sample was kept in the water thermostat bath until it reaches the equilibrium temperature of 40 °C. After maintaining the equilibrium temperature, the Vibro viscometer tip was inserted to the sample to measure the dynamic viscosity and the reading was taken from the controller

Determination of Acid Value (AV): The AV of the oil was determined using the method described by IUPAC (1979) and modified by Egan et al. (1981).

25ml of diethyl ether and ethanol mixture was added to 5gm of oil in a 250ml conical flask and the solution was titrated with 0.1N ethanolic KOH solution in the presence of 5 drops of phenolphthalein as indicator until the endpoint (colorless to pink) is recognized with consistent shaking. The volume of 0.1 N ethanolic KOH (V) for the sample titration was recorded.

The total acidity of oil in mg KOH/ gram was calculated using the following equation:

$$AV = \frac{56.1 * N * V}{W} \qquad [2.7]$$

Where, V = the volume expressed in milliliter of 0.1N solution of ethanolic KOH.

W = the weight of oil sample (the mass in gram of the test portion)

N = concentration of ethanolic KOH

Then, the % FFA value was calculated from the acid value using the following relationship:

$$\%FFA = \frac{AV}{2} \qquad [2.7a]$$

The acid value of biodiesel was determined by applying the ASTMD 664.

Determination of Saponification Number (SN): Indicator method was used as specified by ISO 3657 (1988). The SN determination was conducted by dissolving the oil in an ethanolic KOH solution. 2g of the sample was weighed into a conical flask then 25ml of 0.1N ethanolic potassium hydroxide solution was added. The content was constantly stirred, and allowed to boil gently for 60min. A reflux condenser was placed on the flask containing the mixture. Few drops of phenolphthalein indicator was added to the warm solution and then titrated with 0.5M HCl (volume Va was recorded) to the endpoint until the pink color of the indicator just disappeared. Then a blank determination was carried out upon the same quantity of potassium hydroxide solution at the same time and under the same conditions and (volume Vb was recorded). The result was calculated using equation:

$$SN = \frac{56.1*N*(Vb-Va)}{W} \qquad [2.8]$$

Where W= weight of oil taken in gram, N= normality of HCL solution,
Va= volume of HCL solution used in test in ml,
Vb= volume of HCL solution used in blank in ml.

The same procedure was used to determine the SN of biodiesel as discussed above.

Determination of Iodine Value (IV): The method specified by ISO 3961 (1989) was used. 0.4g of the sample was weighed into a conical flask and 20ml of carbon tetra chloride was added to dissolve the oil. Then 25ml of Dam's (Iodine monochloride) reagent was added to the flask using a safety pipette in fume chamber. Stopper was then inserted and the content of the flask was vigorously swirled. The flask was then placed in the dark place for 2.5 hours. At the end of this period, 20ml of 10% aqueous potassium iodide and 125ml of water were added using a measuring cylinder. The content was titrated with 0.1N sodium-thiosuphate solutions until the yellow color almost disappeared.

Few drops of 1% starch indicator was added and the titration continued by adding sodium thiosuphate drop wise until blue coloration disappeared after vigorous shaking. The same procedure was used for blank test and other samples. The iodine value (IV) is given by the expression:

$$IV = 12.69 * N * \left(\frac{V2-V1}{M}\right) \qquad [2.9]$$

Where, N = normality of sodium thiosuphate, V1 = Volume of sodium thiosuphate
V2 = Volume of sodium thiosuphate used for blank, M = Mass of the sample

The same procedure was used to determine the Iodine value of biodiesel.

Determination of Heating Value (Calorific Value): The HHV of the castor oil and its biodiesel was determined using the empirical formula suggested by Demirbas (1998).

$$HHV = 49.43 - [0.041(SN) + 0.015(IV)] \qquad [2.10]$$

Determination of Cetane Number (CN), ASTMD 613: The Cetane number of the biodiesel was determined using

the empirical formula suggested by (Kalayasiri et al., 1996), using the result of SN and IV of the biodiesel.

$$CN = 46.3 + \left(\frac{5458}{SN}\right) - 0.225(IV) \qquad [2.11]$$

Determination of Flash Point, ASTMD 93: The FP of the biodiesel was determined using empirical formula by Ayhan Demirbas (2008)[9]. The equation between FP and HHV for biodiesel is:

$$HHV = 0.021FP + 32.12 \qquad [2.12]$$

2.4. Experimental Design for Base Catalyzed Biodiesel Production

In order to optimize the reaction factors, a five-level three-factor central composite design (CCD) was utilized in this study. In order to gain information regarding the interior of the experimental region and to evaluate the curvature, this study was conducted in 20 experiments in accordance with a 2^3 complete factorial design, six central points and six axial points (star points). The distance of the star points from the center point is provided by $\alpha = (2^n)^{1/4}$, in which n is the number of independent factors, for three factors α=1.68 [10]. The variable ranges adopted, as provided in Table 2.1. Table 2.1 describes the coded and un coded independent factors. The methanol to oil molar ratio, catalyst concentration and reaction temperature were the independent variables selected to optimize the conditions for FAME production using KOH catalyst. The reaction period and rotational speed was set at optimum point where maximum conversion could be achieved based on literature data at atmospheric pressure for all runs. The responses measured were the yield of FAMEs. These independent variables were assigned as (−1, 1) interval where the low and high levels were −1 and +1, respectively. The axial points was located at (±α, 0, 0), (0, ±α, 0) and (0, 0, ±α) where α is the distance of the axial point from center and makes the design rotatable

Twenty experiments were carried out & data was statistically analyzed by Design-Expert 8 program to find suitable model for the percentage of FAME as a function of the above three variables. The central values (zero level) chosen for experimental design were the Methanol to oil molar ratio of 6:1, Catalyst concentration of 1% (w/w), and Temperature 55°C.

Table 2.1. Independent variables and levels used in CCD for base-catalyzed transesterification process.

Variable (Factors)	Factor Coding	Unit	*Levels				
			-1.68	-1	0	+1	+1.68
Reaction temperature(T)	A	°C	46.9	50	55	60	63.4
Methanol to Oil ratio(M)	B	-	0.96	3	6	9	11.04
Amount of Catalyst(C)	C	Wt%	0.16	0.5	1	1.5	1.84

2.5. Base-Catalyzed Transesterification Reaction

Initially, pretreated Castor seed oil was poured into a

three-necked 500ml glass reactor and then preheated at 120°C to remove the moisture content using temperature controlled hot plate for 30 minutes as shown in fig 2.2. In order to maintain the catalytic activity, the solution of KOH in methanol was freshly prepared so that prolonged contact with the air was not diminishing the effectiveness of the catalyst through interaction with moisture and carbon dioxide. The catalyst solution was added slowly to the preheated oil until the reaction was completed. After the reaction was accomplished, the mixture was allowed to settle under gravity for 24 hrs in the separatory funnel at room temperature. During separation, two layers were formed in such a manner that the crude ester phase present at the top and the glycerol phase at the bottom. The upper layer consists of ME, methanol traces, residual catalyst and other impurities, whereas the lower layer consists of glycerin, excess methanol, catalyst and other impurities. The glycerin and other impurities were removed from biodiesel by opening the tap provided at the bottom.

2.6. Purification of Biodiesel

After separated from the glycerin layer, the MEs layer were purified by washing with warm distilled water by adding 1-2 drops of acetic acid at 60°C until the washing water have a neutral pH value. The gentle washing action of hot distilled water to crude ME ratios were 3:1. Gentle washing prevents the possibility of losing the ME due to the formation of emulsions and results in a rapid and complete phase separation [11]. Then, the excess methanol and any remaining water was removed from the ME layer by heating the product at 120 ℃ [7].

The primary purpose of biodiesel washing step was to remove any soap formed during transesterification reaction. In addition, warm water with acetic acid provides neutralization of the remaining catalyst and removes the formed salts. The use of warm water prevents precipitation of saturated fatty acid esters and retards the formation of emulsions with the use of gentle washing action. Finally, Biodiesel properties such as density, viscosity, FP, CN, AV, SV, IV, and calorific value was determined and compared with ASTM6751 and EN14214 standards

3. Results and Discussion

3.1. Castor Bean Oil Extraction and Purification Process

Moisture Content Determination
The amount of sample was weighted using a sensitive balance for each experiment. Then, it was dried in digital drying oven of model 202-1AB at 80°C for 8hours. Again, the weight of the sample after drying was measured. Five experiments was conducted and the moisture content was determined for each of them and the averaged value of castor kernel seed moisture content was found to be 3.82%.This result varied for literature findings of 4.15% [12] and (5-7)%[13].

3.2. Pretreatment of Crude Castor Oil Processes

Acid Pretreatment Process
Based on the method discussed in chapter 2: 3% (v/v) of hot distilled water and 2% (v/v) phosphoric acid is required for degumming crude Castor bean oil in order to remove phosphatides, gums and other complex compounds, which enhance the hydrolysis of FFA. Therefore, the crude oil was degumming using 40ml of phosphoric acid and 60ml of distilled hot water. After treatment, the amount of oil loss was determined and obtained as 4% (80ml). This shows that crude castor oil contains water insoluble impurities which increases FFA and phosphoric acid have a power to remove these impurities.

Caustic Pretreatment Process
Norris (1982) points out that during the caustic pretreatment of the high FFA oils, a loss of oil is normally three times the amount of FFA. This has been observed in soybean and cottonseed oil with high FFA of more than 5%. In this study, the caustic pretreatment reduces the FFA of crude castor oil from 3.52 to 0.932%. From 1.92liter (1.84kg) of acid pretreated castor oil, 14% oil was loosed during caustic pretreatment .This is higher than that of acid pretreatment oil losses due to saponification and occlusion of oil in the soap stock. Moreover, the amount of AV present in crude castor oil before treatment was 7.04mgKOH/g whereas after neutralization using caustic soda, the acid value minimize to 1.86mgKOH/g that shows caustic soda have a power to neutralize the FFA found in the oil.

3.2.1. Characterization of Pretreated Castor Bean Oil

Using the various formulae as indicated in the experimental procedure, the physico-chemical properties of the pretreated oil were evaluated. The density, viscosity, AV, percentage of FFA, SN, IV and HHV of the purified Castor bean oil were determined and the results are presented in Table 3.1

Table 3.1. Physico-chemical properties of crude and pretreated castor oil

Property	Experimental Result		Unit
	Crude Castor oil	Pretreated castor oil	
Specific Gravity	0.9628	0.9618	-
Density at 15°C	0.9628	0.9618	g/ml
Kinematic viscosity at 40°C	-	208.96	mm²/s
Acid Value	7.049	1.862	mg KOH/g oil
Composition of Free Fatty Acid	3.52	0.931	%
Saponification Number	-	185.3645	mg KOH/g oil
Higher Heating Value	-	39.7	MJ/kg
Iodine value	-	87.9	gI2/100g

Specific gravity: The Specific gravity values for both crude and refined oil were obtained nearly the same (0.9618). Hence, the density of the oil is determined using the specific gravity. Therefore, the density of oil was 0.9618g/ml that are in agreement with the reported in literature [14].

Kinematic viscosity: The viscosity of oil was measured using Vibro viscometer. The device detects the dynamic viscosity, which is the resistance to flow with vibration. The observed kinematic viscosity was 208.96mm^2/sec that is in agreement with literature data [13].

Acid Value: The chemical properties analysis shown in Table 3.1 indicates that the acid value of crude and pretreated castor oil is 7.04 mgKOH and 1.862mg KOH/g of oil respectively. The value is higher in crude oil due to FFA (3.52%) present; while it is less for degummed and neutralized oil because of 0.5N of NaOH used in the treatment of the crude oil, which must have neutralized some of the free fatty acid present in it. The result agrees within the range specified in literature.

The FFA value was also calculated from the AV relation using Eq (2.5a) and determined as 3.52% and 0.931% for crude and purified oil respectively. Therefore, the percentage of FFA value in purified oil was in range to use alkali–catalyzed transesterification process.

Saponification Value/Number (SN): The SN was calculated using Equation (2.6) and the observed value was taken for three trials and the average value obtained as 185.3645 mg of KOH/g of oil, which is in agreement with the result specified for quality castor oil.

Higher Heating Value: It was determined using empirical formula given in Equation (2.10).The calculated value of HHV in the oil was equal to 39.7 MJ/Kg

Iodine Value: It is the measure of the degree of unsaturation of a particular oil or fat. It was determined using titration. The observed value of iodine in the oil was equal to 87.9g I2/100g which is in the range(82-88I2/100g) reported in literature[15].

3.3. Transesterification Reaction

The yield and characteristics of biodiesel is depending on the type of oil used due to variation in the fatty acid composition and other characteristics of oil. Taking into consideration this aspect, the castor oils from inedible sources have been taken as a raw material for the preparation of biodiesel using KOH catalyst and methanol alcohol.

Various reaction parameters such as alcohol to oil molar ratio, concentration of catalyst and temperature have been taken for the study to analyze their effect on the yield and the characteristics of biodiesel. The results obtained are discussed as follows:

3.3.1. Effect of Operating Conditions on Biodiesel Yield
Effect of Methanol to Oil Molar Ratio

After selecting methanol for transesterification reaction, the effect of its concentration on yield and characteristics of biodiesel from castor oils pretreated with phosphoric acid

and caustic soda was investigated. Biodiesel was prepared from this oil at different molar ratio of methanol to oil. As shown in Figure 3.1, the methanol to oil molar ratio is one of the most factors that affect the conversion of triglyceride to FAME. The Stoichiometry of the transesterification reaction requires three mol of alcohol per one mol of triglyceride to yield three mol of fatty esters and one moles of glycerol [6]. However, to shift the transesterification reaction to the right and to achieve equilibrium, it is necessary to either use more than 100% excess alcohols or remove one of the products from the reaction mixture continuously in order to produce more FAME products.

Several researchers studied the effect of molar ratio (from 1:1 to 6:1) on ester conversion with vegetable oils(Soybean, sunflower, peanut and cottonseed) behaved similarly and achieved highest conversions (93–98%) at a 6:1 molar ratio[7] where as experiment conducted by Dennis Y.C leung(2008) has recorded a yield of 93.1% for a methanol-to-oil ratio of 8:1 and 10:1.

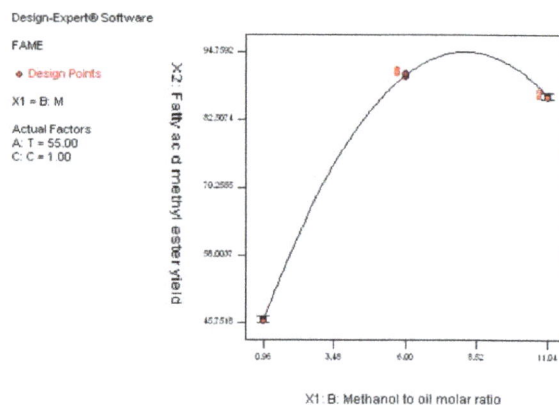

Figure 3.1. *The effect of methanol to oil ratio verses methyl ester yield*

Similarly, the result obtained in this study as shown in Figure 3.1, the methanol-to-oil ratio has a great influence to the yield of methyl ester. When the methanol to oil molar ration increased from 0.96:1 to 8.10:1, the methyl ester yield is increased and the saponification value decreases. However, the yield started to decrease when the molar ratio increase beyond 1:8.10. This is due to separation problem resulted from excessive methanol, minimize the contact of access of triglyceride molecules on the catalyst's active sites, which could decrease the catalyst activity, losses during washing step and interference the separation of glycerol because of increasing glycerol solubility. Moreover, methanol with one polar hydroxyl group can work as an emulsifier that enhances emulsion causing separation of ester layer difficult from the water layer (Leung and Guo, 2006) and when glycerin remains in solution, it will drive the equilibrium back to the left, which lowering the yield of esters. Therefore, the optimum operating condition for biodiesel production using KOH catalyst is obtained at 8.10:1 methanol to oil molar ratio in this study.

The Effect of catalyst concentration: To study the effect

of catalyst concentration on yield and characteristics of biodiesel obtained from castor oil, the biodiesel was prepared with different amount of catalyst (0.16 to 1.84% by weight of oil). The trend of yield with respect to catalyst concentration is shown in Figure 3.2. As we could understand from the figure below, the yield of biodiesel increases with increase in amount of catalyst up to 1.22% and then decreases. At lower concentration of catalyst, the reaction is incomplete as a result lower yield was obtained where as at higher concentration of catalyst ,the yield decreases due to the enhancement of saponification reaction causing triglyceride to form soap faster than ester. Dorado et al. (2004) and Encinar et al. (2005) have reported that the formation of soap in presence of high amount of catalyst increases the viscosity of the reactants and thus lowers the rate of biodiesel production

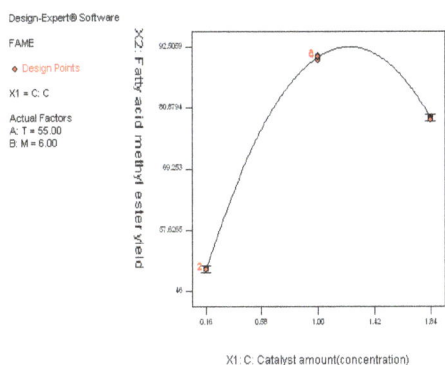

Figure 3.2. The effect of catalyst amount verses methyl ester yield

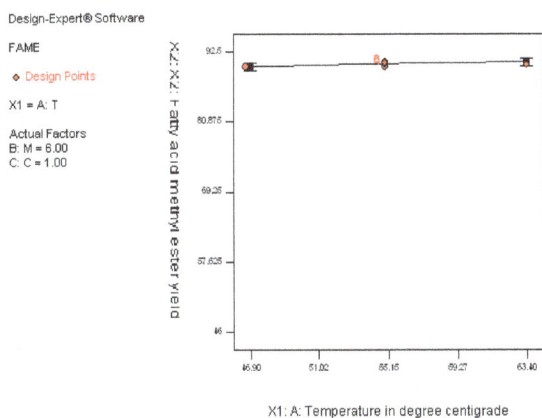

Figure 3.3. The effect of reaction temperature verses methyl ester yield.

Effect of reaction temperature: Several researchers have studied the effect of temperature on conversion of oils and fats into biodiesel. Their results indicate that as the temperature increase, the methyl ester content increase due to the viscosity of the oil decreases, which resulted in an increase in the solubility of the oil in the methanol, leading to an improvement in the contact between oil and methanol. However, in this experiment, the temperature increment effect was not significant on biodiesel yield but there is a little change on biodiesel yield as the temperature increases and decreases as show in Figure 3.3. The increase in the

yield of FAME at higher reaction temperature is due to higher rate of reaction and molecular collision. Moreover, from the experimental model analysis and ANOVA, the p-value of the temperature term in both liner and quadratic model was greater than the p-value limit. Hence, the result showed that increases in reaction temperature did not significantly affect the fatty acid ester content at any of the tested oil-to-methanol molar ratios in the process of castor oil transesterification. In a similar case, researchers reported that the reaction temperature did not affect the castor oil alcoholysis reaction because castor oil is soluble in ethanol at room temperature [17]. However, other studies show that the reaction temperature exerted a significant degree of influence on the rate of biodiesel synthesis using vegetable oils and fats as feedstock [10].

3.4. Optimization of Biodiesel Production from Castor Oil in Base-Catalyzed Transesterification Process Using Response Surface Method

The selected independent reaction parameters (reaction temperature, methanol to oil molar ratio and weight of catalyst) were optimized using design expert software. In contrast with the classical optimization process, this may lack to account the effectiveness of different combination of parameters. Response surface method (RSM) provides elaborate vision over various combinations of parameters.

In order to optimize the reaction factors for castor oil biodiesel production, a CCD with a five-level three-factor design was employed. Table 3.2 describes these experimental parameters and their results based on the CCD experimental design. Twenty designed experiments were conducted and analyzed with multiple regressions using Design-Expert 8.0.7.1 software. Regression analysis yield three linear coefficients (A, B, C), three quadratic coefficients (A^2, B^2, C^2) and three cross product coefficients (AB, AC, BC) for the full model (Table 3.3). Table 3.3 also describes the ANOVA for the response surface quadratic model. The transesterification was carried out using the previously shown experimental setup at Figure2.1. The reaction was carried out using a 500ml capacity three necked glass reactor, which is equipped with a magnetic stirrer in a temperature-controlled hot plate. The statistical analysis of the biodiesel was discussed below.

Statistical analysis on Factors Affecting Biodiesel Yield

The Design Expert 8.0.7.1 program was used in the regression analysis and analysis of variance (ANOVA). The Statistical software program was used to generate surface and counter plots using the fitted equation obtained from the regression analysis, holding one of the independent variables constant. Experimental as well as predicted values of percentage conversion of the oil to biodiesel at the design points are shown in Table 3.2. The actual yields of biodiesel produced at different process parameters were calculated and obtained ranged from 46% to 92.5%. The yield of the transesterification processes were calculated as the sum of weight of FAME produced to weight of oil used multiplied by 100.The formula is given as:

$$\text{Yield of FAME} = \frac{\text{weight of FAME}}{\text{weight of oil used}} * 100\ \% \quad [3.1]$$

Table 3.2. CCD arrangement and response for alkali transesterification reaction processes

Std	Run	Coded factor			Actual factors			FAME (%)		
		A	B	C	T: Temperture	M: Methanol	C: Catalys	Actual value	Predicted value	Residual
6	1	+1	-1	+1	60.0	3.00	1.50	72	71.36	0.64
15	2	0	0	0	55.00	6.00	1.00	90.5	90.50	-1.194E-003
5	3	-1	-1	+1	50.00	3.00	1.50	70	70.12	-0.12
9	4	-1.68	0	0	46.59	6.00	1.00	90	89.96	0.040
12	5	0	+1.68	0	55.00	11.05	1.00	86.5	86.73	-0.23
13	6	0	0	-1.68	55.00	6.00	0.16	50	50.15	-0.15
11	7	0	-1.68	0	55.00	0.95	1.00	46	46.31	-0.31
14	8	0	0	+1.68	55.00	6.00	1.84	78.6	78.99	-0.39
16	9	0	0	0	55.00	6.00	1.00	90	90.50	-0.50
17	10	0	0	0	55.00	6.00	1.00	90.6	90.50	0.099
8	11	+1	+1	+1	60.00	9.00	1.50	92.5	92.39	0.11
18	12	0	0	0	55.00	6.00	1.00	90.5	90.50	0.099
1	13	-1	-1	-1.68	50.00	3.00	0.50	51	50.72	0.28
4	14	+1	+1	-1	60.00	9.00	0.50	78	77.50	0.50
2	15	+1	-1	-1	60.00	3.00	0.50	51	50.96	0.038
19	16	0	0	0	55.00	6.00	1.00	90.8	90.50	0.30
7	17	-1	+1	+1	50.00	9.00	1.50	92	91.66	0.34
10	18	+1.68	0	0	63.41	6.00	1.00	90.2	90.78	-0.58
3	19	-1	+1	-1	50.00	9.00	0.50	77.5	77.76	-0.26
20	20	0	0	0	55.00	6.00	1.00	90.7	90.50	0.20

Development of Regression Model Equation

The model equation that correlates the response (yield of the castor oil to FAME) of the transesterification process variables in terms of actual value after excluding the insignificant terms was given below. The predicted model for percentage of FAME content (FAME %) in terms of the coded factors is shown below.

Final Equation in Terms of Coded Factors:

$$FAME = +90.50 + 0.24 * A + 12.02 * B + 8.57 * C - 0.12 * A * B + 0.25 * A * C - 1.37 * B * C - 0.046 * A^2 - 8.48 * B^2 - 9.17 * C^2 \quad [3.2]$$

Where, A = Reaction temperature, B= Molar ratio of methanol to oil, C=weight of catalyst

The statistical analysis of the ANOVA is given in .Table 3.3. The multiple regression coefficients were obtained by employing a least square technique to predict quadratic polynomial model for the FAME content (Table 3.4). Hence, the best fitting model was determined. The model was selected based on the highest order polynomial where the additional terms were significant and the model was not aliased as suggested by the software. The coefficients of the response surface model as provided by the above quadratic model equation was also evaluated. From the ANOVA of response surface quadratic model for FAME conversion, the Model F-value of 2808.95 and Prob > F of <0.0001 implied that the model was significant. For the model terms, values of Prob>F less than 0.0500 indicate that the model terms are significant. In this case B, C, BC, B^2 and C^2 are significant model terms (all have Prob > F less than 0.050).

This tells us the methanol to oil ratio, catalyst, and their quadratic terms affect the yield much significantly. However, the interaction terms were found to be insignificant except BC Since the values greater than 0.1000 indicates the model terms were insignificant (Table 3.3).

As we observe from p-values of the model coefficients in Table 3.3, the value of the methanol to oil molar ratio and catalyst in both linear and quadratic model are much less than 0.0001. This indicated that they are the most significant in determining the model than the rest. However, in order to minimize error, all of the coefficients were considered in the design. The lack of fit from the ANOVA analysis indicated that the model does indeed represent the actual relationships of reaction parameters, which are well within the selected ranges. The Lack of Fit F-value of 4.17 implies its insignificant relative to the pure error. Non-significant lack of fit is good because we want the model to fit.

Table 3.3. Analysis of variance (ANOVA) for response surface quadratic model of alkali transesterification process

Source	Sum of Squares	Difference	Mean Square	F -Value	P-value, Prob > F	Significance
Model	5072.81	9	563.65	2808.95	< 0.0001	Significant
A	0.82	1	0.82	4.06	0.0715	Not Significant
B	1972.12	1	1972.12	9828.14	< 0.0001	Significant
C	1004.06	1	1004.06	5003.75	< 0.0001	Significant
AB	0.13	1	0.13	0.62	0.4483	Not Significant
AC	0.50	1	0.50	2.49	0.1455	Not Significant
BC	15.13	1	15.13	75.38	< 0.0001	Significant
A^2	0.031	1	0.031	0.15	0.7037	Not Significant
B^2	1035.94	1	1035.94	5162.65	< 0.0001	Significant
C^2	1211.27	1	1211.27	6036.40	< 0.0001	Significant
Residual	2.01	10	0.20	-	-	-
Lack of Fit	1.62	5	0.32	4.17	0.0717	Not Significant
Pure Error	0.39	5	0.078	-		
Cor Total	5074.81	19				

Table 3.4. Regression coefficients and significance of response surface quadratic model for the base catalyzed

Factor	Coefficient estimate	Difference	Standard error	95%CI low	95% CI high	VIF
Intercept	90.50	1	0.18	90.09	90.91	
A-T	0.24	1	0.12	-0.026	0.51	1.00
B-M	12.02	1	0.12	11.75	12.29	1.00
C-C	8.57	1	0.12	8.30	8.84	1.00
AB	-0.12	1	0.16	-0.48	0.23	1.00
AC	0.25	1	0.16	-0.10	0.60	1.00
BC	-1.37	1	0.16	-1.73	-1.02	1.00
A^2	-0.046	1	0.12	-0.31	0.22	1.02
B^2	-8.48	1	0.12	-8.74	-8.22	1.02
C^2	9.17	1	0.12	-9.43	-8.90	1.02

Final Equation in Terms of Actual Factors:

$$FAME = 32.29297 + 0.20210 * T + 16.68521 * M + 90.49174 * C - 8.33333E - 003 * T * M + 0.10000 * T * C - 0.91667 * M * C - 1.84764E - 003 * T^2 - 0.94205 * M^2 - 36.67147 * C^2 \quad [3.1a]$$

Table 3.5. Model adequacy of quadratic model for alkali catalyzed transesterification

Std. Dev	Mean	C.V. %	R-Squared	AdjR-Squared	Pred R Squared	Adeq Precision
0.45	78.42	0.57	0.9996	0.9992	0.9973	145.489

Where, T = reaction temperature, M= molar ratio of methanol to oil, C= weight of catalyst

Model Adequacy Check

The quality of the model developed was evaluated based

on the correlation coefficient value, R square (R^2). The R^2 value for Equation (3.1) was 0.9996. This indicated that 99.96 % of the total variation in the biodiesel yield was attributed to the experimental variables studied. The closer the R^2 value to unity, the better the model will be, as it will give predicted values, which are closer to the actual values for the response

The Pred R-Squared" of 0.9973 is in reasonable agreement with the "Adj R-Squared" of 0.9992. "Adeq Precision" measures the signal to noise ratio. A ratio greater than 4 is desirable. The model ratio of 145.489 indicates an adequate signal. This model can be used to navigate the design space. The value of the adjusted coefficient of determination (Adj R^2 =0.9992) is also high, thus indicating the significance of the model as well as the value of coefficient of variation (CV) is low (0.57%), thereby indicating the reliability of the results of the fitted model.

From the ANOVA and regression analysis on Table3.3 and Table3.4 respectively, it can be seen that the linear terms (B, C), the quadratic term (B^2, C^2) and the cross product BC were significant (because Prob > F less than 0.05), but the interactions (cross products) AB, AC and A, A^2 were insignificant.

The graph of the predicted values obtained using the developed correlation versus actual values forms a line of unit slope, i.e. the line of perfect fit with points corresponding to zero error between predicted values and

actual values as shown in Figure 3.4. The results in Figure 3.4 demonstrated that the regression model equation provided a very accurate description of the experimental data, in which all the points are very close to the line of perfect fit. This result indicates that it was successful in capturing the correlation between the three-transesterification process variables to the yield of FFA.

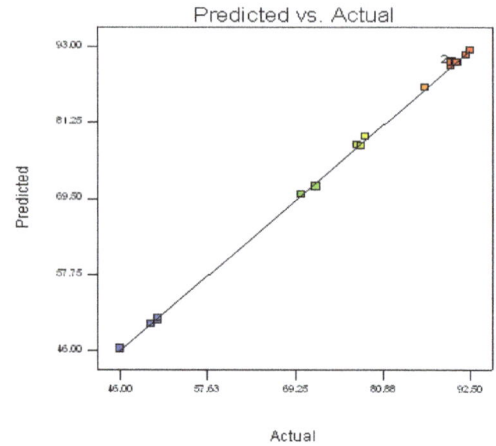

Figure 3.4. *Plot for actual vs. predicted value of FAME yield*

Effect of interactive operating conditions on biodiesel yield

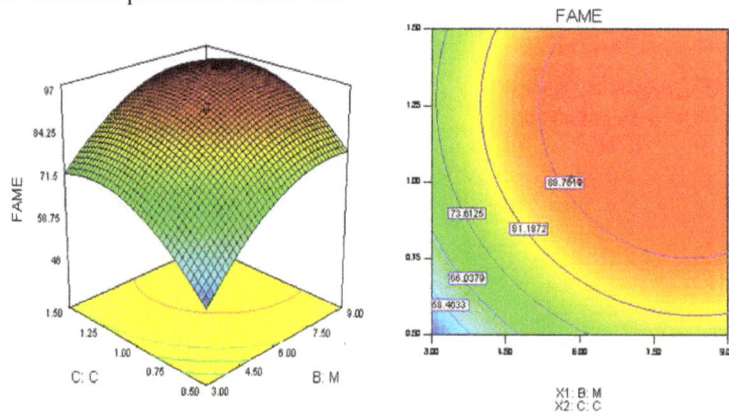

(a) Methanol to oil ratio verses catalyst amount when reaction temperature @ 55^0C

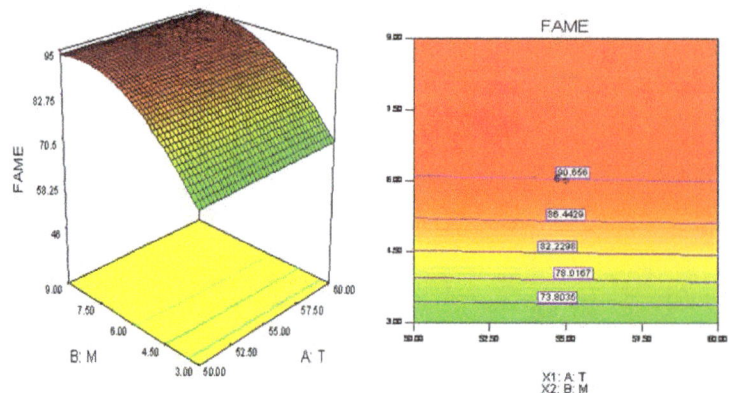

(b)Methanol to oil ratio verses Reaction temperature when the catalyst weight is 1%

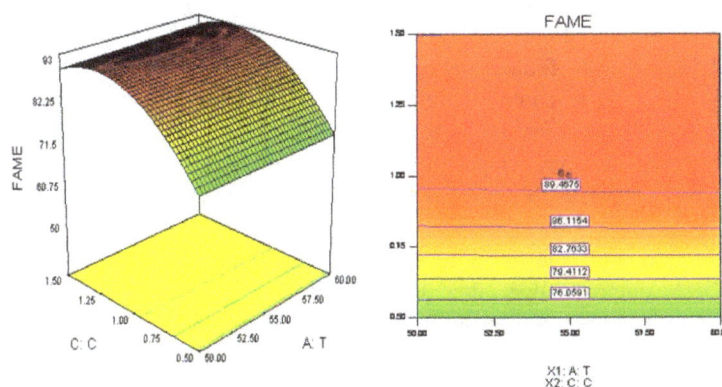

(c). Catalyst amount verses reaction temperature when the molar ratio is 6:1.

Figure 3.5. Response surface (to the left) & Contour (to the right) plot of FAME yield (%) in terms of coded factors (a, b & c)

Surface and Contour plots (Figure 3.5a-c) were drawn to show the relationships between dependent and independent variables of the developed model. Each contour curve presented the effect of two variables on the methyl ester yield, holding the third variable at constant level. The third variable is held at the selected zero level. However, the interaction factor also must be considered since the individual effect plot does not give information regarding the significant interaction involved. Remarkable interaction between the independent variables could be observed, if the contour plots had an elliptical profile. The relationship between independent and dependent variables of the developed model in the response surface plots at the stationary value of 6:1 methanol-to-oil molar ratio, 1% of catalyst concentration and 55^0C Reaction temperature is shown in Figure 3.5.

From Equation (3.1), it was clearly shown that, all the linear terms had positive coefficients, whereas the quadratic terms and the interaction terms had negative coefficients except AC. Therefore, an increase in temperature, KOH and methanol to oil molar ratio to a certain extent could result in a higher percentage of FAME. However, a reduction in the percentage of FAME could be obtained when using too high KOH, and methanol to oil molar ratio. Figure 3.5a showed that, strong interaction between methanol to oil molar ratio (M) and KOH catalyst concentration (C). This can also be confirmed by the high p-values of the interaction parameters. It could also be seen from Figure 3.5a, the FAME yield increased with increasing catalyst concentration at first. However, when the catalyst concentration reached 1.22%by weight of batch oil, the reverse trend was observed. Similar pattern was followed when increasing methanol to oil molar ratio. This is due the positive coefficient for the linear parameters (A B, and C) played the main role when the KOH catalyst concentration and methanol/oil molar ratio were at lower level. While at higher level, the interaction as well as the quadratic terms shows negative significant effect that leads to decrease the yield since the methanol and triglyceride in the oil are immiscible. Addition of catalyst can facilitate the transesterification reaction and rapidly increase the yield.

However, when the catalyst concentration was too high, soap could be quickly formed which made the separation of glycerol from biodiesel more difficult, thus reducing the yield. Similarly, the increase of the methanol amount, on one side, it will drive the reaction to the right since the reaction is an equilibrium process; on the other hand, excess methanol will help to increase the solubility of glycerol , which favors the backward reaction to the left. Therefore, the yield of FAME is decreasing.

Figure 3.5b showed that, the effect of methanol to oil ratio and the reaction temperature when the level of catalyst concentration was fixed. At low methanol to oil ratio, the percentage of FAME increased with reaction temperature increase. In addition, the FAME yield increases with increased molar ratio at a certain level.

Figure 3.5c showed that, the effect of reaction temperature and catalyst concentration on the methyl ester yield when the level of methanol/oil molar ratio was fixed. At a certain level of catalyst concentration, increase in reaction temperature (T), increases the methyl ester yield. An explanation to this has been attributed to the fact that, higher initial temperature helps in faster settlement of glycerol. However, the increments of temperature affect the FAME yield in a positive manner until 60^0C. After that, the effect was negligible. This could be explained by the higher p-value and the negative coefficient for the reactive and quadratic term in the model, which indicates non-significant effect.

Optimization of Process Variables

The results above have shown that three-transesterification process variables and the interaction among the variables that affect the yield of FAME. Therefore, the next step is to optimize the process variables in order to obtain the highest yield using the model regression developed. The methanol to oil molar ration, catalyst weight and the interaction between them are highly and significantly affect the transesterification process. From optimization function in Design Expert 8.0.7.1, it was predicted that at the following conditions (8.10:1 methanol to oil molar ratio, 1.22% catalyst concentration and 59.89 ^0C of reaction temperature) an optimum FAME yield of

94.5% could be obtained. In order to verify this prediction, experiments were conducted and the results were comparable with the prediction. It was found that the experimental value of 93.5% of FAME content, which agreed well with the predicted value. Therefore, this study shows that KOH is a potential catalyst for the production of biodiesel from castor bean oil via homogeneous transesterification process.

The optimization result also tells the same result as the ANOVA output. The ANOVA output shows that the methanol to oil molar ratio, catalyst weight and the interaction between them are highly and significantly affect the transesterification process.

3.5. Physicochemical Properties of Biodiesel

Specific gravity: It was found and observed in the range of 0.920 to 0.932. Therefore, the density of the biodiesel was measured and values were found to be in the range of 920 to 932kg/m^3. When the result is compared with the EN14214, which is 860–900 kg/m^3 for biodiesel, the value is out of the range. The change in the density shows that the density of the biodiesel decreased with increasing molar ratio. This was probably due to a decrease in residual triglycerides. Moreover, the density of the biodiesel also decreased with increasing reaction temperature and catalyst amount. Therefore, further reduction on the density of the biodiesel is required to satisfy the international biodiesel standards.

Kinematic Viscosity: The viscosities of the biodiesel produced at lower temperature are higher than that of the corresponding experiments conducted with the same feed ratio at higher temperature. This is due to the effects of operating parameters that affect the transesterification reaction. On the other hand, the viscosity of the biodiesel increased slightly with decreases in reaction temperature.

Increase in molar ratio decreases the viscosity to some extent. This is probably because of the free area created for the triglycerides to convert to biodiesel as the molar ratio increased. However, as the molar ratio increases it inhibits the contact between the triglycerides and the catalyst. Hence, no change in viscosity is observed when excess molar ratio was used. Viscosity decreased up to optimal catalyst concentration then it was almost constant.

Transesterification reaction is responsible for minimizing the viscosity of vegetable oil in order to apply it as a fuel for engines yet it is significantly affected by temperature. This experimental result shows a viscosity of (12.5 to 20 mm^2/sec) which is out of both the ASTM (1.9 to 6mm^2/sec) and EN14214 (3.5 to 5mm^2/sec) range for the requirement of biodiesel viscosity.

According to the results, it has been determined that pure RC biodiesel usage can cause problems in the injection system because of its high viscosity. In order to solve the viscosity problem it can be suggested that RC biodiesel may use a mixture of others either diesel or biodiesels. Therefore, in this paper further reduction of viscosity is done by blending with diesel fuel, which is the best

solution for RC biodiesel usage in compression ignition engines.

Acid Value: The Acid value of the biodiesel was found to be in the range of 0.324 to 0.784 mgKOH/g and 0.567 mgKOH/g at optimum conditions. The result indicates that the acid value of the oil (1.86mgKOH/g) decreased significantly after transesterification reaction. Furthermore, higher acid value resulted in low yield of biodiesel. Acid value affects storage ability of biodiesel by Contact with air and water, which is the major factors affecting storage stability. Oxidation is usually accompanied by an increase in the acid value and viscosity of the fuel. In the presence of water, the ester can hydrolyze to long-chain FFA, which also causes the acid value to increase. The castor bean oil biodiesel has acid values within the standard specification limit of max 0.8 in ASTM D664.

Heating Value: The heating value of biodiesel depends on the composition of the fuel. Since all the oils have very nearly the same carbon, hydrogen and oxygen contents, the gross and net heating values of each fuel per unit mass will be close to each other. Biodiesel has lower energy content (lower heating value) than conventional diesel fuel. The result obtained (40.5MJ/kg) is nearly the in the range of ASTMD6751 for diesel oil.

Iodine value: All of the measured IVs value falls in the En14124 standard. Higher IV indicates a higher quantity of double bonds in the sample and greater potential to polymerize in engine and hence lesser stability. The process of transesterification reduces the iodine value to a small extent. The EN14214 requirement is a maximum of 120 where as the result shows a maximum value of 86 I$_2$/100I$_2$.

Cetane Number: Although the viscosity and the density of RC biodiesel were noted to be greater than that of diesel fuel, the Cetane number was found in the range of EN 14214. Cetane number is known as a measurement of the combustion quality of diesel fuel. It has been observed that Ricinus Communis biodiesel has a higher Cetane number, which causes shorter ignition delays, and thus, higher efficiency in engine. CN was determined using empirical formula and obtained an average of 57.11 for nine selected samples based on their higher percentage of FAME (>90%) and 57.7 at optimization condition. The results showed that most of them have increased the CN within the permissible minimum limit. In general, diesel engines will operate on fuels with CN > 47(ASTM D613).

Flash Point: Equations were developed for the calculation of the Higher Heating Value of vegetable oils and biodiesel from their viscosity (v), density (ρ) and flash point (FP) [9]. The FP was determined and the values are ranged from 131.2 to 135.0^{0}C. Hence, the FP of the castor bean oil biodiesel lies within the ASTM6751 (>130^{0}C) and EN14214 (>101^{0}C) permissible range.

3.6. Determination of Diesel and Biodiesel Blending Proportion

When biodiesel is blended with petro diesel, the concentration of biodiesel is always written as BXX. The

'XX' refers to the percentage volume of Biodiesel. For example, pure 100 % biodiesel will be named as B100 and B20 is 20% Biodiesel and 80% petroleum diesel. Biodiesel can be used as B100 (neat). However neat biodiesel; because of having a narrow range of boiling points, slightly higher viscosity and density requires blending with petroleum diesel. Hence, blends with a petroleum diesel, at different proportions as B5, B10, & B20 were used in different literatures. As a result, Cloud point and pour point are adjusted by blending. Blending up to 5% is also useful for lubricating purpose [15].

In this study, the properties of B100 and its B5, B10 ,B15, B20, B25, B30 , B35, B40, and B45 mixtures were tested and compared to those of petroleum diesel and acceptable value is obtained within the specified for biodiesel in the ASTM D 6751 standard (with the exception of viscosity and density for B100). The results obtained are shown in Table3.6. It was found that viscosity was higher as the proportion of biodiesel in the mixtures increased. However, this event does not affect the atomization characteristics.

Table 3.6. Mixing proportion of diesel fuel and castor bean biodiesel

Properties	Units	Diesel	B5	B10	B15	B20	B25	B30	B35	B40	B45	B50	B100
SG	-	0.835	0.84	0.843	0.845	0.847	0.853	0.862	0.865	0.869	0.876	0.88	0.920
Density	Kg/m3	835	840	843	845	847	853	862	865	869	876	880	920
viscosity	mm^2/s	3.81	3.87	3.98	4.00	4.38	4.89	5.50	5.87	5.98	6.05	7.54	12.5

4. Conclusions

Based on the forgoing discussion, the conclusions of this study are summarized as follows:

- RC oil can be used as a biodiesel raw material with its high oil content and its non-edible characteristics.
- Castor oil has very high kinematic viscosity and density, which was reduced by using high molar ratio during transesterification but still needed to be blended with diesel fuel to bring it to the limits for biodiesel.
- Of all the variables studied, the interaction between Methanol to oil ratio& amount of catalyst had more influence on the yield of fatty acid methyl ester.
- The ester yield obtained from the transesterification process ranged from 46 to 92.5%.
- The optimum FAME yield of 94.5% was obtained at a catalyst concentration of 1.22wt%, methanol to oil molar ratio of 8.10:1 and Reaction temperature of 59.89°C at a reaction time of 2hr and 600rpm
- HHV of castor biodiesel is slightly lower than that of diesel but has a higher calculated Cetane number.
- In this study, pure RC biodiesel usage can cause problems in injection system because of its high viscosity and density. Therefore, further reducing the viscosity and density of biodiesel is performed by blending with diesels oil up to B45 to use as alternative fuel for diesel oil in the existing conventional diesel engine.

References

[1] Demirbas A., (2003), Biodiesel fuels from vegetable oils via catalytic and non-catalytic supercritical alcohol transesterification and other methods: Energy Converse Manag.44, 2093-2109.

[2] Emission Standards, European Union. Cars and light trucks [online]. Available from: http://www.dieselnet.com/standards/eu/ld.php [Accessed 2008].

[3] Lapuerta M., Armas O., and Rodriguez-Fernandez J., Effect of biodiesel fuels on diesel emissions: Progress in Energy and Combustion Science, 34 (2008) 198–223.

[4] Agarwal and Das, (2001), Agarwal, A.K., and L.M. Das., (2001), Biodiesel development and characterization for use as a fuel in compression ignition engines, Journal of Engineering for Gas, Turbines and Power 123: 440-447.

[5] Mulugetta Y., (2008), Evaluating the economics of biodiesel in Africa. Renew Sust Energy Rev., 13, 1592-15989.

[6] Fangrui, M., Milford A. Hanna, "'biodiesel Production: A review", Bioresource Technology (1999), vol. 70, p. 1-15.3.

[7] Freedman B., Pryde E.H., Mounts T.L., (1984). Variables affecting the yields of fatty esters from transesterified vegetable oils: J Am Oil Chem Soc 61:1638-1984.

[8] Canoira, L., Galean, J.G., Alcantara, R., Lapuerta, M., Contreras, R.Y. (2010), Fatty acid methyl esters (FAMEs) from castor oil: Production process assessment and synergistic effects in its properties. Renewable Energy, vol. 35, p. 208-217.

[9] Ayhan Demirbas, Biodiesel: A Realistic Fuel Alternative for Diesel Engines, Energy Technology Sila Science and Energy Trabzon Turkey, Springer, (2008).

[10] Jeong, G. T., Kim, D. H., & Park, D. H. (2007), Applied Biochemistry and Biotechnology, 136–140, 583–594. Doi: 10.1007/s12010-007-9081.

[11] Akpan, U. G., Jimoh A.and Mohammed, A. D., (2006). Extraction, Characterization and Modification of Castor Seed Oil, Leonardo Journal of Science

[12] Salunke D. K., Desai B. B., Post-harvest Biotechnology of Oil Seeds, CRC Press, p. 161-170, 1941

[13] Lew Kowitseh J. I., Chemical Technology and Analysis of oils, Vol. 2: Fats and waxes, Macmillan, (1909)

[14] Marter A. D., Castor: Markets, Utilization and Prospects, Tropical Product Institute, G152, p. 55-78, 1981 and Weise E. A., Oil seed crops, Tropical Agriculture Series, p. 31-53, Longman, 1983

[15] Ayhan Demirbas. 2008. Relationships derived from physical properties of vegetable oil and biodiesel fuels. Fuel. 87: 1743-1748

[16] Encinar, J. M.; Juan, F.; Gonzalez, J. F.; Rodriguez-Reinares, A., (2005), Biodiesel from used frying oil: Variables affecting the yields and characteristics of the biodiesel. Ind. Eng. Chem.Res, 44 (15), 5491-5499.

[17] Silva, N. D. L. D., Maciel, M. R. W. M., Batistella, C. B., & Filho, R. M. (2006). Applied Biochemistry & Biotechnology, 129–132,405–414. Doi: 10.1385/ABAB: 130:1:405.

12

Fibre, physical and mechanical properties of Ghanaian hardwoods

Emmanuel Tete Okoh

Department of Furniture Design and Production, Accra Polytechnic, P O Box GP 561, Accra, Ghana

Email address:
etokoh@apoly.edu.gh

Abstract: Wood fibre properties (fiber length, fiber width, cell wall thickness and lumen diameter), physical (oven-dry density) and mechanical properties (modulus of rupture, modulus of elasticity, compression parallel to the grain) of four tropical hardwood species (*Terminalia superba* (*Ofram*) and *Terminalia ivorensis* (*Emere*), as currently threatened timber species and *Quassia undulata* (*Hotrohotro*) and *Recinodendron heudelotii*(*Wama*) as lesser used timber species were investigated to measure and compare their timber properties as potential substitutes. Tree normal trees of each tree species were selected and log samples were cut at the middle portion of stem height to determine the properties. The study revealed that, the densities, compression parallel to grain, modulus of rapture and modulus of elasticity of *Ofram* and *Hortrohotro* were not significant, but that of *Emere* and *Wama* were significant. The modulus of elasticity of Emere was however not significant. Based on these findings *Hortrohotro* could be substituted for *Ofram* and *Emere* with *Wama*.

Keywords: Fiber, Hardwood, Mechanical Properties, Lumen

1. Introduction

The cells that make up the anatomical structure of tropical hardwoods are the vessels, fibres, parenchyma and the wood rays. Fibres are the most important element that is responsible for the strength of the wood [1]. According to [2], wood density is an important wood property for both solid wood and fibre products. [3], also reported that factors that determine wood density are cell wall thickness, the cell diameter, the ratio of early wood to latewood and the chemical content of the wood. [4], indicated that density is a general indicator of cell size and is a good predictor of strength, stiffness, ease of drying, machining, hardness and various paper making properties. Many of the density variations within a tree can be ascribed to the anatomical structure of wood, such as characteristics of vessels and fibres [5]. Wood density also serves as an indicator of wood quality due to its strong positive correlation with, for example, mechanical strength properties [6]. According to [7], density is one of the most important properties that influences the use of a timber. Also [8], stressed the fact that wood density affects the technical performance of wood and in particular the strength and processing behavior of sawn wood and veneer, and the yields of wood fibre in pulp production. [9] reported that wood density is a measure of the cell wall material per unit volume and as such gives a very good indication of the strength properties and expected pulp yields of timber [10]. According to [11], basic density is closely related to end-use quality parameters such as pulp yield and structural timber strength. [12] stated that the density of wood is recognized as the key factor influencing wood strength. [13] agreed that much of the variation in wood strength, both between and within species, can be attributed to differences in wood density.

Wood density is, therefore, the single most important single factor determining pulp yield and quality and is also reasonably closely related to various wood properties, for example timber strength, and properties of sawing, machining, glueing, shrinkage, seasoning, peeling and, preservation [14]. Density has been shown to be positively correlated with the strength and stiffness of small clear samples of wood [15], and consequently high density timber is generally associated with superior mechanical performance. In structural size samples, however, the presence of other strength reducing factors mean that density alone is not always a good predictor of mechanical properties. Although for sawn timber, variations in tracheid length per se are not generally considered to have a significant impact, short tracheids are associated with high microfibril angles which do reduce timber strength, stiffness and dimensional stability. Wood density is a measure of the amount of cell wall material present but gives no indication of the anatomy of the cell wall nor of its properties. For example, compression wood is denser than normal wood but is weaker. [16] concluded that while density was of significance in affecting wood strength in Sitka spruce, it was not as important as other factors, such as grain angle and the presence of juvenile wood , which lower performance. This has been supported by [17] who found that variations in wood density only explained part of the variations in mechanical properties observed in trees of differing growth rates, and that this was particularly evident for Abies and Picea species. A relatively small change in wood density can be accompanied by a considerably larger change in mechanical properties, with the result that estimates of structural performance based solely on evaluation of wood density may not be reliable [18]. [19] reported that density was highly significant, although not the most important, influence on Sitka spruce batten stiffness. Hence, density clearly has an important influence on timber strength and stiffness, but the impact on utilization depends on the integration of other factors such as knots, grain angle and juvenile wood.

The Ghanaian forest is a continuum of the tropical forest with fast growing timber species which are used in a wide variety of applications. The current well known primary timber species in Ghana have been exploited selectively by millers, but mostly without permission by illegal chain saw operators, resulting in their reduction both in number and quantum of each of them and the urgent need for finding alternatives for use by both local and the export industry as well as their contribution to the local economy.

Furthermore, there is also ample evidence that timber production in Ghana is not proportional to its potential. Because, its under-utilization is partly as a result of the lack of general information about the wood properties and the great number of timber species. Consequently, *Terminalia superba (Ofram)* and *Terminalia ivorensis (Emere)*, as currently threatened timber species and *Quassis undulata (Hotrohotro)* and *Recinodendron heudelotii (Wama)* as lesser used timber species were selected for the study to investigate and compare the physical, anatomical and mechanical properties as these properties form the basis for specifying timber for any structural application.

The objective of this work is to measure and compare the variation of physical, anatomical and mechanical properties of *Emere* and *Ofram* which are currently threatened timber species to that of *Wama* and *Hortrohotro* which are lesser used species as potential substitutes for utilization.

2. Materials and Methods

The study area is Kajease Forest reserve at Afosu (06°22´N, 00°57´W) and an elevation of 217m in the Eastern Region. The vegetation is characterized by the moist semi-deciduous forest. The total land area is 1557km^2 and an altitude ranging from 152-610m above sea level. It is characterized by a wet semi-equatorial climate with annual rainfall ranging from 1,500-1800mm.The forest floor is closed with tree species of the *Celtic-Triplochiton* association, dominated by *Celtic mildbraebii (esa)*, *Triplochiton scleroxylon* (wawa), *Ceiba pentandra* (silk cotton), *Ricinodendron heudelotii(Wama)*, *Hannoa klaineana (Hotrohotro))*, *Melicia exelsa (Odum)*, *Khaya ivorensis* (African Mahogany), *Terminalia ivorensis (emere)*, *Terminalia superba (Ofram)* and *Entandrophragma cylindricum (Sapele)*. The timbers extracted for the research were taken from the yield allocated to the contractor by the Forest Services Division (FSD) of the Forestry Commission. Each timber felled had a merchantable diameter of least 60cm. Inventory records from FSD was used to determine the age of the trees. From each of the species, three normal tress were selected. Logs were cut at the middle portion of stem of tree height. All testing samples were taken from mature wood for the determination of the different wood properties.

Sample Preparation: Defect free boards of, *Ricinodendron heudelotii (Wama)*, *Hannoa klaineana (Hotrohotro))*, *Terminalia ivorensis.,* (Emeri) and *Terminalia superba (*Ofram) were cut from the middle portion of trees into 15mm thick boards with multiple rip saws. All boards were prepared with the same equipment. Maceration: Fibres were separated by maceration of match stick sized wood pieces originating from 5 arbitrary chosen samples per species in Jeffrey's solution at 40°C for 4h. The resulting cell suspension was washed thoroughly with distilled water. Fibres were spread from this suspension onto a glass slide and left to dry for 12h.

A Leica EZ 4D light microscope was used to determine the diameter and ratio of vessels, as well as the fibre length and amount of parenchyma cells. Fifteen images were acquired per section for cell analysis. To determine the number of vessels per mm^2 and the vessel diameter, images were acquired with a magnification of 20x and all visible vessels counted and their diameter recorded. Fifteen images of the cell suspension were acquired for each species at a magnification of 35x, and the length of fibres, and parenchyma cells, as well as their amount per mm^2 were determined.

A Leica EZ 4D light microscope was used to determine the fibre diameter and cell wall thickness. The fiber diameter and cell wall thickness were determined from fifteen images acquired with a magnification of 4x. Fiber diameter and cell wall thickness were measured on all visible cells in the image.

Density: Twenty samples were dried for 24 h at 105°C before being tested for density. The density of wood was determined on a dry-mass basis. A digital caliper was used to measure the dimensions of the samples of oven-dried wood at a moisture content of 12% in order to determine their volumes. The samples were then weighed using an electronic balance. The calculated volume was divided by the mass to obtain the density (ρ), using the formula below:

$$\rho = m/v \tag{1}$$

where ρ is density (g·cm^3),m is mass (g), and v is volume (cm^3).

Mechanical properties: The Flexure Testing Machine was used to determine the Modulus of rupture and the Modulus of elasticity. The sample dimensions for determination of mechanical properties were 450 × 50 × 15 mm for static bending strength tests, such as modulus of rupture (MOR) and modulus of elasticity (MOE) and the compression parallel to grain[CPG(бcpl)].

The prepared samples (N= 5 for each species) were then conditioned in a room at a temperature of 20°C and 65 ±5% relative humidity until the specimens reached an equilibrium moisture content of 12%. The load was applied in the tangential direction. The mechanical strength properties were calculated using the following equation;

$$MOR= 3PL/2db^2 \qquad (2)$$

$$MOE= P'L^3/4\Delta'bd^3 \qquad (3)$$

where P' = load at the limit of proportionality (kN); P = maximum load (KN), L = span of the test specimen (mm), b= breadth of the test specimen (mm), d = depth of the test specimen (mm) and Δ' = deflection at the limit of proportionality (mm).

$$\sigma cpl = Pmax /A \qquad (4)$$

Where σcpl = MCS (MPa),

Pmax = maximum crushing load at break point (KN) and A = area of cross section of the specimen on which force was applied (mm^2).

Statistical analysis to determine the effect of hardwood species on anatomical (fiber length, fiber width, cell wall thickness and lumen diameter), physical (oven-dry density) and mechanical properties (modulus of rupture and modulus of elasticity), was conducted using the analysis of variance (ANOVA) techniques. Duncan's multiple range test (DMRT) was used to test the statistical significance at the α = 0.05 level. The Pearson correlation was used to analyze the relationship among the wood's various properties.

3. Results and Discussions

Fibre cell dimensions: The analysis of variance (ANOVA) shows that there is significant difference between the wood species and their fibre cell dimensions. *Wama* has by far the highest values for fibre length, cell wall thickness and rankle ratio or wood fraction, but the lowest fibre diameter. Although *Hotrohotro* has the highest fibre diameter it recorded the lowest cell wall thickness and rankle ratio or wood fraction. *Emere* and *Ofram* however recorded intermediate fibre cell values. All these anatomical properties are displayed in Table 1.

Table 1. Fibre cell mean values (m ± SE) of the wood species(n=5)

Species	Fibre diameter/μm	Fibre length/mm	Cell wall thickness/μm	Rankle ratio
Hotrohotro	29.984±5.119 B	1.588±0.221 A	6.695±1.099 A	0.207± 0.035 A
Ofram	28.481±6.937 B	1.314± 0.239 A	7.258±1.389 AB	0.283±0.142 A
Emere	29.241± 5.931 B	1.314±0.244 A	8.359±2.625 AB	0.303±0.146 A
Wama	20.256±2.804 A	1.727±0.528 A	8.744±946 B	0.432±0.111 B

Note: columns with same letters are not significantly different at p=0.05
columns with different letters are significantly different at p=0.05

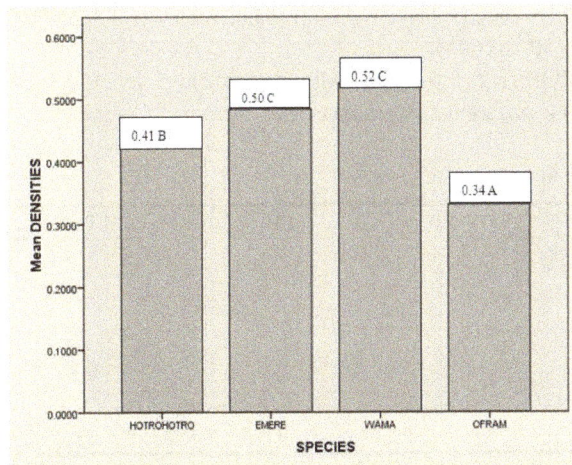

Fig. 1. Oven dry density values of the four hardwood species(g/cm^3)

Oven dry density: The oven dry density values for the four the tropical hardwood species are displayed in fig.1 below. It is evident from fig.1 that *Wama* had the highest density of 0.524g/cm^3, whilst *Ofram* had the lowest of 0.331g/cm^3. The analysis of variance (ANOVA) shows that there is significant difference between the types of species and oven dry density value. [20] puts the wood density value of *Ofram* between 0.37 – 0.73g/cm^3 which is a little higher than the one determined by this work.

Also [21] Dudek,, Förster, and Klissenbauer (1981) quote the wood density value of *Hotrohotro* between 0.29-0.45g/cm^3, and this is consistent with the wood density value of 0.41g/cm^3 determined by this study. *Wama* had a wood density value of 0.524g/cm^3 which is little higher than the one determined by [22](Richter and Dallwitz, 2000). The differences in wood density values for *Wama* and *Ofram* may be due to differences in soil and climatic conditions.

Mechanical Properties: The mechanical properties

[Compression parallel to grain (CPG), Modulus of Elasticity (MOE) and Modulus of Rapture (MOR)] of the wood species are measured to serve as the basis for timber specification and utilization. The results are shown in the figures below.

Fig. 2 shows the compression parallel to grain of the four hardwood species (e.g *Ofram, Emere, Wama* and *Hotrohotro*) used for the study. *Emere* had the highest of compression strength of 25.37 MPa with *Ofram*. The analysis of variance (ANOVA) shows that there is no significant difference between the type of species and the compression parallel to grain.

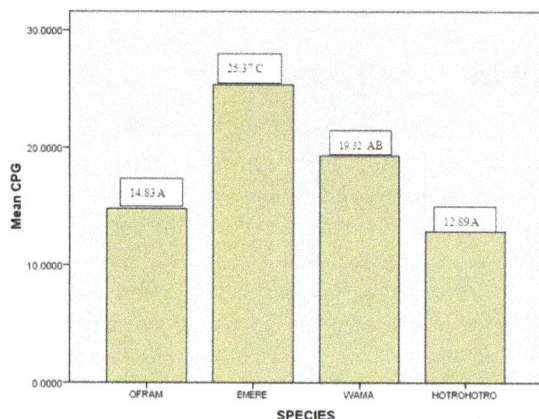

Fig. 2. *shows the compression parallel to grain of the four hardwood species .*

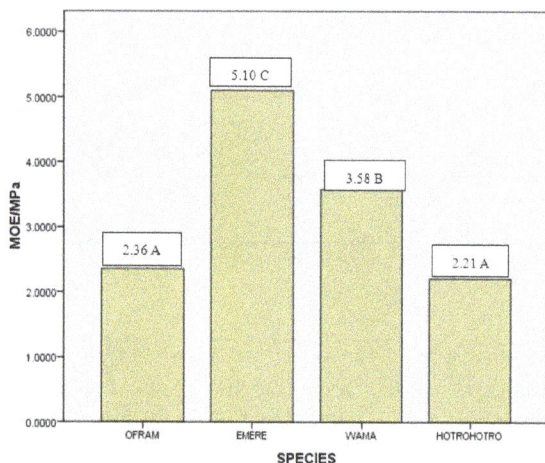

Fig. 3. *Modulus of elasticity(MOE) of the four hardwood species*

Modulus of Elasticity: The modulus of elasticity (MOE) values for the four hardwood species such as *Ofram, Emere, Wama* and *Hotrohotro* are depicted in fig. 3 above. The MOE values for the species are 2.361, 5.099, 3.579 and 2.206 MPa respectively. The analysis of variance (ANOVA) shows that there is significant difference between the types of species and the modulus of elasticity values. The highest value of 5.099 was recorded in *Emere* with *Hotrohotro* having the least value of 2.206 MPa

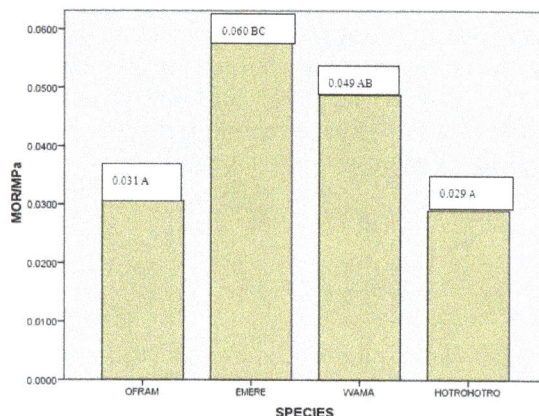

Fig. 4. *Modulus of Rapture values of the four hardwood species(MPa)*

Modulus of rapture: The modulus of rapture (MOR) values for the four hardwood species are shown in fig. 4. *Emere* recorded the highest rapture value of 0.060 MPa, whilst Hotrohotro had the least rapture value of 0.029 MPa. *Ofram* and *Wama* however recorded intermediate values. The analysis of variance (ANOVA) shows that there is significant difference between the type of species and the modulus of rapture. The relationship between oven dry density and mechanical properties are shown in Table 1. Results show that there are positive correlation between wood density and CPG (R^2=0.644), (ANOVA) shows that there is significant difference between the type of species and the modulus of rapture.

The relationship between oven dry density and mechanical properties are shown in Table 3. Results show that there are positive correlation between wood density and CPG (R^2=0.644),

MOR (R^2= 0.680) and Modulus of Elasticity (R^2= 0.646) at four different species level.

Table 2. *The relationship between different wood properties (p= 0.01)*

	Do	CPG	MOR	MOE	FD	FL	CWT
Do	1						
CPG	0.644**	1					
MOR	0.680**	0.878**	1				
MOE	0.646**	0.910**	0.764**	1			
FD	-0.281	-0.022	-0.110	0.101	1		
FL	0.251	-0.092	-0.133	0.017	-0.074	1	
CWT	0.375	0.084	0.168	0.049	-0.260	-0.089	1

Do: Oven dry density, CPG: Compression parallel to Grain, MOR: Modulus of Rapture, MOE: Modulus of Elasticity, FD: Fibre Diameter, FL: Fibre Length, FCWT: Fibre Cell Wall thickness

The relationship between wood density and mechanical strength properties within a species have been investigated tremendously by researchers. The relationship between wood density and mechanical properties within a species has been studied by many researchers. [23] observed that a significant linear relationship exists between wood density and mechanical properties of timber. According to [24] the modulus of rupture and the maximum crushing strength in compression parallel to the grain are most closely and almost linearly related to wood density, whereas modulus of elasticity is poorly and least linearly related to wood density [25]. The density of timber is a function of both cell wall thickness and lumen diameter and there exists correlation between strength and density of timber. The results of this study show a significant linear relationship between wood density and mechanical strength properties of timber.

Although there are positive relationship between wood density and mechanical strength properties, their biometric features (fibre diameter and fibre length) are weak and negatively correlated. Interestingly, the relationship between wood density and mechanical properties with fibre cell wall thickness is positive though weak at four species level. There are also positive relationship between MOR and CPG (R^2=0.878), MOE and CPG (R^2= 0.910) and MOE and MOR (R^2=0.764).

4. Conclusion

The study revealed that, the densities, compression parallel to grain, modulus of rapture and modulus of elasticity of *Ofram* and *Hortrohotro* were not significant, but that of *Emere* and *Wama* were significant. The modulus of elasticity of Emere was however not significant. There were however positive relationships between wood density and MOE, CPG and MOR, but not fibre cell dimensions. Based on these findings *Hortrohotro* could be substituted for *Ofram* and *Emere* with *Wama*.

References

[1] A. J. Pan,shin, and C. de Zeeuw, Textbook of wood technology(4th ed.).1980. New York: McGraw-Hill

[2] De Guth, E.B., (1980), Relationship between wood density and tree diameter in Pinus elliottii of Missiones Argentina. IUFRO Conf. Div. 5 Oxford, England. 1p (Summary).

[3] ID Cave and JCF Walker, Stiffness of wood in fast grown plantation softwoods: the influence of microfibril angle. Forest Prod. J. 44(5), 43-48, 1994.

[4] R.M. Roque, M.T. Filho, Relationships between anatomical features and intra-ring wood density profiles in Gmelina arborea applying x-ray densitometry. Cerne 13:384–392, 2007.

[5] G. Nepveu, Croissance et qualite′ du bois de framire′. Evolution de la largeur de cerne et des composantes densitome′triques en fonction de l'age. Bois et Fore′ts des Tropiques 165: 39–58. 1976.

[6] J.D. Brazier and R.S. Howell, The use of a breast height core for estimating selected whole tree properties of Sitka spruce. Forestry. 52(2), 177-185, 1979..

[7] J.D. Brazier and R.S. Howell, The use of a breast height core for estimating selected whole tree properties of Sitka spruce. Forestry. 52(2), 177-185, 1979Cown, D.J, (1992) Core wood (Juvenile Wood) in Pinus radiata- should we be concerned? New Zealand J. Forestry Sci. 22(1), 87-95.

[8] E.W.J. Philips, The inclination of the fibrils in the cell wall and its relation to the compression strength of timber. Empire Forestry J. 20, 74-78, 1941.

[9] E.W.J. Philips, The inclination of the fibrils in the cell wall and its relation to the compression strength of timber. Empire Forestry J. 20, 74-78, 1941.

[10] C, Harvald and P.O. Olesen, The variation of the basic density within the juvenile wood of Sitka spruce (Piceasitchensis). Scand. J. Forest Res. 2, 525-537, 1987.

[11] Cown, D.J, (1992) Core wood (Juvenile Wood) in Pinusradiata- should we be concerned? New Zealand J. Forestry Sci. 22(1), 87-95.

[12] Schniewind, A.P., (1989). Concise encyclopedia of wood and wood-based materials. Pergamon Press. pp: 248.

[13] G. Nepveu, Croissance et qualite′ du bois de framire′. Evolution de la largeur de cerne et des composantes densitome′triques en fonction de l'age. Bois et Fore′ts des Tropiques 165: 39–58, 1976.

[14] A. J. Pan,shin, and C. de Zeeuw , C Desch, H.E. and Dinwoodie, J.M., (1996), Timber Structure, Properties, Conversion and Use. MacMillan Press, London.

[15] J.D. Brazier and R.S. Howell, The use of a breast height core for estimating selected whole tree properties of Sitka spruce. Forestry. 52(2), 177-185, 1979

[16] Zhang, S.Y., (1997) Wood quality: its definition, impact and implications for value-added timber management and end uses. In Timber Management Toward Wood Quality and End-Product Value.

[17] S.Y. Zhang, R. Gosselin and G. Chauret (eds). Proceedings of the CTIA/IUFRO International Wood Quality Workshop, Quebec City. Part I, pp. 17–39

[18] K.W. Maun (1992), Sitka spruce for construction timber: the relationship between wood growth characteristics and machine grade yields of Sitka spruce. Forestry Commission Research Information Note No. 212. Forestry Commission, Edinburgh

[19] Phongphaew, P., 2003. The commercial woods of Africa. Linden Publishing, Fresno, California, United States. 206 pp.

[20] S. Dudek,, B. Förster, and, K. Klissenbauer, 1981. Lesser known Liberian timber species. Description of physical and mechanical properties, natural durability, treatability, workability and suggested uses. GTZ, Eschborn, Germany. 168 pp

[21] H.G. Richter, and M.J. Dallwitz, Commercial timbers: descriptions, illustrations, identification, and information retrieval. 2000. [Internet]. Version 18th October 2002. http://delta-intkey.com/wood/index.htm. Accessed May 2005.

[22] S.Y. Zhang, R. Gosselin and G. Chauret (eds). *Proceedings of the CTIA/IUFRO International Wood Quality Workshop*, Quebec City. Part I, pp. 17–39.

[23] S.Y. Zhang,, Effect of growth rate on wood specific gravity and selected mechanical properties from distinct wood categories. *Wood Sci. Technol*. 29, 451–465, 1995.

[24] J.M. Dinwoodie, Timber Structure, Properties, Conversion and Use. 1996. MacMillan Press, London.

13

Modelling the competition for forest resources: The case of Sweden

Anna Olsson, Robert Lundmark

Economics Unit, Luleå University of Technology, Luleå, Sweden

Email address:
Anna.Olsson@ltu.se (A. Olsson), Robert.Lundmark@ltu.se (R. Lundmark)

Abstract: Past decades increasing shares of forest resources have been diverted from the forest sector to the energy sector. The increasing utilization of forest fuel is, to a large extent, caused by economic policies introduced to reduce the emission of greenhouse gases. Since the energy sector is believed to continue to increase its use of forest fuel in the energy production in Sweden, it is of interest to investigate the effects of this. The purpose of this study is to analyses the extent and degree of forest resources competition in the presence of climate policy by accounting for the inter-linkages of forest resources utilization between the energy sector and the forest industries. A partial equilibrium model was thus constructed and applied to the Swedish forest sector and energy sector. A baseline scenario is calibrated using the GAMS software. Four scenarios with alternative development paths are then simulated and compared to the baseline scenario. The results indicate that the impacts on the procurement competition between the forest sector and the energy sector are relatively moderate also in situations of expanding production in the forest industries. An increase in the competition between industries in the energy sector can however be observed.

Keywords: Partial Equilibrium Model, Forestry, Forest Industries, Energy Sector, Bioenergy, Forest Fuel

1. Introduction

This study focuses on procurement and market issues of forest products used by the energy sector and forest industries. The specific purpose is to analyze the extent and degree of the competition for forest products in the presence of climate and energy policies. This is done by developing a partial equilibrium model capturing the forest and the energy industries and the product flows to and between them. The model is applied on Sweden, which is a good case-study since it has both a large forest industry and a significant utilization of forest products in the energy sector.

Historically, the bulk of forest products have been used by the forest industries. Lately however, increasing shares are diverted to the energy sector. The increasing utilization by the energy sector is, to a large extent, caused by changes in the relative price between biofuels and alternative fuels. In turn, the changing relative prices are partly a consequence of the economic instruments and policies introduced to e.g., reduce the emission of greenhouse gases.

In countries where forests already are highly utilized, such instruments might result in an increase in the procurement competition between the forest industries and the energy sector (e.g., Hammarlund et al., 2010; Lundmark and Söderholm, 2004; Brännlund et al., 2010).In this context, it is important to note that an intensified competition, as a result of market interventions, is not a problem in itself from an economic perspective if the policies are implemented to correct for market failures. For example, if a carbon tax, which changes the relative price between fossil fuels and biofuels, is reflecting the social cost of the negative external effect.

Forest products are normally allocated on the basis of users' willingness to pay. Economic policies can change the willingness to pay for specific sectors both positively and negatively depending on the specific policy implemented. For example, a policy promoting the use of bioenergy will increase the energy sector's willingness to pay for forest products. Moreover, from an international perspective, national policies promoting the use of bioenergy can affect the competition of forest products between countries – assuming that there are differences in the national policies (Brännlund et al, 2010; Hammarlund et al, 2010).

It has been estimated that the emission of greenhouse gases needs to be reduced by approximately 85 percent until 2050 to achieve the 2-degree target (Åkerman et al.,

2007). This is a central assumption in the forthcoming analysis. To achieve the emission target, substantial technological efficiency improvements have to be realized as well as changes in the current trends of both production and consumption of energy. From a production perspective it is important to stimulate the development of silvicultural – and agricultural – practices. An increasing supply of forest fuels can offset an increasing demand without causing price changes. But it is more likely that the increases in supply will not keep up with the increases in demand, causing price changes and an intensified competition.

2. Modelling the Swedish Forest Sector

The model is designed as an analytical tool for assessing the consequences of external changes. It has explicitly been designed to analyze how policies aimed to reduce emission of greenhouse gases will affect the forestry, forest industries and the energy sector.

2.1. Model Overview and Assumptions

The model is designed as a static partial equilibrium model for a closed economy. Equilibrium models have been used to analyze changes in market conditions for the forest sectors (e.g., Buongiorno et al., 2003; Lundmark, 2006). Input demand and supply functions as well as production technologies are defined for each industry under the assumption of profit maximizing behavior (Kallio et al., 2004).Assuming constant returns to scale, a CES production function for the *n*-factor case in the standard coefficient form is specified as:

$$y_i^f = \gamma_i \left[\sum_j \alpha_i^j \left(x_i^j\right)^{\rho_i} \right]^{1/\rho_i} \quad (1)$$

Where y_i^f is output f of industry i, γ_i is a scale parameter, α_i^j is a distribution parameter and x_i^j is the demand of industry I for input j, and $\rho_i = 1/(1-\sigma_i)$ where σ_i is the elasticity of substitution (Varian, 1992). A list of all notations used in the model can be found in the Appendix. The production function is assumed to be weakly separable between forest products and other production factors, e.g., labor and capital (Varian, 1992).

The conditional input demand is derived by maximizing profits (or equivalently minimizing the cost for each level of output) and is expressed as:

$$x_i^j = \left(\alpha_i^j\right)^{\sigma_i} \left(w^j\right)^{-\sigma_i} \gamma_i^{-1} \left[\sum_{j=1}^n \left(\alpha_i^j\right)^{\sigma_i} \left(w^j\right)^{(1-\sigma_i)} \right]^{\sigma_i / (1-\sigma_i)} y_i^f \quad (2)$$

where w^j is the price of input j. The inverse supply of forest resources is written as:

$$w^j = \beta^j + \delta^j \left(\frac{x_i^j}{k^j} \right)^{1/\phi^j} \quad (3)$$

where β^j is an exogenous cost component in forest product prices that is independent of harvesting levels, δ^j is a shift parameter accounting for factors other than price affecting the supply, k^j is a capacity restriction, and the ϕ^j is the supply elasticity of input j.

The profit function (π)is specified as:

$$\pi_i = \sum p^f y_i^f - \sum w^j x_i^j \quad (4)$$

The objective of the model is to maximize the sum of the individual sector's profit defined as:

$$\pi_{TOT} = \sum \pi_i \quad (5)$$

The sectors included are the forestry (*F*),sawmill industry (*S*), pulp and paper industry (*P*) and the heating industry (*H*). In total six forest products are included. The forestry sector supply roundwood (*RW*) and logging residues (*LG*). The production of roundwood is roughly equally divided between pulpwood and timber. Pulpwood is used by the pulp and paper industry and timber is used by the sawmill industry. Depending on its price, pulpwood can be used by the energy sector. The extraction of harvesting residues is conditional on a harvesting operation and is considered a by-product of the harvesting operation. Harvesting residues is not used by the forest industries, thus no immediate competition for it exists. It has been estimated that the extraction of logging residues could potentially increase before the competition between the forest industries and energy sector intensified (Swedish Forest Agency, 2009; Lundmark and Söderholm, 2004). The sawmill industry is producing sawdust and woodchips as a by-products (*BY*) from their main production of sawn wood products. The energy sector's choice between by-products and logging residues is determined by the relative price. Since sawdust and woodchips are by-products, the supplied volume will not respond to changes in the price level. Only an increase in the production of sawn wood products will affect the supply of the by-products. The by-products are mainly used by the pulp and paper industry, but are also relevant for the energy sector. The endogenous flow of forest products and the connection between the sectors are illustrated in Figure 1.

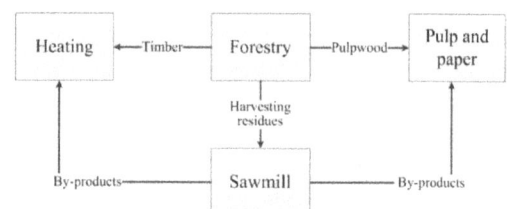

Figure 1. Actors and flows of forest resources in Sweden

2.2. Empirical Specification of the Forestry

The specification of the forestry sector mainly follows the specification in Kallio et al. (2004). The inverse supply function of roundwood is specified as:

$$w^{RW} = \beta^{RW} + \delta^{RW}\left(\frac{y_F^{RW}}{k^{RW}}\right)^{1/\phi^{RW}} \qquad (6)$$

where w^{RW} is the price of roundwood and y_F^{RW} is the harvesting level of roundwood. β^{RW} is an exogenous cost component of the roundwood price that is independent of harvesting levels (following Kallio et al. (2004) it is set at zero), δ^{RW} is a shift parameter accounting for factors other than price affecting the supply, k^{RW} is a capacity restriction, and the ϕ^{RW} is the supply elasticity of roundwood. The baseline values of w^{RW}, y_F^{RW}, β^{RW}, along with the k^{RW} and ϕ^{RW}, is substituted into equation (4) in order to calibrate the value of δ^{RW}. In order to ensure a sustainable forestry a restriction on the harvesting level of roundwood is applied.

$$y_F^{RW} < \psi_{FGROWTH} \qquad (7)$$

The restriction requires the harvesting level of roundwood to be less than the annual growth rate of the forest ($\psi_{FGROWTH}$). The inverse supply of logging residues is specified as:

$$w^{LG} = \beta^{LG} + \delta^{LG}\left(\frac{y_F^{LG}}{k^{LG}}\right)^{1/\phi^{LG}} \qquad (8)$$

Where w^{LG} is the price of logging residues and y_F^{LG} is the output level of logging residues from the forestry. β^{LG} is an exogenous cost component in logging residue price that is independent of output level (following Kallio et al. (2004) it is set at zero), δ^{LG} is a shift parameter accounting for factors other than price affecting the supply, k^{LG} is a capacity restriction, and the ϕ^{LG} is the supply elasticity of logging residues. The baseline values of w^{LG}, y_F^{LG}, β^{LG}, along with the k^{LG} and ϕ^{LG}, is substituted into equation (8) in order to calibrate the value of δ^{LG}. The output level of logging residues, y_F^{LG}, is assumed not to exceed a fixed proportion (μ_{LG}) and is consequently conditional on the actual level harvested roundwood.

$$y_F^{LG} \le \mu^{LG} y_F^{RW} \qquad (9)$$

The value of μ^{LG} is calibrated by substituting the baseline values of y_F^{LG} and y_F^{RW} into equation (9). Following this the forestry's profit function is specified as:

$$\pi_F = (p^{RW}y_F^{RW} + p^{LG}y_F^{LG}) - \tau_{FCOST}x_F^{RW} \qquad (10)$$

The forestry's costs are assumed to be represented by τ_{FCOST}, which is the annual logging cost per cubic meter harvested roundwood.

2.3. Empirical Specification of the Sawmill Industry

The sawmill industry produces sawn wood products, sawdust and woodchips using roundwood as a single input. Thus, no production function is specified. The production of sawn wood products is assumed to be exogenously fixed. The inverse supply of by-products is specified as:

$$w^{BY} = \beta^{BY} + \delta^{BY}\left(\frac{y_S^{BY}}{k^{BY}}\right)^{1/\phi^{BY}} \qquad (11)$$

Where w^{BY} is the roundwood price and y_S^{BY} is the output level. β^{BY} is an exogenous cost component in the by-product price that is independent of output level, δ^{BY} is a shift parameter accounting for factors other than price affecting the output, k^{BY} is a capacity restriction, and the ϕ^{BY} is the supply elasticity of by-products. The baseline values of w^{BY}, y_S^{BY}, β^{BY}, along with the k^{BY} and ϕ^{BY}, is substituted into equation (11) in order to calibrate the value of δ^{BY}.

The output level of by-products is proportional to sawmills' demand of roundwood. That is, the output level of by-products cannot exceed a fixed proportion (μ_{BY}) of the actual quantity of roundwood used by sawmills.

$$y_S^{BY} \le \mu^{BY} x_S^{RW} \qquad (12)$$

The value of μ^{BY} is calibrated by substituting baseline values of y_S^{BY} and x_S^{RW} into equation (12).

2.4. Empirical Specification of the Pulp and Paper Industry

The pulp and paper industry is assumed to use roundwood and by-products in their production. The CES production function is specified as:

$$y_P^{PP} = \gamma_P\left[\alpha_P^{RW}\left(x_P^{RW}\right)^{-\rho_P} + \alpha_P^{BY}\left(x_P^{BY}\right)^{-\rho_P}\right]^{-1/\rho_P} \qquad (13)$$

where y_P^{PP} is the pulp and paper industry's output level of pulp, x_P^{RW} and x_P^{BY} are the demanded input quantities of roundwood and by-products. γ_P is a scale parameter for the pulp and paper industry, α_P^{RW} and α_P^{BY} (where $\alpha_P^{BY} = (1 - \alpha_P^{RW})$) are distribution parameters for roundwood and by-products respectively. These are calibrated using benchmark values. ρ_P is the pulp and paper industry's

specific ρ-value based on the elasticity of substitution (σ_P) between the inputs. The pulp and paper industry's associated input demand functions of roundwood and by-products are specified as:

$$x_P^{RW} = (\alpha_P^{RW})^{\sigma_P}(w^{RW})^{-\sigma_P}\gamma_P^{-1}\left[(\alpha_P^{RW})^{\sigma_P}(w^{RW})^{\rho_P\sigma_P} + (\alpha_P^{BY})^{\sigma_P}(w^{BY})^{\rho_P\sigma_P}\right]^{\frac{1}{\rho_P}} y_P^{PP} \tag{14}$$

and

$$x_P^{BY} = (\alpha_P^{BY})^{\sigma_P}(w^{BY})^{-\sigma_P}\gamma_P^{-1}\left[(\alpha_P^{RW})^{\sigma_P}(w^{RW})^{\rho_P\sigma_P} + (\alpha_P^{BY})^{\sigma_P}(w^{BY})^{\rho_P\sigma_P}\right]^{\frac{1}{\rho_P}} y_P^{PP} \tag{15}$$

where w^{RW} and w^{BY} are the input prices of roundwood and by-products respectively.

The profit function of the pulp and paper industry is written as:

$$\pi_P = p^{PP}y_P^{PP} - \left(w^{RW}x_P^{RW} + w^{BY}x_P^{BY}\right) \tag{16}$$

where p^{PP} is the price of pulp.

2.5. Empirical Specification of the Heating Industry

The heating industry is assumed to use logging residues and by-products. The production function is specified as:

$$y_H^{HE} = \gamma_H\left[\alpha_H^{LG}\left(x_H^{LG}\right)^{-\rho_H} + \alpha_H^{BY}\left(x_H^{BY}\right)^{-\rho_H}\right]^{-1/\rho_H} \tag{17}$$

where y_H^{HE} is the output of heat, x_H^{LG} and x_H^{BY} are the demanded inputs of logging residues and by-products. γ_H is a scale parameter for the heating industry, α_H^{LG} and α_H^{BY} (where $\alpha_H^{BY} = (1 - \alpha_H^{LG})$) are distribution parameters for logging residues and by-products respectively. As in the case of the pulp and paper industry these parameters are calibrated using benchmark values. ρ_H is the heating industry's specific ρ-value based on the elasticity of substitution (σ_H) between the inputs. The heating industry's associated demand functions of logging residues and by-products are specified as:

$$x_H^{LG} = (\alpha_H^{LG})^{\sigma_H}(w^{LG})^{-\sigma_H}\gamma_H^{-1}\left[(\alpha_H^{LG})^{\sigma_H}(w^{LG})^{\rho_H\sigma_H} + (\alpha_H^{BY})^{\sigma_H}(w^{BY})^{\rho_H\sigma_H}\right]^{\frac{1}{\rho_H}} y_H^{HE} \tag{18}$$

and

$$x_H^{BY} = (\alpha_H^{BY})^{\sigma_H}(w^{BY})^{-\sigma_H}\gamma_H^{-1}\left[(\alpha_H^{LG})^{\sigma_H}(w^{LG})^{\rho_H\sigma_H} + (\alpha_H^{BY})^{\sigma_H}(w^{BY})^{\rho_H\sigma_H}\right]^{\frac{1}{\rho_H}} y_H^{HE} \tag{19}$$

where w^{LG} and w^{BY} are the input prices of logging residues and by-products respectively.

The profit function of the heating industry is written as

$$\pi_H = p^{HE}y_H^{HE} - \left(w^{LG}x_H^{LG} + w^{BY}x_H^{BY}\right) \tag{20}$$

where p^{HE} is the price of heat.

As the quantity of logging residues demanded by the heating industry only accounted for a small share of the total supplied quantity, a slack variable is included to account for the remaining quantity. x_E^{LG} is the demanded quantity of logging residues by the other industries in the energy sector. Logging residues are also utilized in e.g., the combined power and heating industry and the refined woodfuel industry. The quantity demanded of logging residues by the other industries in the energy sector is assumed to be exogenously fixed.

It is assumed that all markets clears where the market clearing conditions are defined as:

$$y_F^{RW} = x_S^{RW} + x_P^{RW} \tag{21}$$

$$y_F^{LG} = x_H^{LG} + x_E^{LG} \tag{22}$$

$$y_S^{BY} = x_P^{BY} + x_H^{BY} \tag{23}$$

The industries' objective is to maximize their profits. The optimal input combination is where the technical rate of substitution equals the input price ratio. The input decision by the industries is hence determined by the marginal product of the inputs and input prices. The model endogenously determines the prices and quantities for the included markets, given a constant exogenously determined output supply. The objective function of the model is defined as:

$$\pi_{TOT} = \pi_F + \pi_P + \pi_H \tag{24}$$

where π_{TOT} is the sum of the profits of the forestry, the pulp and paper industry and the heating industry, respectively.

3. Data, Parameters and Scenario Description

3.1. Data Sources

The output and input volumes and prices are presented in Table 1. All observations are for year 2007. Output consumed internally by the included industries is not considered, i.e., only the net output production is considered. Data on the forestry's output of roundwood was obtained from FAOSTAT (2010). Following Lundmark (2007) the production of logging residues are approximated to be 15 percent of the harvested roundwood level. The output of by-products was calculated by adding the output levels of woodchips and sawdust produced by the sawmill industry and collected from SDC (2009). The output of sawn wood is obtained from SDC (2009) and the output of pulp from the Swedish Forest Agency (2009). The output of the heating industry is based on heat produced by forest fuel. The production of heat originating from by-products and logging residues is approximated using the average efficiency (87 percent) and the input levels. The required data was collected from Statistics Sweden (2009). Data on the sawmill industry's input of roundwood (timber) is collected from SDC (2009). The input of roundwood (pulpwood) and by-products to the pulp and paper industry is also obtained from SDC (2009). The input level of by-products is calculated using the input levels of chips and sawdust. The heating industry's input of by-products and logging residues is collected from Statistics Sweden (2009).

The price of roundwood, by-products and logging residues was collected from the Swedish Forest Agency (2009). The price of roundwood is the weighted average price of timber and pulpwood. The price paid for by-products differs between the forest industry and the energy sector, with the price paid by the forest industry generally being higher. The price presented here is a weighted average of the two. The price of sawn wood and pulp are export prices obtained from the Swedish Forest Agency (2009). The price of heat is collected from Statistics Sweden (2008).

Table 1. Output and input volumes and prices

	Output level (million m3s)	Input level (million m3s)	Price (SEK/m3s)
Forestry			
Roundwood	72.30	-	335
Logging residues	10.85	-	304
Sawmill			
Roundwood	-	38.19	335
By-products	21.55	-	306
Sawn wood	18.74	-	2354
Pulp and paper			
Roundwood	-	36.20	-

	Output level (million m3s)	Input level (million m3s)	Price (SEK/m3s)
By-products	-	11.66	306
Pulp	12.40 a		4439b
Heating			
Logging residues	-	0.04	304
By-products	-	6.34	306
Heat	5.90c	-	1357
Other energy industries			
Logging residues	-	7.39	-

[a]Measured in (million) tonnes. [b]Measured in SEK/tonnes. [c]Converted from TWh to (million) m^{3}s using 1TWh = $10^6*0.47$ m^{3}s.
Sources: FAOSTAT (2010); SDC (2009); Statistics Sweden (2008, 2009); Swedish Forest Agency (2009)

In addition, the harvesting costs and the annual growth rate of the forest are included in the model. The harvesting costs are the costs for thinning and cutting incurred by the forestry and set to 253 SEK per m^3 (Swedish Forest Agency, 2009). The growth rate is the mean annual volume increment for all forest types and set to 101.6 million m^3 (Swedish Forest Agency, 2009).

3.2. Parametric Assumptions

Generally, empirical estimates of substitution elasticities for forest products are difficult to obtain. It is commonly assumed that it is more difficult to substitute between forest products in the pulp and paper industry than in the heating industry. That is, the elasticities of substitution are generally assumed to be lower in the pulp and paper industry than in the heating industry. This assumption is confirmed by Olsson and Lundmark (2013), which estimates elasticities of substitution for forest products. The elasticity of substitution between roundwood and by-products in the pulp and paper industry is estimated to 0.19. The substitution elasticity between logging residues and by-products in the heating industry is estimated to 6.93.

There are several econometric studies on the relationship between roundwood supply and price. For example, Toppinen and Kuuluvainen (1997) estimate the pulpwood price elasticity of supply to 0.41 in Finland (which has a similar market structure to Sweden). The price elasticity of pulpwood supply and timber supply is estimated to 0.23 and 0.5, respectively by Ankarhem et al., (1999) in Sweden. The same elasticities are estimated to 0.37 and 0.68 by SLU (2004). The elasticity of roundwood supply is set to 0.5. The elasticity of supply for logging residues and by-products are assumed to be relatively more elastic and set to ten.

3.3. Scenario Descriptions

The baseline scenario is the norm to which the other scenarios are compared to and calibrated to the 2007 situation. Four scenarios with different development paths are considered. The policies implemented to achieve the 2-degree target are assumed to result in an increase in the demand for forest products from the heating industry. In addition, different development paths for the industries are considered. An overview of the scenarios is given in Table 2.

Table 2. Development paths in the scenarios

	Heating industry	Pulp and paper industry	Sawmill industry	Other industries in energy sector
	------Output level-------		---------Input level--------	
Scenario 1	50 % increase			
Scenario 2	50 % increase	20% increase		
Scenario 3	50 % increase	20% increase	20% decrease	
Scenario 4	50 % increase			50 % increase

4. Scenario Results and Implications

Table 3present the results for the baseline scenario and for the four scenarios, reported as percentage changes compared to the baseline scenario. In the baseline scenario the forestry supplies 74.4 million m3 of roundwood and 7.4 million m3 of logging residues, while he sawmill industry supplies 18 million m3 of by-products. The pulp and paper industry consumes 36.2 million m3 of roundwood and 11.7 million m3 of by-products. The heating industry consumes 0.04 million m3 of logging residues and 6.3 million m3 of by-products. The prices are 335, 306 and 304 SEK per m3 for roundwood, logging residues and by-products, respectively.

In scenario 1 the heat production is assumed to increase by 50 percent. The results suggest that the heating industry's demand of by-products will increase by 50 percent and the demand for logging residues by 58.3 percent relative the baseline scenario. However, the increase in logging residues is from a low level thus the increase does only slightly affect the price. The price of by-products increase by 0.8 percent inducing the pulp and

paper industry to substitute by-products (decreases by 0.1 percent) in favor of roundwood (increases by 0.03 percent).In total, the heating industry's increase in the demand of by-products is larger than the decrease in demand by the pulp and paper industry. The sawmills increase of by-products by 8.1 percent imply that the maximum level is reached. Overall, the competition for by-products may become relatively intense given an increase of 50 percent in the heat production. The pulp and paper industry's ability to substitute, at least small quantities, of by-products for roundwood may help reduce the price of by-products and the competition.

Scenario 2 combines an increase in heat production by 50 percent with an increase in the pulp production by the pulp and paper industry by 20 percent. The results indicate that demand for roundwood by the pulp and paper industry increases by 19.2 percent and the demand for by-products increases by 23.0 percent. The price of roundwood increases by 19.5 percent and the price of by-products by 0.8 percent. The heating industry increases their demand for logging residues by 58.3 percent while their demand for by-products decreases by 19.3 percent. The output of roundwood increases by 9.3 percent while the output of by-products increases by 8.1 percent. The scenario was expected to result in an increase in the input demand of the heating and pulp and paper industries. While the pulp and paper industry increase their demand for its inputs, the heating industry decrease their demand for by-products. The reason behind this behavior is likely the limited supply of by-products. The price of by-products does however only increase slightly and the heating industry seems to decrease their by-product consumption in favor for logging residues. In addition, the relative price between by-products and logging residues is lower than between by-products and roundwood. This could be reasons why the competition seems to be relatively limited also in this case.

Table 3. Simulated output and input volumes and prices of forest products for baseline and policy scenarios

Output level	Baseline	Scenario 1	Scenario 2	Scenario 3	Scenario 4
Forestry	(million m3)		(%)		
Roundwood	74.4	0.01	9.3	-0.5	0.01
Logging residues	7.4	0.3	0.3	0.2	49.9
Sawmill					
By-products	18.0	8.1	8.1	-13.5	8.1
Input level					
Pulp and paper					
Roundwood	36.2	0.03	19.2	20.0	0.03
By-products	11.7	-0.1	23.0	20.1	-0.1
Heating					
Logging residues	0.04	58.3	58.3	36.1	19.4
By-products	6.3	50.0	-19.3	-75.3	23.2
Price	(SEK/m3)				
Roundwood	335	0.03	19.5	-1.1	0.03
Logging residues	304	0.03	0.03	0.02	4.1
By-products	306	0.8	0.8	-1.4	0.8
Profit	(million SEK)				
Forestry	6,100	0.1	96.6	-5.0	0.1
Sawmill	36,822	1.3	-5.5	5.0	1.3
Pulp and paper	39,358	-0.1	12.7	20.6	-0.1
Heating	6,050	58.7	72.5	90.8	58.7

Scenario 3 combines the assumptions of the two previous scenarios with a decrease in the demand for roundwood by the sawmill industry by 20 percent. Not surprisingly, the results show that the decrease in the demand for roundwood by sawmills will reduce their output of by-products negatively by 13.5 percent. This output level is at the limit of the restriction set by sawmills' roundwood demand. The pulp and paper industry increase their demand of roundwood and by-products by 20.0 and 20.1 percent, respectively. The heating industry increase their input demand of logging residues but decrease their input demand of by-products by a relatively large amount. The increase in the pulp and paper industry's demand, in percentage terms, is larger for by-products than for roundwood, this is explained by the relatively larger fall in the price of by-products than in price of roundwood.

Finally, the results for scenario 4 indicate that the pulp and paper industry increases their demand of roundwood slightly (0.03 percent) and decrease their demand of by-products slightly (0.10 percent). The heat industry increases their demand of by-products by 50.2 percent and logging

residues by 19.4 percent. The price of logging resides increases by 4.1 percent. Overall, the pulp and paper industry is relatively unaffected in this scenario. However, the logging residue price does increase suggesting an intensified competition for logging residues between the industries within the energy sector.

4.1. Sensitivity Analysis

As the elasticities of substitution are important parameters, these have been changed to test the robustness of the results. The elasticities of substitution for the pulp and paper industry and the heating industry are increased respectively decreased by 50 percent. The elasticity of substitution for the heating industry is thus set to 10.39 respectively 3.46 and for the pulp and paper industry it is set to 0.28 and 0.09 respectively. The results from the sensitivity check of input and output quantities and prices are reported in Table 4. The results from the sensitivity analysis indicate that the model is relatively robust.

Table 4. Percentage Change in Output and Input Quantities and Prices with a ± 50 % Change in the Elasticity of Substitution

Output level	Scenario 1		Scenario2		Scenario3		Scenario4	
	+50%	-50%	+50%	-50%	+50%	-50%	+50%	-50%
Forestry				---(%)---				
Roundwood	0.01	-0.01	-0.17	0.18	0.00	0.00	0.01	-0.01
Logging residues	0.01	-0.03	0.01	-0.03	-0.04	0.03	-0.04	0.04
Sawmill								
By-products	0.00	0.00	0.00	0.00	0.00	0.00	0.00	0.00
Input level								
Pulp and paper								
Roundwood	0.02	-0.01	-0.33	0.34	-0.01	0.01	0.02	-0.01
By-products	-0.05	0.05	1.16	-1.21	0.02	-0.03	-0.05	0.05
Heating								
Logging residues	1.75	-3.51	1.75	-3.51	-6.12	4.08	-9.30	11.63
By-products	-0.02	0.01	-0.02	0.01	-0.19	0.26	0.05	-0.05
Price								
Roundwood	0.01	-0.02	-0.35	0.36	-0.01	0.01	0.01	-0.02
Logging residues	0.00	0.00	0.00	0.00	0.00	0.00	0.00	0.00
By-products	0.00	0.00	0.00	0.00	0.00	0.00	0.00	0.00

5. Conclusions

The energy sector has in the past decades increased its utilization of forest products. This is, to a large extent, caused by economic policies introduced to reduce the emission of greenhouse gases. Since the increasing utilization is believed to continue, it is of interest to analyze the affects it might have on the competition between the forest industries and the energy sector. Thus, the purpose is to analyze the extent and degree of the competition for forest products in the presence of climate and energy policies.

The results indicate that climate policies that would increases the utilization of forest products in the energy sector would have rather modest effects on the pulp and paper industry. The forest product prices would slightly increase and the pulp and paper industry would be able to alleviate this

effect by substituting between different feed-stocks.

Climate policies could however affect the possibilities for the pulp and paper industry to expand its production. In such a situation both the heat industry and the pulp and paper industry will demand larger volumes of forest products compared to the current situation. The limited supply of by-products from sawmills suggests that the competition in this case would be relatively intense. The price of by-products does however only increase slightly and the heat industry seems to decrease their by-product consumption in favor for logging residues.

In the event of an expanding pulp and paper industry, simultaneously as sawmills experiences a fall in their production, the effects of a climate policy is rather unintuitive. Despite the lower supply and higher demand for by-products, the price does not increase. Instead the price decreases. Therefore, the competition does not intensify in this situation either.

An increase in the demand for forest products from other industries in the energy sector does seem to give rise to a more intense competition. That is, a climate policies that only affects the increase in the heating industry's utilization of forest products in combination with a policy that increases the rest of the energy sector's demand for forest products will result to a higher price of logging residues and increase the competition.

In summary, the results do not indicate an intensified competition between the forest industries and the energy sector. An increase in the competition between industries in the energy sector can however be observed.

Acknowledgement

Financial support from the LETS research program and its financiers: the Swedish Environmental Protection Agency, the Swedish Energy Agency, the Swedish Transport Administration, and Vinnova are gratefully acknowledged. We also thank Bio4Energy, a strategic research environment appointed by the Swedish government, for supporting this work.

Appendix

Table A1. List of indices, parameters and variables.

Indices	
i	industry (i = F, S, P, H)
j	input (j = RW, BY, LG)
f	output (f = RW, BY, LG, SW, PP, HE)
Parameters	
γ_i	scale parameter of industry i
α_i^j	distribution parameter of industry i for input j
$\rho_i = \frac{1}{1-\sigma_i}$	function of the elasticity of substitution of industry i
σ_i	elasticity of substitution in industry i
β^j	exogenous cost component in input prices independent of harvesting levels
δ^j	shift parameter accounting for factors other than price affecting the supply of input j
ϕ^j	elasticity of supply for input j
k^j	capacity restriction (maximum production level) for input j
τ_{FCOST}	annual logging cost per cubic meter harvested roundwood
$\psi_{FGROWTH}$	annual growth rate of the forest
Variables	
y_i^f	output of f of industry i
x_i^j	demand of industry i for input j
w^j	price of input j
p^f	price of output f
π_i	profit of industry i
π_{TOT}	sum of the profits of all industries

References

[1] Ankarhem, M., R. Brännlund and M. Sjöström.(1999). Biofuels and the forest sector. In Yoshimoto, A. and K. Yukutake (eds.). Global concerns for forest resource utilization; sustainable use and management. Kluwer Academic Publishers, Dordrecht.

[2] Bolkesjo T.F., E. Trømberg and B. Solberg. (2006). Bioenergy from the forest sector: Economic potential and interactions with timber and forest products markets in Norway. Scandinavian Journal of Forest Research, 21(2): 175-185.

[3] Buongiorno, J., S. Zhu, D. Zhang J. Turner and D. Tomberlin. (2003). The global forest products model: Structure, estimation and applications. Academic Press, Elsevier Sci., New York, USA.

[4] Brännlund, R., R. Lundmark and P. Söderholm. (2010). Kampen om skogen – Koka, såga, bränna eller bevara? SNS Förlag, Stockholm, Sweden.

[5] Böhringer, C., T.F. Rutherford and W. Wiegard. (2003). Computable general equilibrium analysis: Opening a black box. ZEW Discussion Paper 03-56, Mannheim, Germany.

[6] European Union (EU). (2009). Directive 2009/28/EC of the European Parliament and Council. Official Journal of the European Union, 140:16-62.

[7] FAO – Food and agriculture organization of the United Nation. (2010). FAOSTAT Forestry (www.fao.org)

[8] Hammarlund, C., K. Ericsson, H. Johansson, R. Lundmark, A. Olsson, E. Pavlovskaia and F. Wilhelmsson. (2010). Bränsle för ett bättre klimat. Rapport 2010:5, AgriFood. Lund, Sweden.

[9] Haynes, R.W. (1993). Forestry sector analysis for developing countries: Issues and methods. PNW-GTR-314. USDA, Forestry Service, Pacific Northwest Research Station, Portland, USA.

[10] Kallio, M., D.P. Dykstra and C.S. Binkley.(1987). The global forest sector: An analytical perspective. Wiley, New York, USA.

[11] Kallio, M., A. Moiseyev, and B. Solberg, (2004). The global forest sector model EFI-GTM – The model structure. Internal report No.15. European Forest Institute, Finland.

[12] Lundmark, R.(2007). Dependencies between forest products sectors: a partial equilibrium analysis. Forest Products Journal, vol. 57(9): 79-86.

[13] Lundmark, R and P. Söderholm. (2004). Brännhett om svensk skog. SNS Förlag, Stockholm, Sweden.

[14] Olsson, A. and R. Lundmark.(2011). Factor substitution and procurement competition for forest resources in Sweden. Economics Unit, Luleå University of Technology.

[15] Mansikkasalo, A. (2007). The European forest trade model – EU policy and the impact on forest raw material use. Licentiate Thesis Economics Unit, Luleå University of Technology.

[16] SDC. (2009). Skogsindustrins virkesförbrukning samt produktion av skogsprodukter 2004-2008. Uppsala, Sweden.

Modelling the competition for forest resources: The case of Sweden   99

[17] SLU. (2004). SLU:s bioenergiutredning. The Swedish University of Agricultural Sciences, Uppsala, Sweden.

[18] Statistics Sweden. (2008). Energy prices and electricity suppliers switching, 4th Quarter 2007. EN 24 SM 0801, Örebro, Sweden.

[19] Statistics Sweden .(2009). Electricity supply, district heating and supply of natural and gasworks gas 2007. EN 11 SM 0901, Örebro, Sweden.

[20] Swedish Forest Agency.(2009). Swedish Statistical Yearbook of Forestry 2009. Jönköping, Sweden.

[21] Toppinen, A. and J. Kuuluvainen.(1997). Structural changes in sawlog and pulpwood markets in Finland. Scandinavian Journal of Forest Research, 12(4): 382-389.

[22] Trømberg E. and B. Solberg. (2010). Forest sector impacts of the increased use of wood in energy production in Norway. Forest Policy and Economics, 12(1):39-47.

[23] United Nations (UN). (2007). United Nations Forum on Forests – Report of the seventh session. E/2007/42.

[24] United Nations Programme to Reduce Emissions from Deforestation and Degradation (UN-REDD) (2010). (www.un-redd.org).

[25] Varian, H.R. (1992). Microeconomic Analysis (3rd ed). Norton & Company, USA.

[26] Åkerman, J., K. Isaksson, J. Johansson and L. Hedberg. (2007). Tvågradersmålet i sikte? Scenarier för det svenska energi- och transportsystemet till år 2050. Rapport 5754, Naturvårdsverket. Stockholm, Sweden.

14

Does a windfall lead to a downfall? A study of mineral rents and genuine growth in selected African countries

Ishmael Ackah[1], Dankwa Kankam[2], Kwaku Appiah- Adu[3]

[1]Department of Economics, Portsmouth Business School, University of Portsmouth, Portsmouth, UK
[2]Opoku, Andoh& Co, Chartered Accountants and Management Consultants, Accra, Ghana
[3]Central Business School, Central University College, Accra, Ghana

Email address:
Ackish85@yahoo.com (I. Ackah), okyeamek@yahoo.com (D. Kankam), kwaku_appiahadu2001@yahoo.com (K. Appiah- Adu)

Abstract: The mining industry helps governments to increase revenues resulting in job creation, infrastructural development and enhancing the standard of living of the local communities. However, it is also a potential source of environmental pollution and rent-seeking. This study examines the relationship between mineral rents and genuine income in a multivariate panel data analysis in Botswana, Egypt, Ghana, Morocco, South Africa and Zambia at the aggregate and industrial levels. The findings suggest that mineral revenues have been a blessing to these countries at the aggregate level but affected industrial growth negatively. These findings support evidence in the literature that mineral resource abundance slows growth in industrial output. The study also reveals that growth in mineral rich countries is principally driven by investment in capital and energy consumption. It is therefore recommended that mineral revenues should be invested in capital and alternative energy sources to boost aggregate and industrial growth.

Keywords: Mineral Rents, Genuine Income, Energy Consumption, Resource Curse

1. Background

Resource abundance should lead to growth and institutional quality (Brunnschweiler and Bulte, 2008). This is because; the discovery of resources such as gold or diamond brings windfalls which can be invested in other sectors of the economy. According to Limi (2010), natural resource abundance should lead to development since resource revenues can be invested in economic infrastructure and human capital development. However, this has always not been the case. Resource discoveries have led to slow growth, corruption, stagnated manufacturing sector and local currency appreciation in many countries (Ross, 2012). The reasons for the poor performance of oil-rich countries have been varied in the mineral resource literature.

According to the literature on natural resource economics, there are four channels through which natural resources become a curse. Firstly, the Dutch disease (Corden and Neary, 1982): secondly, rent seeking (Collier and Hoeffler, 2004):thirdly, institutional failures (Mehlum et al, 2006) and finally, the type of natural resource (Boschini et al 2007). The summary of these literatures indicates that resource

scarce economies often outperform their resource rich counterparts with higher growth rates, thus, natural resource abundance inhibits growth. This paradoxical hypothesis suggests a shift from the classical conception of the growth enhancing effect of rich natural resource endowments to a growth inhibiting effect termed as the 'Resource Curse'.

The Dutch disease has been found to be a major determinant of the resource curse. It describes an economy where a booming sector and a lagging one co-exist due to natural resource discovery. For instance, in an open economy, when a windfall such as diamond or gold find occurs, it leads to current account surplus which intend leads to the appreciation of the local currency under a flexible interest rate regime. This makes the non- natural resource sector uncompetitive and leads to high prices (Sachs and Warner, 2001). In addition, the natural resource earnings are absorbed by the domestic non-tradable sector which leads to real appreciation of the local currency (Corden and Neary, 1982). These high prices of local produce leads not only to a stagnated export sector which deprives the economy of the benefits of export-led growth but also increase unemployment.

Gylfason et al (1999) identifies several main channels of transmission of abundant resources into a stunted economy. The first channel is through the surge in primary raw material exports and upward adjustment of the exchange rate which reduces the exports from other productive sectors such as service, high-tech and manufacturing sectors. This leads to overvaluation of a country's currency or real wages. Again, since mineral prices are volatile, the recurrent booms and busts put pressure on the exchange rate regime (Corden, 1984). These challenges coupled with mass exodus of workers from other productive sectors to the natural resource sector have been referred to as the Dutch disease. In addition to currency appreciation, resource abundance leads to reallocation of natural resources and deindustrialization. Other factors include the lack of incentive to collect taxes in mineral rich countries. According to Harford and Klein (2005), natural resources such as gold and diamond export remove incentives to establish a well-functioning tax system, reform institutions or even improve infrastructure development due to increased rent-seeking and corruption.

On Africa, Aryeetey et al., (2012) suggest that the major challenge to growth has been the inability of African economies to translate mineral resource revenues into development. The reasons for the inability of natural resource to play a major role in economic growth have been partially attributed to the different sides of mineral revenues. On the positive side, resource revenues have the distinctive qualities of scale, stability, source (Ross, 2012) and superiority. These qualities should translate into growth in the form of job creation, access to credit facilities by the companies and individuals, technological transfer from major foreign mineral companies to local partners, training and capacity building and increased government revenue mineral related taxes and resource rent. Aryeetey et al (2012) suggest that resource revenues can be used to finance diversification as was the case of Malaysia which used resource revenues to fund the expansion of light manufacturers that now dominate its export.

On the negative side however, mineral revenues are volatile, exhaustible, and uncertain and traced to international market conditions (Barnett and Ossowski, 2002).Owing to the exhaustible nature of mineral revenues, issues of intergenerational and intragenerational development have become central in policy design in mineral producing countries. Since long term plans require near stable prices and conditions for predictions, volatility makes economic forecasts challenging which subsequently creates loopholes for rent seeking.

The resource abundance-growth hypothesis has attracted increased attention in Africa. For instance, Sala-i-Martin and Subramanian (2003) find a negative and nonlinear association between natural resource abundance and growth in Nigeria and attribute this inverse relationship to corruption and poor institutional quality. Iimi (2006) explores the case of Botswana and concludes that natural resources have growth impact and attributes this finding to good governance.

Minerals such as gold, diamond, bauxite and steel abound

in Africa. Figure 1.0 shows the mineral rent of Ghana and Botswana over the estimated period.

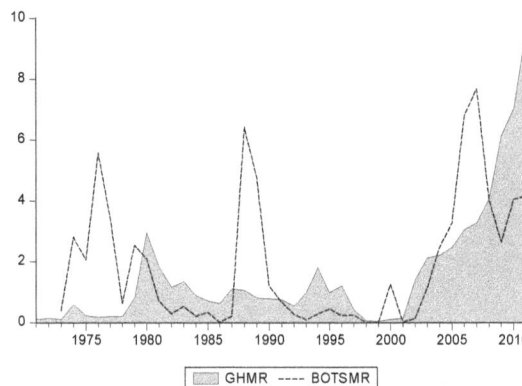

Fig1.Mineral rent for Ghana and Botswana

Often, the mineral revenue-economic growth hypothesis has been investigated by modelling minerals abundance as a function of GDP. However, Aryee et al, (2002) suggest that mining leads to siltation, coloration and chemical pollution of streams and rivers that provide drinking water for mining communities. In Africa, the mining sector is generally thought to be the second largest source of pollution after agriculture. The sector is resource intensive and generates high concentrations of waste and effluents. For example Kuhndt et al (2008) argue that during the extraction process and smelting process, a ton of copper generates about: 100-350 tons of residues, 50-250 tons of extraction waste and 300 kg of sulphur dioxide.

It would therefore be prudent if environmental damages due to mining are considered when estimating the resource curse hypothesis. This study contributes to the resource literature by using the genuine growth as a proxy for GDP. The reason is that, genuine growth captures the environmental effect of growth. That is, the mining related pollution is quantified and subtracted from the GDP. In addition, a multivariate fixed effect model that controls for country specific effect is used. Apart from mineral rents, the impact of variables such as energy consumption, human capital development, investment and inflation on growth are also measured. Finally, an industrial model is estimated to ascertain the impact of mineral rents on industrial output. Six countries: Botswana, Egypt, Ghana, Morocco, South Africa and Zambia are used for this study due to their relatively higher mineral (gold, diamond, copper) output in Africa. For instance, Botswana produced 27.9% of the African diamond production in 2009. Africa is the world's largest producer of diamonds, producing as much as 52.4% of global production in 2009 (Brown et al., 2010), and reserving 60% of the global diamond. Ghana is the second largest producer of gold in Africa and accounted to for 20% of total gold production in Africa in 2009 (Sharaky, 2010). In addition, it produces diamond, manganese, and bauxite just as South Africa.

2. Method

The purpose of this study is to examine the relationship between mineral revenues and economic growth in mineral producing African countries at both the aggregate and industrial levels. The panel fixed effect is used since it has the ability to control for individual country effects such as culture, education, corruption and institutional quality. According to Ackerberg et al., (2007) the panel fixed effects model has the advantage of overcoming simultaneity bias and gives consistent estimates. The equation proceeds as:

$$Y_{it} = e^{\beta} MR_{it}^{\alpha} K_{it}^{\psi} H_{it}^{\phi} EN_{it}^{\delta} IN_{it}^{\gamma} e^{\varepsilon_{it}} \qquad (1)$$

Where Y_{it} is genuine income, MR_{it} is mineral rent, K_{it} is investment, H_{it} is human capital development, EN_{it} is energy consumption and IN_{it} is inflation. $\alpha, \psi, \phi, \delta$ and γ represent the elasticities of mineral rents, investment, human capital, energy consumption and inflation respectively.

Applying logs to equation 1, For instance mr = log MR

$$y_{it} = \beta + \alpha mr_{it} + \psi k_{it} + \phi h_{it} + \delta en_{it} + \gamma in_{it} + \varepsilon_{it} \qquad (2)$$

The same procedure is followed to estimate the relationship between industrial output and mineral rents.

2.1. Data Sources

The study relied on secondary data obtained from the World Bank Development Indicators (WDI). Instead of GDP, the study used genuine income. This is because; the value of the genuine income is GDP minus environmental damages such as pollution and degradation due to mining and other commercial activities. Energy consumption is measured in tonnes of oil equivalent. Government expenditure on education is used as a proxy for human capital development and is measured in US dollars. In addition, fixed capital formation is used as a proxy for investment in capital. The estimation period is from 1971 to 2011. Export, industrial output, and all other data are in current US dollars.

3. Data Analysis

3.1. Panel Unit Root

To estimate the long relationship among the variables, a panel unit root test is carried out to examine their stationarity. Four different tests are applied. These are Levin, Lin and Chu test, the lm, Pesaran and Shin, the ADF Fisher test and the PP Fisher test. The panel root equation is given as:

$$Y_{it} = \chi_t Y_{it-1} + \varpi_t F_{it} + \varepsilon_{it}$$

Where if $\chi_t = 1$, then Y_{it} contains a unit root but if $\chi_t < 1$, then Y_{it} is considered a weakly trend stationary. F_{it} represents the fixed effects, individual time trends and other exogenous variables. All four tests rejected the null of unit root 1%, 5% and 10% confidence level at level but were significant after the first difference as shown in Table 1. Table 1. presents the output of the unit root test.

Table 1. *panel unit root results*

Test	Results
Levin, Lin and Chu	-29.687***
lm, Pesaran and Shin-W stat	-29.956***
ADF Fisher	787.004***
PP Fisher	862.484***

Table 2. *Panel data output - Aggregate estimation*

Regressors	Coefficients
C	1.945***
MR	0.021**
H	0.191***
IN	-0.002
EN	0.363***
K	0.533***

Table 2 shows the estimated output of the panel estimation. All the regressors have a positive relationship with genuine income apart from inflation which is not significant even at 10% confidence level. The results suggest that genuine growth is mainly driven by investment in capital and energy consumption. Table 3 presents the preferred model after insignificant variables are removed from the estimation.

Table 3. *Panel data output- preferred Aggregate model*

Regressors	Co-efficients
C	1.911***
MR	0.017*
H	0.183***
EN	0.384***
K	0.536***

Table 2 reports the output of the of the panel data estimation at the aggregate level. The results indicate a positive relation between mineral revenues and genuine income. Similar findings are reported in Cavalcantiet al (2009) van der Ploeg (2011). This implies that at the aggregate level, mineral resources contribute to increased genuine income to the extent that any 1% increase in mineral revenues leads to 0.017% increase in genuine income. This finding may be due to three reasons. First, mineral companies contribute immensely to government revenues through the payment of tax and royalties. The mining sector has been consistently contributed to Africa's growth. To illustrate this point with examples, on the average mining accounted for 12.1 per cent of government revenue from 1990 to 2004 in Ghana, and 48.3 per cent from the 1998/1999 to the 2008/2009 fiscal year in Botswana. The high contribution of mining revenue to total revenue in Botswana is due to the high value addition to diamonds. In addition, minerals such as gold, diamond, and bauxite form a vital part of the export of these countries. Finally, the introduction of environmental regulations seems to have reduced the impact of mining activities on the environment. Investment in capital is the main driver of genuine income. This finding buttress Hartwick's (1977) assertion that investment in man-made capitalis the key to sustainable growth in resource rich countries. A change of 1% percentage in investment leads to a 0.536% change in genuine growth in the same direction. This may be due to the capitalintensive nature of the mining industry and the trickledown effect of capital investments to other sectors of the economy. Energy consumption plays a vital role in every economy. Energy is an important resource in the production process of firms and source of comfort in households. Energy consumption explains 0.384of variations in genuine income.

Table 4. Industrial sector

Regressors	Co-efficients
C	-6.724***
EX	0.561***
MR	-0.055**
EN	2.541***

Table 4 shows the preferred model of the industrial estimation. The results confirms the findings of Sachs and Warner (1997, 1999 and 2001) and Gylfason et al (1999) who reported a negative relationship between resource revenues and output .The results show that the main growth determinants in the industrial sector are energy consumption and export. This may be due to the fact theindustrial sector relies on energy to power its machines for production. The study reports that, any 1% increase in mineral revenues reduces industrial output by 0.055% which confirms the Dutch disease hypothesis. The results further indicate that energy consumption is very vital to industrial performance. There is a need to enhance energy access and security since reductions in energy supply can have a negative effect on industrial output. Finally, mineral rich countries should open their economies since exports have a direct relationship with industrial growth.

4. Conclusion and Policy Recommendations

This study examines the relationship between mineral revenues and genuine income in 6 mineral-rich African countries. These are Botswana, Egypt, Ghana, Morocco, South Africa and Zambia. Multivariate panel data estimation with controls for country specific effects is used. The results indicate that mineral revenues contribute positively to genuine income in these countries. In addition, the study reveals that genuine growth at the aggregate level is principally driven by investment in capital and energy consumption. On the contrary, the resource curse hypothesis is confirmed in the industrial sector. Increased mineral

revenues lead to a reduction in industrial output. This may be attributed to the Dutch disease or movement of skilled labour from the industrial sector to the mining sector. This could alsobe due to the fact that increased natural resource revenues lead to increased growth in the service and agriculture sectors at the expense of the industrial sector. Besides, growth in industrial output is influenced positively by export and energy consumption. Ross (2012) suggests that the best means of avoiding or minimizing the impact of the mineral curse is to avoid export-oriented extractive industries and develop the agriculture and the manufacturing sectors. Furthermore, the role natural resources play in economic development is highly dependent on policy choices (Auty, 1998) and therefore policy makers should make choices that are optimal for long term growth. Atyeh and Al-Rashed (2013) suggest that the diversification of the economy with emphasis on modernization of agriculture can lead to growth in mineral rich countries.

Since minerals are exhaustible resources, mineral revenues should be invested in man-made capital to enhance sustainable development. Therefore, investment in physical capital such as schools, roads, hospitals and other forms of capital such as bonds and equity holdings should be encouraged. This will help to spread the benefits of the resources to the present as well as future generations.

Additionally, since energy consumption has a direct and significant relationship with both genuine income and industrial output, there should be investment in alternative sources of energy. This is because, renewable energy sources are not depletable and abounds in Africa. Investment in renewable energy will enhance security of supply and minimize the impact of energy consumption on the environment.

Our study recommends that mineral revenues should be invested in human capital development such as education and health infrastructure and training, since it is a major determinant of growth. Finally, policies which seek to promote export should be encouraged to enhance aggregate and industrial growth.

Regarding future research areas, subsequent studies can compare the effect of oil and mineral revenues on growth to ascertain if the curse is resource specific. This study was limited by lack of data on corruption, institutional quality and other factors that can affect the mineral resource-growth relationship.

References

[1] Ackerberg, D., Benkard, C.L., Berry, S. and Pakes, A. (2007) Econometric tools for analyzing market outcomes In J. Heckman and E. Leamer (eds), Handbook of Econometrics (Vol. 6(1), pp. 4171–4276) Amsterdam: North-Holland

[2] Aryee, B. N. A., Ntibery, B. K., Atorkui, E. (2002) Trends in small-scale mining of precious minerals in Ghana; a perspective on its environmental impacts. Journal of Clean production, vol. 11, pp 131-140.

[3]   Aryeetey, E., Devarajan, S., Kanbur, R., &Kasekende, L. (Eds.) (2012) The Oxford companion to the economics of Africa. Oxford University Press

[4]   Atyeh, M and Al-Rashed, W. (2013) Testing the existence of integration: Kuwait and Jordanian financial markets, International Journal of Economics, Financial and Management Sciences 2013; 1(2): 89-94

[5]   Auty, R. M., & Mikesell, R. F. (1998). Sustainable development in mineral economies. Oxford University Press.

[6]   Boschini, A. D., Pettersson, J., & Roine, J. (2007). Resource Curse or Not: A Question of Appropriability. The Scandinavian Journal of Economics, 109(3), 593-617.

[7]   Brunnschweiler, C. N., &Bulte, E. H. (2008). The resource curse revisited and revised: A tale of paradoxes and red herrings. Journal of Environmental Economics and Management, 55(3), 248-264.

[8]   Cavalcanti, Tiago V. de V., Kamiar Mohaddes, and Mehdi Raissi (2009), "Growth, Development and Natural Resources: New Evidence Using a Heterogeneous Panel Analysis," Faculty of Economics, University of Cambridge (mimeo).

[9]   Collier, P., & Hoeffler, A. (2004). Greed and grievance in civil war. Oxford economic papers, 56(4), 563-595.

[10] Corden, Max W. and J. Peter Neary (1982), "Booming Sector and De-Industrialization in a Small Open Economy," The Economic Journal 92 (368), pp. 825–848.

[11] Gylfason, Thorvaldur, Tryggvi T. Herbertson, and Gylfi Zoega (1999), "A Mixed Blessing: Natural Resources and Economic Growth," Macroeconomic Dynamics 3, pp. 204–25.

[12] Hartwick, J. M. (1977). Intergenerational equity and the investing of rents from exhaustible resources. The American economic review, 67(5), 972-974.

[13] Harford, T. M. Klein, Aid and the Resource Curse, The World Bank Group, Private Sector Development Vice Presidency, Note #291, Washington, DC, 2005.

[14] Iimi, Atsushi (2006), "Did Botswana Escape from the Resource Curse?" IMF Working Paper06/138, Washington, D.C.

[15] Mehlum, H., Moene, K., &Torvik, R. (2006). Institutions and the Resource Curse*. The Economic Journal, 116(508), 1-20.

[16] SHARAKY, A. M. Mineral Resources and Exploration in Africa. Egypt: Cairo University. 2010

[17] Kuhndt M, Tessema F, and Martin H. (2008) Global Value Chain Governance for Resource Efficiency Building Sustainable Consumption and Production Bridges across the Global Sustainability Divides. Environmental Research, Engineering and Management, 2008. No. 3(45), P. 33-41.

[18] Ross, M. L. (2012). The oil curse: how petroleum wealth shapes the development of nations. Princeton University Press.

[19] Sachs, Jeffrey D. and Andrew M. Warner (1997), "Natural Resource Abundance and Economic Growth," Center for International Development and Harvard Institute for International Development, Harvard University, Cambridge, Mass.

[20] Sachs, Jeffrey D. and Andrew M. Warner (1999), "The Big Push, Natural Resource Booms and Growth," Journal of Development Economics (59), pp 43–76.

[21] Sachs, Jeffrey D. and Andrew M. Warner (2001), "Natural Resources and Economic Development. The curse of natural resources," European Economic Review 45(2001), pp. 827–838.

[22] Sala-i- Martin, Xavier and Arvind Subramanian (2003), "Addressing the Natural Resource Curse: An Illustration from Nigeria," NBER Working Paper 9804, Cambridge, Mass.

[23] Van der Ploeg, F. (2011). Natural resources: Curse or blessing? Journal of Economic Literature, 49(2), 366-420.

15

Water absorption properties of some tropical timber species

Emmanuel Tete Okoh

Department of Furniture Design and Production, Accra Polytechnic, P O Box GP 561, Accra, Ghana

Email address:
etokoh@apoly.edu.gh

Abstract: The water absorption characteristics during soaking of *Terminalia superba* (Ofram), *Terminalia ivorensis* (*Emere*) as currently threatened timber species and *Quassia undulata* (Hotrohotro) and *Recinodendron heudelotii*.(*Wama*) as lesser used timber species were studied to determine and compare their absorption and diffusion coefficients as potential substitutes for utilization. Water soaking was carried out for nineteen days and the data were fitted into the Fick's model to determine both the water absorption and diffusion coefficients. The study showed that, the mean values of the water absorption coefficient at initial stages of moisture sorption for *Hortrohotro*, *Ofram*, *Emere* and *Wama* were 3.51x 10^{-3}, 4.31x10^{-3}, 1.67x10^{-3} and 8.27x10^{-4} (kg/m^2/s) respectively. The corresponding mean values of this parameter for the entire soaking process were also determined for the timber species viz; *Ofram* (2.91x 10^{-3}), *Hortrohotro* (2.58x 10^{-3}), *Emere* (1.14x 10^{-3}) and *Wama* (6.11x 10^{-4}) kg/m^{2}s respectively. The measured diffusion coefficient for *Wama*, *Emere*, *Ofram* and *Hotrohotro* timber species were 9.637x10,$^{-4}$ 6.694x10^{-3} 4.185x10^{-2} and 2.899x10^{-2} kg/m^2/s respectively. *Emere* and *Wama* had lower absorption and diffusion characteristics than *Hortrohotro* and *Ofram*. Based on this study, *Wama* could be substituted for *Emere* and *Hortrohotro* with *Ofram*.

Keywords: Absorption, Diffusion, Fick's Law, Sorption

1. Introduction

[1], reported two types of liquid movement in wood: diffusion through the cell walls and flow in the cell lumens. The latter is considerably more prevalent during wood processing. The anatomical features affect non-steady liquid flow in radial and longitudinal directions. The wetting rate through capillary action is much faster than that through diffusion; therefore, the present study aimed to observe free liquid soaking through capillary and diffusion action only.

Although fiber often constitutes the majority of woody tissue, general fiber is not considered as important as vessels in primary liquid flow [2]. However, fiber permeability may influence the subsequent spreading of liquid from vessels or other cells connecting them to pits. Comparing fibers to vessels, non perforated ones are thick walled with relatively small pits that are not adapted for efficient liquid conduction. Apart from this, interconnecting pits provide one of the main pathways for the flow of liquid between cells, and their structure and distribution affects the penetration of liquid in wood [3]. Also, the air that is compressed during liquid penetration lowers the permeability of wood [4]. For this [5],

reported that air is compressed during water soaking and additional counter pressure is formed that substantially reduces permeability.

Knowledge of the capillary system is very important for studying the movement of fluids and vapors through a porous material. This fact influences its sorption properties, especially at high relative humidity (RH), where equilibrium is mainly controlled by the capillary forces and consequently by the microstructure of wood species. The capillary system of wood consists of cavities interconnected by narrow channels. The variation in dimensions between the different types of cavities connected in series suggests that desorption tends to be governed by a lower water potential, which is determined by the narrower sections of the pores. In contrast, adsorption tends to be governed by a higher water potential, which depends on the larger sections of the pores; thus, the desorption isotherm will depend on the size of channels connecting the lumina, whereas the adsorption isotherm will depend on the size of these lumina [6]. The permeability of wood is strongly dependent on its moisture content [7], as well as the principal direction of the grain [8] and various physical and chemical properties [9].

In exterior applications, the wood-moisture content fluctuates roughly between 8 and 40% (mass) which causes dimensional changes between 2 to 10% depending on wood species [10]. The physical processes that control the uptake and release of moisture are: adsorption of water to the hydroxyl groups in the cell walls, diffusion of water molecules through the air inside the wood or the cell wall and capillary flow of liquid water into the pores of wood. Logically, the last process is only involved in the uptake of water as a liquid [11].

Wood is a heterogeneous material and wood pore structure varies greatly among species, logs, and different parts within the same log, resulting in large differences in location and quantity of the penetrated resin. [12] observed that the resin penetration was greater in the early wood than in the latewood. Penetration of UF adhesive in the tangential direction was greater than in the radial direction for beech veneer [13]. The water-swollen wood cell wall contains water amounting to 30% or more of its dry weight[14]. This might be expected to make the cell wall permeable to water and therefore to provide an alternative to the pathway through the pits for water flowing between adjacent wood cell cavities. The bulk of the axial flow of water in the living tree is known to pass through the pits, but the cell wall pathway could be important when pit closure has occurred or when pits are absent [15].

Wood is biologically degraded in exposed conditions. The uptake of moisture by wood above the fiber saturation point is responsible for wood-decaying fungi to germinate and grow[16]. Another disadvantage is that an increased wood's ability to absorb moisture, affect dimensional stability of wooden materials in service. The uptake and release of moisture and the subsequent changes in dimensions are involved almost in all physical and biological degradation process of the wood and strongly influence the degradation of the coating as well.

Wood in storage is exposed to both periodic water absorption and desorption processes. The water absorption by wood frequently assumes great importance, especially in the structural uses of wood [17]. In residential buildings and in industrial applications, some components are often wood or wood-based [18]. These components are exposed to liquid water, for example wetting by rain or by water infiltration. Thus, wood is always undergoing changes in moisture content. Understanding water absorption by wood during soaking is of practical importance, since it affects the mechanical properties of the product. The effects of moisture content on the mechanical properties of wood have been the subject of an intense investigation worldwide [19]. All strength properties decrease as wood adsorbs moisture in the hygroscopic range. Important properties such as modulus of rupture and compressive strength parallel to grain may decrease up to 4 and 6 percent, respectively, for each percent increase in moisture content [20]. The periodic water absorption has also a negative effect on wood quality. The ability of microorganisms to attack wood depends on the moisture content of the wood cell wall [21]. Hence evaluating water

transfer in wood during soaking has attracted considerable attention. The amount of absorbed water in wood is dependent on the density and water diffusivity of wood. The water diffusivity coefficient describes the rate at which water moves from surface to the interior of products. These effects are caused by the porous structure of wood and the reactivity of its chemical components

The current well known primary timber species in Ghana have been exploited selectively by millers, but mostly without permission by illegal chain saw operators, resulting in their reduction both in number and quantum of each of them and the urgent need for finding suitable alternatives for use by both local and the export industry.

Furthermore, there is also ample evidence that timber production in Ghana is not proportional to its potential. Because, its under-utilization is partly as a result of the lack of general information about the wood properties and the great number of timber species. Consequently, *Terminalia superba (Ofram)* and *Terminalia ivorensis (Emere),* as currently threatened timber species and *Quassis undulata (Hotrohotro)* and *Recinodendron heudelotii (Wama)* as lesser used timber species were selected for the study to determine and compare their water absorption and diffusion characteristics. Knowledge of moisture uptake and transport properties are essential for predicting the moisture content and utilization of wood.

The primary objective of this research was to measure and compare the water absorption and diffusion coefficients of the four tropical timber species.

2. Materials and Methods

Four wood species were selected for the experiments: Wood samples were cut with dimensions L × W ×T = 70 × 50 × 35 mm³. The initial moisture content of samples was 0%. Water absorption data were obtained by placing the wood samples between screw-clamps and immersed in a water bath. Experiments were conducted at 25°C and for immersion periods, from several minutes to 19 days. After soaking, the moisture content of samples was calculated based on the increase in the sample weight at corresponding times. For this purpose, at regular time intervals, ranging from 60 mins at the beginning to 12 hours during the last stages of the process, the samples were rapidly removed from the water bath and superficially dried with filter paper to eliminate the surface water. The samples were then weighed to determine the moisture uptake. The samples were subsequently returned into water, and the process was repeated until the moisture content attained a range of 109-115%. Five experiments were conducted for each wood species and the mean results were used for further analysis. Finally, curves showing the cumulative weight gain versus the square root of time were plotted, and linear regression curves were computed for each wood sample. The water absorption coefficient of the wood samples was determined by using the following equation [22]

$$M_w = A\sqrt{t} \qquad (1)$$

where, m_w is the amount of water absorbed in kg/m², and A is the water absorption coefficient (kg/m² s½). Following the definition, the water absorption coefficient A is given by the slope of the fitted curve divided by the contact area.

In this study, the Fick's second law of diffusion was used to determine the diffusion coefficients of water in the wood samples. It has been demonstrated that for a short period of soaking time, the following mathematical model may be used to correlate the water uptake ratio (Mt-Mo)/(Ms-Mo) with diffusion of water in solids of arbitrary shape during soaking in water [23]

$$\frac{Mt-Mo}{Ms-Mo} = 2/\sqrt{\pi}(\frac{S}{V})\sqrt{D_e}t = (\alpha_b)^2 \qquad (2)$$

where, Ms and Mo are constants for wood samples, depending on the physical properties, and the ratio of volume-to-surface area (V/S) may be taken as constant, irrespective of moisture content. To determine the diffusion coefficient, data were plotted as water uptake data against the square root of time, t. If the initial part of the curve was linear, it would be possible to determine its slope, α_b, and the coefficient of diffusion, D_e, by the following relation:

$$D_e = \frac{\pi}{4}(\frac{V}{S})^2(\alpha_b)^2 \qquad (3)$$

3. Results and Observations

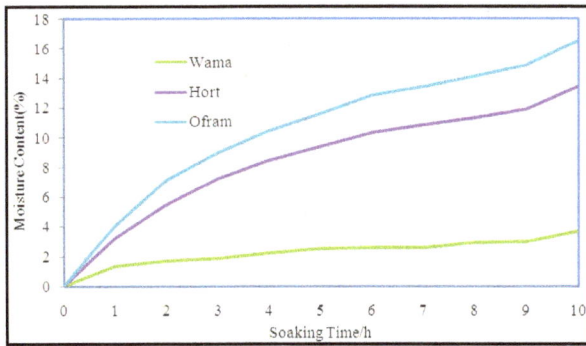

Figure 1a. *Water absorption graphs during initial soaking of timber species in plain water.*

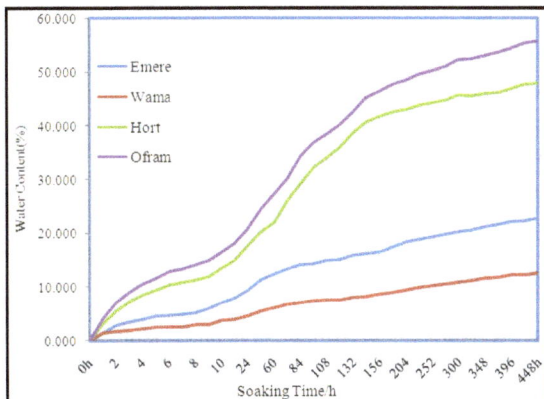

Figure 1b. *Water absorption graphs during the entire soaking period of timber species in plain water.*

Figure 1a shows the initial water absorption curves for a

period of ten hours by the timber species in plain water. By the end of the tenth hour, *Ofram* had absorbed 16.512 percent of moisture followed by *Hortrohotro* with 13.448 percent of moisture. During the same absorption period under review, *Emere* and *Wama* absorbed 7.05 and 3.70 percent moisture respectively. It is clear from the figure 1 that, both *Ofram* and *Hotrohotro* had high initial uptake of moisture *Emere* and *Wama*.

Figure 1b shows the entire water absorption curves for a period of nineteen days of soaking of the timber species in plain water. By the end of the nineteenth day, *Ofram* had absorbed 55.733 percent of moisture followed by *Hortrohotro* with 47.802 percent of moisture. *Emere* and *Wama* absorbed 22.796 and 12.545 percent moisture respectively. It can be noted that for the entire soaking period of nineteen days (figure 1b), both *Ofram* and *Hotrohotro* had the highest uptake of moisture as compared to *Emere* and *Wama*.

Figure 2a displays the variation in the initial amount of water absorbed against soaking time of the timber species. After four hours of initial soaking, the water absorption coefficients of the timber species were determined according to equation 1. The mean values of the water absorption coefficient (A) at initial stages of moisture sorption for *Hortrohotro*, *Ofram*, *Emere* and *Wama* were 3.51×10^{-3}, 4.31×10^{-3}, 1.67×10^{-3} and 8.27×10^{-4} (kg/m²/s) respectively. The corresponding mean values of this parameter for the entire soaking process were determined for the timber species viz; *Ofram* (2.91×10^{-3}), *Hortrohotro* (2.58×10^{-3}), *Emere* (1.14×10^{-3}) and *Wama* (6.11×10^{-4}) kg/m²s and are shown in Fig. 2b.

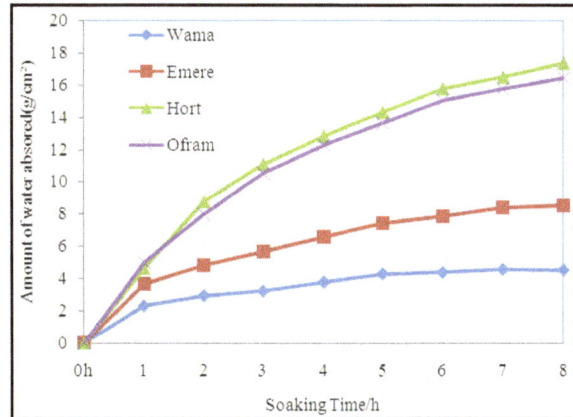

Figure 2a. *Variation in the initial amount of water absorbed against soaking time of the timber species.*

The corresponding mean values of this parameter for the entire soaking process were determined for the timber species viz; *Ofram* (2.91×10^{-3}), *Hortrohotro* (2.58×10^{-3}), *Emere* (1.14×10^{-3}) and *Wama* (6.11×10^{-4}) kg/m²s and are shown in Fig. 2b.

Similarly, the diffusion coefficient (De) of the timber species were also determined. This diffusion coefficient (De) was calculated after neglecting the non-Fickian behaviour and using the slope, the absorption data was fitted into equation 3 after the initial period of soaking. The measured diffusion coefficient(De) for *Wama, Emere, Ofram* and *Hotrohotro* timber species were 9.637×10^{-4}, 6.694×10^{-3}, 4.185×10^{-2} and 2.899×10^{-2} kg/m²/s respectively.

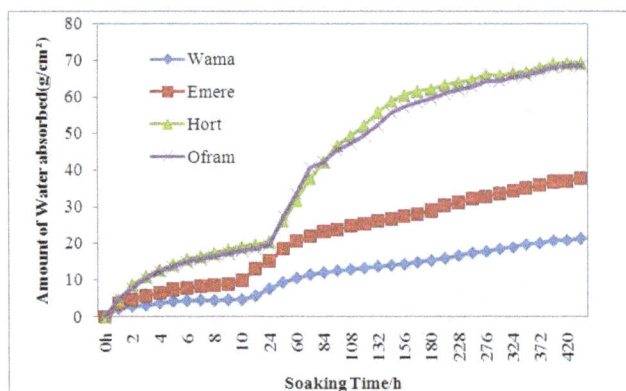

Figure 2b. *Variation in the amount of water absorbed during the entire soaking process of the timber species.*

4. Discussion

The timber species showed an initial rate of high moisture uptake (Fig. 2a) followed by slower absorption in the later stages, also known as the relaxation phase [24]. The nature of water vapour uptake indicates a two-stage process, in which *Ofram* and *Hotrohotro* absorbed 29% and 27% of the total moisture content during the initial 10 hour period (Figure 1a). This was followed by a period of very slow but ongoing water vapour uptake. The moisture content absorbed by the timber species over the entire soaking period of 19days stood at 56% and 48% respectively for *Ofram* and *Hortrohotro* (Figure 1b). [25] measured a similar value of 56.78% moisture content for a hardwood species from Iran. *Emere* and *Wama* reached 31% and 30% of moisture content after 10 hours of soaking as against the entire soaking period's values of 23% and 13% respectively. It can be inferred from this study that, although *Ofram* and *Hotrohotro* had an overall higher moisture uptake over the entire soaking period, *Emere* and *Wama's* initial moisture uptake values of 31 and 30% exceeded that of *Ofram* and *Hotrohotro* i.e. 29% and 27% respectively. This absorption phenomenon can be explained by the presence of capillaries present in wood, which quickly reaches equilibrium with the hydration medium by capillary action. During the early stages of the water absorption process, capillaries and cavities near the surface are filled quickly. Consequently, the water concentration at the surface is raised to saturation level almost instantaneously. The moisture gradient is limited to the inside of the material exclusively. Large wood cell cavities aid unrestricted water flow, but in the small ones, the presence of trapped air bubbles control the interior water gradient. Water absorption measurement is important because of the structural uses of wood [27]. Wood components in residential buildings and industrial applications are periodically exposed to liquid water in the form of rain or water vapour. An increase in 2 or 3% of water vapour can cause reduction in strength properties of timber

structures [28]. Also wood is attacked by microorganisms when the moisture content exceeds the fiber saturation point (FSP). This FSP is between 27 to 30% of moisture content. In this study *Emere* and *Wama* reached moisture content values of 23 and 13% respectively during the entire soaking period of 19 days. These moisture content values were far below the FSP of these timber species for any structural deformation and microbial attack to occur. However, *Hotrohotro* and *Ofram* reached in excess of 48 and 56% respectively of moisture content which were above the FSP during the same period of soaking. *Hotrohotro* and *Ofram* can be used outside only by subjecting them to some form of surface treatments to prevent them from infiltration of rain and water vapour as well as fungal attack. For a prolong exterior utilization of any of these timbers, it will be advisable to apply some form of surface treatments to protect them from structural and microbial damage.

The diffusion coefficient(De) measured for *Wama* (9.637×10^{-4} m²/s) was the lowest among the other timber species(*Hortrohotro*-2.899×10^{-2}, *Ofram*- 4.180×10^{-2} and *Emere*-6.694×10^{-3} m²/s respectively) The amount of absorbed water in wood is also dependent on the density and water diffusivity of wood. In this study *Wama* (>0.5g/cm³) as the densest wood, had the lowest water diffusivity. The water diffusivity coefficient describes the rate at which water moves from surface to the interior of products. These effects are caused by the porous structure of wood and the reactivity of its chemical components. This explains why *Wama* had the lowest absorption and diffusion rates.

5. Conclusion

The study showed that, the mean values of the water absorption coefficient at initial stages of moisture sorption for *Hortrohotro, Ofram, Emere* and *Wama* were 3.51×10^{-3}, 4.31×10^{-3}, 1.67×10^{-3} and 8.27×10^{-4} (kg/m²/s) respectively. The corresponding mean values of this parameter for the entire soaking process were also determined for the timber species viz; *Ofram* (2.91×10^{-3}), *Hortrohotro* (2.58×10^{-3}), *Emere* (1.14×10^{-3}) and *Wama* (6.11×10^{-4}) kg/m²s respectively. The measured diffusion coefficient for *Wama, Emere, Ofram* and *Hotrohotro* timber species were 9.637×10^{-4}, 6.694×10^{-3}, 4.185×10^{-2} and 2.899×10^{-2} kg/m²/s respectively. *Emere* and *Wama* had a lower absorption and diffusion characteristics than *Hortrohotro* and *Ofram*. Based on these findings, *Wama* could be substituted for *Emere* and *Hortrohotro* with *Ofram*.

References

[1] U. Watanabe, Y. Imamura and I. Iida, Liquid penetration of pre-compressed wood VI: Anatomical characterization of pit fractures. *J. Wood Sci.* 44: 158-162, 1998.

[2] S. Leal, V.B. Sousa, and H. Pereira., Radial variation of vessel size and distribution in cork oak wood (*Quercus suber L.*). *Wood Sci. Technol.* 41: 339-350, 2007.

[3]   Y. Sano, Intervascular pitting across the annual ring boundary in *Betula platyphylla var. japonica* and *Fraxinus mandshurica var. japonica*. IAWA J. 25: 129 140, 2004.

[4]   J. Virta,, S. Koponen and I. Absetz., Modeling moisture distribution in wooden cladding board as a result of short-term single-sided water soaking. *Building Environ*. 41: 1593-1599, 2006.

[5]   W.B. Banks, Some factors affecting the permeability of Scots pine and Norway spruce. J. Inst. Wood Sci. 5 , 10 – 17,1970.

[6]   Y. FortinMoisture content-matric potential relationship and water flow properties of wood at high moisture contents. PhD thesis, University of British Columbia, Vancouver. 1979 Strømdahl, K., (2000). Water sorption in wood and plant fibres (Series R, No. 78). *Department of Structural Engineering and Materials,* Technical University of Denmark

[7]   C. Hansmann,, Gindl, W., Wimmer, R. and Teischinger, A., Permeability of wood – a review. *Wood Res. Drevarsky Vysk.* 47, 1 – 16, 2002.

[8]   G. Bramhall, The validity of Darcy's Law in the axial penetration of wood. *Wood Sci. Technol.* 7, 319 – 322, 1971.

[9]   A.B. Wardrop, and Davies, G.W., Morphological factors relating to the penetration of liquids into wood. *Holzforschung* 15, 129 – 141, 1961.

[10] T. Rypstra, Analytical techniques for the evaluation of wood and wood fi nishes during weathering, Ph.D. Thesis, University of Stellenbosch, 1995. and Rijsdijk,J.F, P.B., Laming, (1994). Physical and related properties of 145 timbers, Kluwer Academic Publishers, Dordrecht

[11] C. F Siau, Wood: Influence of moisture on physical properties. Blacksburg, Virginia Polytechnic Institute and State University, 1995.

[12] L.A. Smith, and W. A. Cote., Studies of penetration of phenol-formaldehyde adhesive into beech wood. *Wood Fiber Sci.* 3(1):56–57, 1971.

[13] M. Sernek, J. Resnik, and F. A. Kamke, Penetration of liquid urea-formaldehyde adhesive into beech wood. *Wood Fiber Sci.* 31(1):41–48, 1999.

[14] C. Skaar, Water in wood. Syracuse University Press, 1972. Syracuse, N.Y.

[15] R. D. Preston, The physical biology of plant cell walls. 1974, Chapman and Hall, London

[16] A. Eckeveld van, Homan, W.J. and Militz, H., Increasing the water repellency of Scots pine sapwood by impregnation with undiluted linseed oil, wood oil, coccos oil and tall oil. *Holzforsch. Holzverwert.* 53, 113 – 115, 2001.

[17] R. Baronas, Ivanauskas F., Juodeikienė I., Kajalavicius A.,– Modelling of Moisture Movement in Wood during Outdoor Storage. Nonlinear Analysis: Modelling and Control, 2, 3-1, 2001.

[18] L. Candanedo, Derome, D., - Numerical simulation of water absorption in softwood. Ninth International IBPSA Conference , 2005. Montréal, Canada

[19] C.C. Gerhards, - Effect of moisture content and temperature on the mechanical properties of Wood: an analysis of immediate effects. *Wood and Fiber Science,* 14, 154-163,1998. Obataya, E, Norimoto, M, Gril, J., (1998) - The effects of adsorbed water on dynamic mechanical properties of wood. *Polymer*, 39(14), 3059-3064 and Severa L., Buchar J., Krivanek I., (2003) - The influence of the moisture content on the fracture of the notched wood beam. *8th International IUFRO Wood Drying.*

[20] B.A. Bendtsen, - Sorption and swelling characteristics of salt-treated wood. U. S. Forest Service Research Paper FPL 60, 1966.

[21] R. Baronas, Ivanauskas F., Juodeikienė I., Kajalavicius A.,– Modelling of Moisture Movement in Wood during Outdoor Storage. Nonlinear Analysis: Modelling and Control, 2, 3-1, 2001.

[22] M. Krus. Moisture Transport and Storage Coefficients of Porous Mineral Building Materials. Theoretical Principles and New Test Methods,1996. Fraunhofer IRB Verlag. Germany, 106 p. plus appendix

[23] N.E. Marcovich, M.M. Reboredo, M.I. Aranguren, Moisture diffusion in polyester– wood flour composites. *Polymer* 40, 7313–7320, 1999.

[24] A. Kumar, P.C., Flynn - Uptake of fluids by boreal wood chips: Implications for bioenergy. Fuel Processing Technology 87, 605–608, 2006.

[25] J. Khazaei, Water absorption in three wood varieties. Cercetări Agronomice în Moldova,2008.Vol. XLI , No. 2 (134).

16

Method for enhancement of power quality at point of common coupling of wind energy system

Jyothilal Nayak Bharothu[1],AbduL Arif[2]

[1]Associate professor of Electrical & Electronics Engineering, Sri Vasavi Institute of Engineering & Technology, Nandamuru, A.P.; India
[2]Assistant professor of Electrical & Electronics Engineering, Sri Vasavi Institute of Engineering & Technology, Nandamuru, A.P.; India

Email address:
nayakeee@gmail.com (J. Nayak Bharothu), arifabdul76@gmail.com (A. Arif)

Abstract: In this paper a compensation strategy based on a particular Custom Power System (CUPS) device, the Unified Power Quality Compensator (UPQC) has been proposed. A customized internal control scheme of the UPQC device was developed to regulate the voltage in the WF terminals, and to mitigate voltage fluctuations at grid side. The voltage regulation at WF terminal is conducted using the UPQC series converter, by voltage injection "in phase" with point of common coupling (PCC) voltage. On the other hand, the shunt converter is used to filter the WF generated power to prevent voltage fluctuations, requiring active and reactive power handling capability. The sharing of active power between converters is managed through the common DC link. Therefore the internal control strategy is based on the management of active and reactive power in the series and shunt converters of the UPQC, and the exchange of power between converters through UPQC DC–Link. This approach increases the compensation capability of the UPQC with respect to other custom strategies that use reactive power only. The proposed compensation scheme enhances the system power quality, exploiting fully DC–bus energy storage and active power sharing between UPQC converters, features not present in DVR and D–STATCOM compensators. Simulations results show the effectiveness of the proposed compensation strategy for the enhancement of Power Quality and Wind Farm stability.

Keywords: Cups, Upqc, Pcc, Dc Link, Wind Farm, Power Quality Etc

1. Introduction

The location of generation facilities for wind energy is determined by wind energy resource availability, often far from high voltage (HV) power transmission grids and major consumption centers. In case of facilities with medium power ratings, the WF is connected through medium voltage (MV) distribution headlines. A situation commonly found in such scheme is that the power generated is comparable to the transport power capacity of the power grid to which the WF is connected, also known as weak grid connection. The main feature of this type of connections is the increased voltage regulation sensitivity to changes in load. So, the system's ability to regulate voltage at the point of common coupling (PCC) to the electrical system is a key factor for the successful operation of the WF. Also, is well known that given the random nature of wind resources, the WF generates fluctuating electric power.

These fluctuations have a negative impact on stability and power quality in electric power systems. Moreover, in exploitation of wind resources, turbines employing squirrel cage induction generators (SCIG) have been used since the beginnings. The operation of SCIG demands reactive power, usually provided from the mains and/or by local generation in capacitor banks. In the event that changes occur in its mechanical speed, i.e. due to wind disturbances, so will the WF active (reactive) power injected (demanded) into the power grid, leading to variations of WF terminal voltage because of system impedance.

This power disturbance propagate into the power system, and can produce a phenomenon known as "flicker", which consists of fluctuations in the illumination level caused by voltage variations. Also, the normal operation of WF is impaired due to such disturbances. In particular for the case of "weak grids", the impact is even greater. In order to reduce the voltage fluctuations that may cause "flicker", and improve WF terminal voltage regulation, several solutions have been posed. The most common one is to upgrade the power grid, increasing the short circuit power level at the point of common coupling PCC, thus reducing the impact of power fluctuations and voltage regulation problems.

In recent years, the technological development of high power electronics devices have led to implementation of electronic equipment suited for electric power systems, with fast response compared to the line frequency. These active compensators allow great flexibility in:

Controlling the power flow in transmission systems using Flexible AC Transmission System (FACTS) devices, and

Enhancing the power quality in distribution systems employing Custom Power System (CUPS) devices.

The use of these active compensators to improve integration of wind energy in weak grids is the approach adopted in this work. In this paper we propose and analyze a compensation strategy using an UPQC, for the case of SCIG–based WF, connected to a weak distribution power grid. This system is taken from a real case. The UPQC is controlled to regulate the WF terminal voltage, and to mitigate voltage fluctuations at the point of common coupling (PCC), caused by system load changes and pulsating WF generated power, respectively. The voltage regulation at WF terminal is conducted using the UPQC series converter, by voltage injection "in phase" with PCC voltage. On the other hand, the shunt converter is used to filter the WF generated power to prevent voltage fluctuations, requiring active and reactive power handling capability. The sharing of active power between converters is managed through the common DC link. Simulations were carried out to demonstrate the effectiveness of the proposed compensation approach.

2. Literature Survey

The location of generation facilities for wind energy is determined by wind energy resource availability, often far from high voltage (HV) power transmission grids and major consumption centers [1].In case of facilities with medium power ratings, the WF is connected through medium voltage (MV) distribution headlines. A situation commonly found in such scheme is that the power generated is comparable to the transport power capacity of the power grid to which the WF is connected, also known as weak grid connection. The main feature of this type of connections is the increased voltage regulation sensitivity to changes in load [2].The system's ability to regulate voltage at the point of common coupling (PCC) to the electrical system is a key factor for the successful operation of the WF. Also, is well known that given the random nature of wind resources, the WF generates fluctuating electric power. These fluctuations have a negative impact on stability and power quality in electric power systems [3].

Moreover, in exploitation of wind resources, turbines employing squirrel cage induction generators (SCIG) have been used since the beginnings. The operation of SCIG demands reactive power, usually provided from the mains and/or by local generation in capacitor banks [4].In recent years, the technological development of high power elec-

tronics devices has led to implementation of electronic equipment suited for electric power systems, with fast response compared to the line frequency. These active compensators allow great flexibility in: a) controlling the power flow in transmission systems using Flexible AC Transmission System (FACTS) devices, and b) enhancing the power quality in distribution systems employing Custom Power System CUPS) devices [6].The use of these active compensators to improve integration of wind energy in weak grids is the approach adopted in this work. In this paper we propose and analyze a compensation strategy using an UPQC, for the case of SCIG–based WF, connected to a weak distribution power grid. This system is taken from a real case [7].

The UPQC is controlled to regulate the WF terminal voltage, and to mitigate voltage fluctuations at the point of common coupling (PCC), caused by system load changes and pulsating WF generated power, respectively. The voltage regulation at WF terminal is conducted using the UPQC series converter, by voltage injection "in phase" with PCC voltage. On the other hand, the shunt converter is used to filter the WF generated power to prevent voltage fluctuations, requiring active and reactive power handling capability. The sharing of active power between converters is managed through the common DC link [8]. The dynamic compensation of voltage variations is performed by injecting voltage in series and active–reactive power in the MV6 (PCC) busbar; this is accomplished by using an unified type compensator UPQC [9]. This transformation allows the alignment of a rotating reference frame with the positive sequence of the PCC voltages space vector. To accomplish this, a reference angle _ synchronized with the PCC positive sequence fundamental voltage space vector is calculated using a Phase Locked Loop (PLL) system. In this work, an "instantaneous power theory" based PLL has been implemented [11].

2.1. Wind Energy

Wind power is the conversion of wind energy into a useful form of energy, such as using wind turbines to make electrical power, windmills for mechanical power, wind pumps for water pumping or drainage, or sails to propel ships. A wind farm is a group of wind turbines in the same location used for production of electricity. A large wind farm may consist of several hundred individual wind turbines, and cover an extended area of hundreds of square miles, but the land between the turbines may be used for agricultural or other purposes. A wind farm may also be located offshore.

Fig.2.1 Wind farm

2.2. Wind Power

Wind is abundant almost in any part of the world. Its existence in nature caused by uneven heating on the surface of the earth as well as the earth's rotation means that the wind resources will always be available. The conventional ways of generating electricity using non renewable resources such as coal, natural gas, oil and so on, have great impacts on the environment as it contributes vast quantities of carbon dioxide to the earth's atmosphere which in turn will cause the temperature of the earth's surface to increase, known as the green house effect. Hence, with the advances in science and technology, ways of generating electricity using renewable energy resources such as the wind are developed. Nowadays, the cost of wind power that is connected to the grid is as cheap as the cost of generating electricity using coal and oil. Thus, the increasing popularity of green electricity means the demand of electricity produced by using non renewable energy is also increased accordingly.

2.2.1. Features of Wind Power Systems

There are some distinctive energy end use features of wind power systems

Most wind power sites are in remote rural, island or marine areas. Energy requirements in such places are distinctive and do not require the high electrical power.

A power system with mixed quality supplies can be a good match with total energy end use i.e. the supply of cheap variable voltage power for heating and expensive fixed voltage electricity for lights and motors.

Rural grid systems are likely to be weak (low voltage 33 KV). Interfacing a Wind Energy Conversion System (WECS) in weak grids is difficult and detrimental to the workers' safety.

There are always periods without wind. Thus, WECS must be linked energy storage or parallel generating system if supplies are to be maintained.

2.3. Power from the Wind

Kinetic energy from the wind is used to turn the generator inside the wind turbine to produce electricity. There are several factors that contribute to the efficiency of the wind turbine in extracting the power from the wind. Firstly, the wind speed is one of the important factors in determining how much power can be extracted from the wind. This is because the power produced from the wind turbine is a function of the cubed of the wind speed. Thus, the wind speed if doubled, the power produced will be increased by eight times the original power. Then, location of the wind farm plays an important role in order for the wind turbine to extract the most available power form the wind.

The next important factor of the wind turbine is the rotor blade. The rotor blades length of the wind turbine is one of the important aspects of the wind turbine since the power produced from the wind is also proportional to the swept area of the rotor blades i.e. the square of the diameter of the swept area.

Hence, by doubling the diameter of the swept area, the power produced will be fourfold increased. It is required for the rotor blades to be strong and light and durable. As the blade length increases, these qualities of the rotor blades become more elusive. But with the recent advances in fiberglass and carbon-fiber technology, the production of lightweight and strong rotor blades between 20 to 30 meters long is possible. Wind turbines with the size of these rotor blades are capable to produce up to 1 megawatt of power.

The relationship between the powers produced by the wind source and the velocity of the wind and the rotor blades swept diameter is shown below.

$$P_{wind} = \pi/8 d D^2 V_{wind}^3 \qquad (2.1)$$

The derivation to this formula can be looked up in. It should be noted that some books derived the formula in terms of the swept area of the rotor blades (A) and the air density is denoted as δ.

Thus, in selecting wind turbine available in the market, the best and efficient wind turbine is the one that can make the best use of the available kinetic energy of the wind.

Wind power has the following advantages over the traditional power plants.

Improving price competitiveness
Modular installation
Rapid construction
Complementary generation
Improved system reliability and
Non-polluting

3.

3.1. Weak Grid

The term 'weak grid' is used in many connections both with and without the inclusion of wind energy. It is used without any rigour definition usually just taken to mean the voltage level is not as constant as in a 'stiff grid'. Put this way the definition of a weak grid is a grid where it is necessary to take voltage level and fluctuations into account because there is a probability that the values might exceed the requirements in the standards when load and production cases are considered. In other words, the grid impedance is significant and has to be taken into account in order to have valid conclusions.

Weak grids are usually found in more remote places where the feeders are long and operated at a medium voltage level. The grids in these places are usually designed for relatively small loads. When the design load is exceeded the voltage level will be below the allowed minimum and/or the thermal capacity of the grid will be exceeded. One of the consequences of this is that development in the region with this weak feeder is limited due to the limitation in the maximum power that is available for industry etc.

The problem with weak grids in connection with wind energy is the opposite. Due to the impedance of the grid the amount of wind energy that can be absorbed by the grid at the point of connection is limited because of the upper voltage level limit. So in connection with wind energy a weak grid is a power supply system where the amount of wind energy that can be absorbed is limited by the grid capacity and not e.g. by operating limits of the conventional generation.

3.2. Basic Power Control Idea

The main idea is to increase the amount of wind energy that can be absorbed by the grid at a certain point with minimum extra cost. There exist several options that can be implemented in order to obtain a larger wind energy contribution. These options include:

Grid reinforcement

Voltage dependent disconnection of wind turbines

Voltage dependent wind power production

Inclusion of energy buffer (storage)

Determination of actual voltage distribution instead of worst case and evaluation if real conditions will be problem Grid reinforcement increases the capacity of the grid by increasing the cross section of the cables. This is usually done by erecting a new line parallel to the existing line for some part of the distance. Because of the increased cross section the impedance of the line is reduced and therefore the voltage variations as a result of power variations are reduced.

Grid reinforcement increases both the amount of wind energy that can be connected to the feeder and the maximum consumer load of the feeder. Since the line impedance is reduced the losses of the feeder are also reduced. Grid reinforcement can be very costly and sometimes impossible due to planning restrictions. Since grid reinforcement can be very costly or impossible other options are interesting. The simplest alternative is to stop some of the wind turbines when the voltage level is in danger of being exceeded.

This can e.g. be done by the wind turbine controller monitoring the voltage level at the low voltage side of the connection point. At a certain level the wind turbine is cut off and it is then cut in again when the voltage level is below a certain limit. The limits can be recalculated and depends on transformer settings, line impedance and other loads of the feeder. This is a simple and crude way of ensuring that the voltage limits will not be exceeded. It can be implemented at practically no cost but not all the potentially available wind energy is utilized.

`A method that is slightly more advanced is to continuously control the power output of the wind turbine in such a way that the voltage limit is not exceeded. This can be done on a wind farm level with the voltage measured at the point of common connection. The way of controlling the power output requires that the wind turbine is capable of controlling the output (pitch or variable speed controlled) and a bit more sophisticated measuring and control equip-

ment, but the amount of wind energy that is dumped is reduced compared to the option of switching off complete wind turbines.

The basic power control idea in the current context of this project is based on the combination on wind turbines and some kind of energy storage. The storage is used to buffer the wind energy that cannot be feed to the grid at the point of connection without violating the voltage limits. Usually the current limit of the grid will not be critical. The energy in the storage can then be fed back to the grid at a later time when the voltage level is lower.

The situations where the voltage level will be high will occur when the consumer load of the grid is low and the wind power production is high. If the voltage level will be critically high depends on the characteristics of the grid (e.g. impedance and voltage control), the minimum load of the consumers, the amount of installed wind power and the wind conditions. The critical issues involved in the design of a power control system are the power and energy capacity, the control bandwidth as well as investment, installation and maintenance cost. The various types of power control systems have different characteristics giving different weights on capacity, investment and maintenance.

Different types of storage can be applied. During the project only pumped storage and batteries has been investigated. Other types of storage include flywheel, super conducting magnetic storage, compressed air and capacitors. These types of storage have not been investigated for several reasons among them cost, capacity and availability.

3.3. Control Strategies

Several different control strategies exist for a power controller with storage. The different control strategies place different weights on voltage and power fluctuations and therefore have different impact on the sizing of the storage capacity and of the power rating. The two main types of control strategies are

Ones controlling the voltage at the point of common connection or another point in the grid and

Ones controlling the power for smoothing or capacity increase.

3.4. Basic Problems

3.4.1. Voltage Level

The main problem with wind energy in weak grids is the quasi-static voltage level. In a grid without wind turbines connected the main concern by the utility is the minimum voltage level at the far end of the feeder when the consumer load is at its maximum.

So the normal voltage profile for a feeder without wind energy is that the highest voltage is at the bus bar at the substation and that it drops to reach the minimum at the far end.

The settings of the transformers by the utility are usually so, that the voltage at the consumer closest to the transformer will experience a voltage, that is close to the maximum value especially when the load is low and that the voltage is close to the minimum value at the far end when the load is high. This operation ensures that the capacity of the feeder is utilized to its maximum. When wind turbines are connected to the same feeder as consumers which often will be the case in sparsely populated areas the voltage profile of the feeder will be much different from the no wind case. Due to the power production at the wind turbine the voltage level can and in most cases will be higher than in the no wind case. As is seen on the figure the voltage level can exceed the maximum allowed when the consumer load is low and the power output from the wind turbines is high. This is what limits the capacity of the feeder.

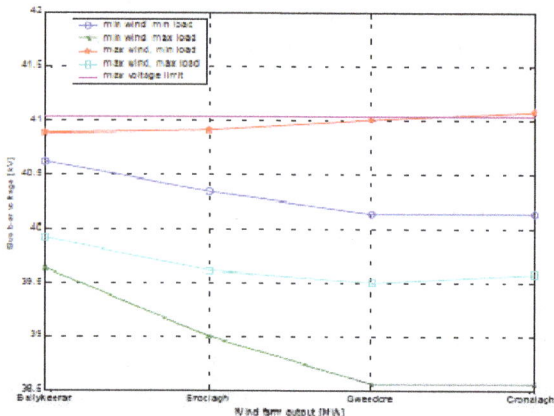

Fig.3.1 Example of voltage for feeder with and without wind power

The voltage profile of the feeder depends on the line impedance, the point of connection of the wind turbines and on the wind power production and the consumer load. For a simple single load case the voltage rise over the grid impedance can be approximated with $\Delta U \cong (R * P + X * Q) / U$ using generator sign convention. This formula indicates some of the possible solutions to the problem with absorption of wind power in weak grids. The main options are either a reduction of the active power or an increase of the reactive power consumption or a reduction of the line impedance.

3.4.2. Power Disturbances

Another possible problem with wind turbines in weak grids are the possible voltage fluctuations as a result of the power fluctuations that comes from the turbulence in the wind and from starts and stops of the wind turbines. As the grids becomes weaker the voltage fluctuations increase given cause to what is termed as flicker. Flicker is visual fluctuations in the light intensity as a result of voltage fluctuations. The human eye is especially sensitive to these fluctuations if they are in the frequency range of 1-10 Hertz. Flicker and flicker levels are defined in IEC1000-3-7.

During normal operation the wind turbulence causes power fluctuations mainly in the frequency range of 1-2 Hertz due to rotational sampling of the turbulence by the blades. This together with the tower shadow and wind shear

are the main contributors to the flicker produced by the wind turbine during normal operation.

The other main contribution to the flicker emission is the cut-in of the wind turbine. During cut-in the generator is connected to the grid via a soft starter. The soft starter limits the current but even with a soft starter the current during cut-in can be very high due to the limited time available for cut-in, especially the magnetization current at cut-in contributes to the flicker emission from a wind turbine.

3.4.2.1. Flicker

Electric power is an essential commodity for most industrial, commercial and domestic processes. As a product, electric power must be of an acceptable quality, to guarantee the correct behavior of the equipment connected to the power distribution system. Low-frequency conducted disturbances are the main factors that can compromise power quality. The IEC 610002-1 standard classifies low-frequency conducted disturbances in the following five groups: harmonics and inter harmonics, voltage dips and short supply interruptions, voltage unbalance, power frequency variations and voltage fluctuations or flicker.

Voltage fluctuations are defined as cyclic variations in voltage with amplitude below 10% of the nominal value. Most of the connected equipment is not affected by voltage fluctuations, but these fluctuations may cause changes in the illumination intensity of light sources, known as flicker.

Main sources of flicker

The main sources of flicker are large industrial loads, such as arc furnaces, or smaller loads with regular duty cycles, such as welding machines or electric boilers. However, from the point of view of power generation, flicker as a result of wind turbines has gained attention in recent years. Rapid variations in wind speed produce fluctuating power, which can lead to voltage fluctuations at the point of common coupling (PCC), which in turn generate flicker.

The IEC 61400-21 standard establishes the procedures for measuring and assessing the power quality characteristics of grid-connected wind turbines. The section dedicated to flicker proposes a complex model for calculating the flicker coefficient that characterizes a wind turbine. This coefficient must be estimated from the current and voltage time series obtained for different wind conditions. The wind turbine being tested is usually connected to a medium-voltage network, having other fluctuating loads that may cause significant voltage fluctuations.

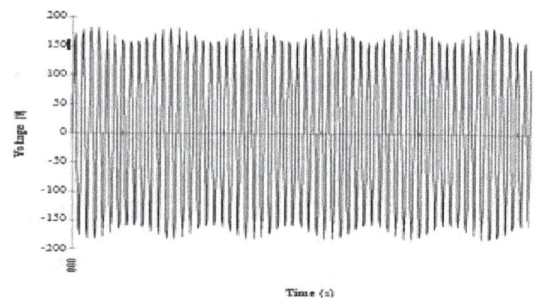

Fig.3.2 Example of voltage flicker

Effects

Flicker is considered the most significant effect of voltage fluctuation because it can affect the production environment by causing personnel fatigue and lower work concentration levels. In addition, voltage fluctuations may subject electrical and electronic equipment to detrimental effects that may disrupt production processes with considerable financial costs.

Other effects of voltage fluctuation include the following:

Nuisance tripping due to misoperation of relays and contactors.

Unwanted triggering of UPS units to switch to battery mode.

Problems with some sensitive electronic equipment, which require a constant voltage (i.e. medical laboratories).

In order to reduce the voltage fluctuations that may cause "flicker" and improve WF terminal voltage regulation, several solutions has been posed. The most common technique is used as custom power device strategy.

3.5. Custom Power Devices

Custom power is a strategy, which is intended principally to convene the requirement of industrial and commercial consumers. The concept of the custom power is tools of application of power electronics controller devices into power distribution system to supply a quality of power, demanded by the sensitive users. These power electronics controller devices are also called custom power devices because through these valuable powers is applied to the customers. They have good performance at medium distribution levels and most are available as commercial products. For the generation of custom power devices VSI is generally used, due to self- supporting of dc bus voltage with a large dc capacitor. The custom power devices are mainly divided into two groups: network reconfiguring type and compensating type.

Custom power devices are classified into two types

Network reconfiguration type cp devices

compensating power devices

4.

4.1. Unified Power Quality Conditioner

A Unified Power Quality Conditioner (UPQC) is a device that is similar in construction to a Unified Power Flow Conditioner (UPFC). The UPQC, just as in a UPFC, employs two voltage source inverters (VSIs) that connected to a D.C. energy storage capacitor. One of these two VSIs is connected in series with a.c. line while the other is connected in shunt with the a.c. system.

A UPQC that combines the operations of a Distribution Static Compensator (DSTATCOM) and Dynamic Voltage Regulator (DVR) together.

The function of UPQC includes

(i) Reactive Power Compensation

(ii)Voltage Regulation

(iii)Compensation for Voltage sag and swell

(iv)Unbalance Compensation for current and voltage (for 3-phase systems)

(v)Neutral Current Compensation (for 3-phase 4-wire systems)

A UPQC is employed in a power transmission system to perform shunt and series compensation at the same time. A power distribution system may contain unbalance, distortion and even D.C. components. Therefore a UPQC operate, better than a UPFC, with all these aspects in order to provide shunt or series compensation. The UPQC is a relatively new device and not much work has yet been reported on it. Sometimes it has been viewed as combination of series and shunt active filters.

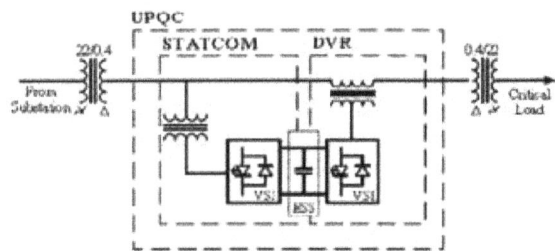

Fig.4.1 Unified power quality conditioner

4.2. Constructional Features of Upqc

UPQC is the integration of series and shunt active filters, connected back-to-back on the dc side, sharing a common DC capacitor as shown in Figure. The series component of the UPQC is responsible for mitigation of the supply side disturbances: voltage sags/swells, flicker, voltage unbalance and harmonics. It inserts voltages so as to maintain the load voltages at a desired level; balanced and distortion free.

The shunt component is responsible for mitigating the current quality problems caused by the consumer: poor power factor, load harmonic currents, load unbalance etc. It injects currents in the ac system such that the source currents become balanced sinusoids and in phase with the source voltages.

This work deals with the review of research work that has been completed so far on this issue. Emphasis has been given on incorporation techniques of UPQC in DG or microgrid system along with their advantages and disadvantages. More DGs such as Photovoltaic or Wind Energy Systems are now penetrating into the grid or microgrid. Again, numbers of nonlinear loads are also increasing. Therefore, current research on capacity enhancement techniques of UPQC to cope up with the expanding DG or microgrid system is also reviewed.

As the UPQC can compensate for almost all existing PQ problems in the transmission and distribution grid, placement of a UPQC in the distributed generation network can be multipurpose. As a part of integration of UPQC in DG systems, research has been done on the following two techniques: DC-Linked and Separated DG-UPQC systems.

4.3. Control Objectives of Upqc

The shunt connected converter has the following control objectives

To balance the source currents by injecting negative and zero sequence components required by the load

The compensate for the harmonics in the load current by injecting the required harmonic currents

To control the power factor by injecting the required reactive current (at fundamental frequency)

To regulate the DC bus voltage.

The series connected converter has the following control objectives

To balance the voltages at the load bus by injecting negative and zero sequence voltages to compensate for those present in the source.

To isolate the load bus from harmonics present in the source voltages, by injecting the harmonic voltages

To regulate the magnitude of the load bus voltage by injecting the required active and reactive components (at fundamental frequency) depending on the power factor on the source side

To control the power factor at the input port of the UPQC (where the source is connected. Note that the power factor at the output port of the UPQC (connected to the load) is controlled by the shunt converter.

5.

5.1. System Description

Fig.5.1 depicts the power system under consideration in this study. The WF is composed by 36 wind turbines using squirrel cage induction generators, adding up to 21.6MW electric power. Each turbine has attached fixed reactive compensation capacitor banks (175kVAr), and is connected to the power grid via 630KVA 0.69/33kV transformer.

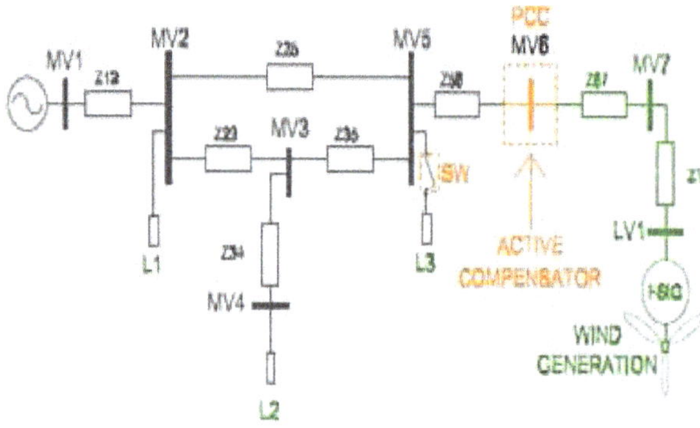

Fig.5.1 Study case power system

This system is taken from and represents a real case. The ratio between short circuit power and rated WF power, give us an idea of the "connection weakness". Thus considering that the value of short circuit power in MV6 is SSC $\simeq$

120MV A this ratio can be calculated

$$r = \frac{S_{sc}}{P_{wf}} \cong 5.5 \tag{5.1}$$

Values of r < 20 are considered as a "weak grid" connection.

5.2. Turbine Rotor and Associated Disturbances Model

The power that can be extracted from a wind turbine is determined by the following expression:

$$p = \frac{1}{2} \rho \pi R^2 V^3 C_p \tag{5.2}$$

Where is ρ air density, R the radius of the swept area, v the wind speed, and Cp the power coefficient. For the considered turbines (600kW) the values are R = 31.2 m, ρ = 1.225 kg/m3 and C_p calculation is taken from. Then, a complete model of the WF is obtained by turbine aggregation; this implies that the whole WF can be modeled by only one equivalent wind turbine, whose power is the arithmetic sum of the power generated by each turbine according to the following equation:

$$P_T = \sum_{i=1\ldots 36} p_i \tag{5.3}$$

Moreover, wind speed v in (1) can vary around its average value due to disturbances in the wind flow. Such disturbances can be classified as deterministic and random. The firsts are caused by the asymmetry in the wind flow "seen" by the turbine blades due to "tower shadow" and/or due to the atmospheric boundary layer, while the latter are random changes known as "turbulence". For our analysis, wind flow disturbance due to support structure (tower) is considered, and modeled by a sinusoidal modulation superimposed to the mean value of v. The frequency for this modulation is $3.N_{rotor}$ for the three–bladed wind turbine, while its amplitude depends on the geometry of the tower. In our case we have considered a mean wind speed of 12m/s and the amplitude modulation of 15%. The effect of the boundary layer can be neglected compared to those produced by the shadow effect of the tower in most cases. It should be noted that while the arithmetic sum of perturbations occurs only when all turbines operate synchronously and in phase, this is the case that has the greatest impact on the power grid (worst case), since the power pulsation has maximum amplitude. So, turbine aggregation method is valid.

5.3. Model of Induction Generator

For the squirrel cage induction generator the model available in Matlab/Simulink Sim Power Systems libraries is used. It consists of a fourth–order state–space electrical model and a second–order mechanical model.

5.4. Dynamic Compensator Model

The dynamic compensation of voltage variations is performed by injecting voltage in series and active–reactive power in the MV6 (PCC) busbar; this is accomplished by using an unified type compensator UPQC. In Fig.5.2 we see the basic outline of this compensator; the busbars and impedances numbering is referred to Fig.5.1. The operation is based on the generation of three phase voltages, using electronic converters either voltage source type (VSI– Voltage Source Inverter) or current source type (CSI– Current Source Inverter). VSI converters are preferred because of lower DC link losses and faster response in the system than CSI. The shunt converter of UPQC is responsible for injecting current at PCC, while the series converter generates voltages between PCC and U1, as illustrated in the phasor diagram of Fig.5.3. An important feature of this compensator is the operation of both VSI converters (series and shunt) sharing.

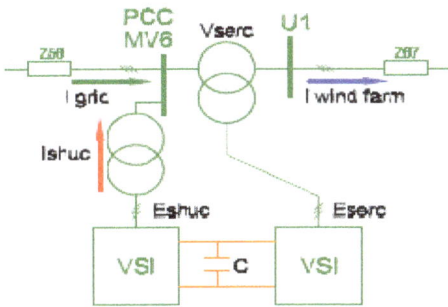

Fig.5.2 Block diagram of UPQC

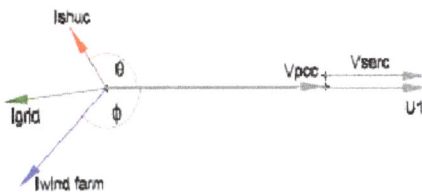

Fig.5.3 Phasor diagram of UPQC

The same DC–bus, which enables the active power exchange between them. We have developed a simulation model for the UPQC based on the ideas taken from. Since switching control of converters is out of the scope of this work, and considering that higher order harmonics generated by VSI converters are outside the bandwidth of significance in the simulation study, the converters are modeled using ideal controlled voltage sources. Fig.5.4 shows the adopted model of power side of UPQC. The control of the UPQC, will be implemented in a rotating frame dq0 using Park's transformation.

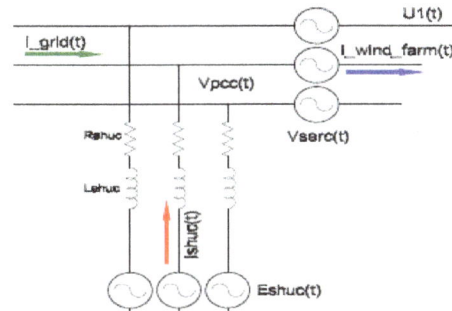

Fig.5.4 Power stage compensator model Ac side

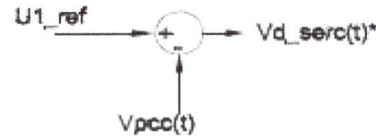

Fig.5.5 Series compensator controller

Where f_i=a, b, c represents either phase voltage or currents, and f_i=d, q, 0 represents that magnitudes transformed to the d_{qo} space. This transformation allows the alignment of a rotating reference frame with the positive sequence of the PCC voltages space vector. To accomplish this, a reference angle synchronized with the PCC positive sequence fundamental voltage space vector is calculated using a Phase Locked Loop (PLL) system. In this work, an "instantaneous power theory" based PLL has been implemented. Under balance steady-state conditions, voltage and currents vectors in this synchronous reference frame are constant quantities. This feature is useful for analysis and decoupled control.

5.5. Upqc Control Strategy

The UPQC serial converter is controlled to maintain the WF terminal voltage at nominal value (see U1 bus-bar in Fig.5.4), thus compensating the PCC voltage variations. In this way, the voltage disturbances coming from the grid cannot spread to the WF facilities. As a side effect, this control action may increase the low voltage ride–through (LVRT) capability in the occurrence of voltage sags in the WF terminals. Fig.5.5 shows a block diagram of the series converter controller. The injected voltage is obtained subtracting the PCC voltage from the reference voltage, and is phase–aligned with the PCC voltage (see Fig.5.3). On the other hand, the shunt converter of UPQC is used to filter the active and reactive power pulsations generated by the WF. Thus, the power injected into the grid from the WF compensator set will be free from pulsations, which are the origin of voltage fluctuation that can propagate into the system. This task is achieved by appropriate electrical currents injection in PCC. Also, the regulation of the DC bus voltage has been assigned to this converter. Fig.5.6 shows a block diagram of the shunt converter controller.

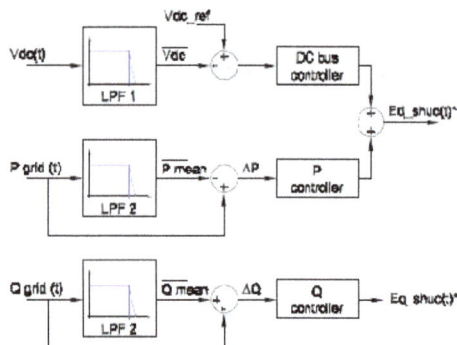

Fig.5.6 Shunt compensator controller

This controller generates both voltages commands $E_{d-shuc*}$ and $E_{q-shuc*}$ based on power fluctuations P and Q, respectively. Such deviations are calculated subtracting the mean power from the instantaneous power measured in PCC.

The mean values of active and reactive power are obtained by low–pass filtering, and the bandwidth of such filters are chosen so that the power fluctuation components selected for compensation, fall into the flicker band as stated in IEC61000- 4-15 standard. In turn, $E_{d-shuc*}$ also contains the control action for the DC–bus voltage loop. This control loop will not interact with the fluctuating power compensation, because its components are lower in frequency than the flicker–band. The powers P_{shuc} and Q_{shuc} are calculated in the rotating reference frame, as follows:

$$P_{shuC(t)} = \frac{3}{2} V_d^{PCC}(t).I_d^{shuC}(t) \qquad (5.4)$$

$$Q_{shuC}(t) = -\frac{3}{2} V_d^{PCC}(t).I_d^{shuC}(t)$$

Ignoring PCC voltage variation, these equations can be written as follows.

$$P_{shuC}(t) = k_p'.I_{d_{shuC(t)}} \qquad (5.5)$$

$$Q_{shuC}(t) = k_q'.I_{q\_shuC}(t)$$

Taking in consideration that the shunt converter is based on a VSI, we need to generate adequate voltages to obtain the currents in (6). This is achieved using the VSI model proposed sin [10], leading to a linear relationship between the generated power and the controller voltages. The resultant equations are:

$$P_{shuC}(t) = k_p''.E_{d-shuC^*(t)} \qquad (5.6)$$

$$Q_{shuC}(t) = k_q''.E_q\text{-}shuC^*(t)$$

P and Q control loops comprise proportional controllers, while DC–bus loop, a PI controller. In summary, in the proposed strategy the UPQC can be seen as a "power buffer", leveling the power injected into the power system grid.

The Fig.5.7 illustrates a conceptual diagram of this mode of operation. It must be remarked that the absence of an

external DC source in the UPQC bus, forces to maintain zero–average power in the storage element installed in that bus. This is accomplished by a proper design of DC voltage controller. Also, it is necessary to note that the proposed strategy cannot be implemented using other CUPS devices like D–STATCOM or DVR. The power buffer concept may be implemented using a D–STATCOM, but not using a DVR. On the other hand, voltage regulation during relatively large disturbances cannot be easily coped using reactive power only from D–STATCOM; in this case, a DVR device is more suitable.

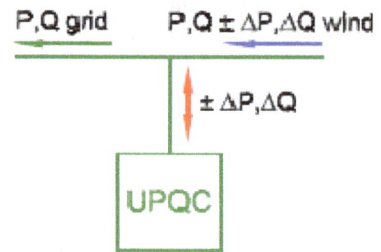

Fig.5.7 Power buffer concept

6. Simulation Results

Fig.6.1 Wind farm terminal voltages

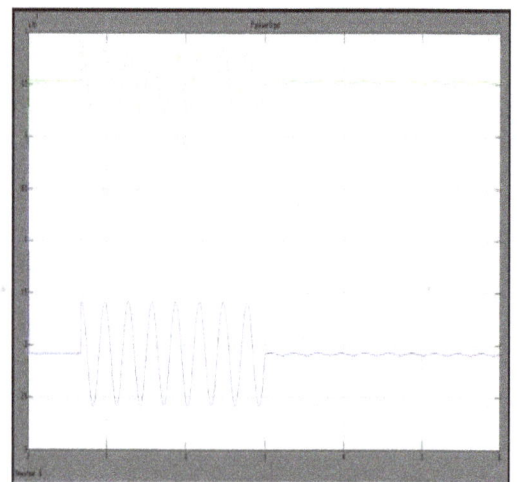

Fig.6.2 Active and reactive power demand at power grid side

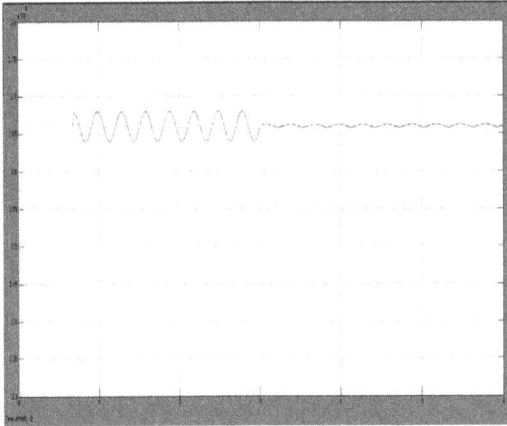

Fig.6.3 Voltage at point of common coupling

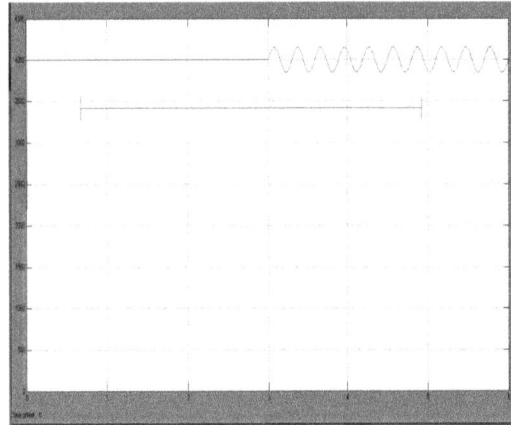

Fig.6.5 Power of the capacitor in the voltage bus

Fig.6.6 DC- Bus voltage

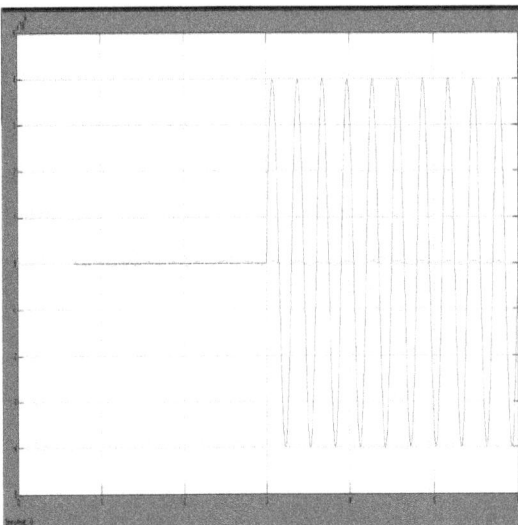

Fig.6.7 Voltage of the capacitor in the DC-Bus

7. Conclusion

The main work of the paper has been carried to show that by using custom power devices like UPQC it is possible to regulate the voltage in the wind farm terminals, and to mitigate voltage fluctuations in distribution systems. The UPQC which can be used at the PCC for improving power quality is modelled and simulated using proposed control strategy and the performance is compared by applying it to a distribution system.

In this paper, a new compensation strategy implemented using an UPQC type compensator was presented, to connect SCIG based wind farms to weak distribution power grid. The proposed compensation scheme enhances the system power quality, exploiting fully DC–bus energy storage and active power sharing between UPQC converters, features not present in DVR and D–STATCOM compensators.

The simulation results show a good performance in the rejection of power fluctuation due to "tower shadow effect" and the regulation of voltage due to a sudden load connection. So, the effectiveness of the proposed compensation approach is demonstrated in the study case. In future work, performance comparison between different compensator types will be made.

Acknowledgements

I Thank to our Institute Executive directors Mr.T Sai kumar & Mr.D Baba for providing creative environment for this work. Also I am very much thankful to our Institute principal Dr. K Ramesh for his kind permission and encouragement to write research paper. I would like to extend my heartfelt thanks to my colleagues. And finally I am very much obliged to my respected parents who inspiring me around the clock.

References

[1]   M. F. McGranaghan, R. C. Dugan, and H. W. Bety, Electrical Power Systems Quality. New York: McGraw-Hill, 1996.

[2]   T. A. Short, Electric Power Distribution Handbook. Boca Raton, FL:CRC, 2004.

[3]   P. Heine, "Voltage sag distributions caused by power system faults," IEEE Trans. Power Syst., vol. 18, no. 4, pp. 1367–1373, Nov. 2003.

[4]   D. S. Dorr, M. B. Hughes, T. M. Gruzs, R. E. Jurewicz, and J. L.McClaine, "Interpreting recent power quality surveys to define the electrical environment," IEEE Trans. Ind. Appl., vol. 33, no. 6, pp.1480–1487, Nov./Dec. 1997.

[5]   T. S. Key and J.-S. Lai, "Comparition of standards and power supply design options for limiting harmonic distorsion in power systems," IEEE Trans. Ind. Appl., vol. 29, no. 4, pp. 688–695, Jul./Aug. 1993.

[6]   A. Domijan, Jr., A. Montenegro, A. J. F. Keri, and K. E. Mattern, "Custom power devices: An interaction study," IEEE Trans. Power Syst., vol. 20, no. 2, pp. 1111–1118, May 2005.

[7]   H. Fujita and H. Akagi, "The unified power quality conditioner: The integration of series- and shunt-active filters," IEEE Trans. Power Electron., vol. 13, no. 1, pp. 315–322, Mar. 1998.

[8]   H. Fujita, Y. Watanabe, and H. Akagi, "Control and analysis of a unified power flow controller," in Proc. IEEE/PELS PESC, 1998, pp.805–811.

[9]   M. Rahman, M. Ahmed, R. Gutman, R. J. O"Keefe, R. J. Nelson, and J.Bian, "UPFC application on the AEP system: Planning

[10]   M.P. P'alsson, K. Uhlen, J.O.G. Tande. "Large-scale Wind Power Integration and Voltage Stability Limits in Regional Networks"; IEEE 2002. p.p. 762–769

[11]   P. Ledesma, J. Usaola, J.L. Rodriguez "Transient stability of a fixed speed wind farm" Renewable Energy 28, 2003 pp.1341–1355

[12]   P. Rosas "Dynamic influences of wind power on the power system". Technical report RISØR-1408. Ørsted Institute. March 2003.

[13]   R.C. Dugan, M.F. McGranahan, S. Santoso, H.W. Beaty "Electrical Power Systems Quality" 2nd Edition McGraw-Hill, 2002. ISBN 0-07- 138622-X

[14]   P. Kundur "Power System Stability and Control" McGraw-Hill, 1994. ISBN 0-07-035958-X

[15]   N. G. Hingorani y L. Gyugyi. "Understanding FACTS". IEEE Press; 2000.

[16]   Z. Saad-Saoud, M.L. Lisboa, J.B. Ekanayake, N. Jenkins and G. Strbac "Application of STATCOM's to wind farms" IEE Proc. Gen. Trans. Distrib. vol. 145, No. 5; Sept. 1998

[17]   T. Burton, D. Sharpe, N. Jenkins, E. Bossanyi "Wind Energy Handbook" John Wiley & Sons, 2001. ISBN 0-471-48997-2.

[18]   Ghosh, G. Ledwich "Power Quality Enhancement Using Custom Power Devices" Kluwer Academic Publisher, 2002. ISBN 1-4020-7180- 9

[19]   Schauder, H. Mehta "Vector analysis and control of advanced static VAR compensators" IEE PROCEEDINGS-C, Vol.140, No.4, July 1993.

[20]   E.M. Sasso, G.G. Sotelo, A.A. Ferreira, E.H. Watanabe, M. Aredes, P.G. Barbosa, "Investigac˛ ˜ao dos Modelos de Circuitos de Sincronismo Trif´asicos Baseados na Teoria das Potˆencias Real e Imagin´aria Instantaneas (p–PLL e q–PLL)", In: Proc. (CDROM) of the CBA 2002 – XIV Congresso Brasileiro de Automtica, pp. 480-485, Natal RN, Brasil, 1-4, Sep. 2002

[21]   International   Electro   technical   Commission"INTERNATIONAL STANDAR IEC 61000-4-15: Electromagnetic compatibility (EMC) Part 4: Testing and measurement techniques Section 15: Flickermeter Functional and design specifications." Edition 1.1 2003

[22]   H.Akagi, E.H.Watanabe, M.Aredes "Instantaneous power theory an applications to power conditioning", John Wiley & Sons, 2007. ISBN 978-0-470-1

17

An evaluation of wind energy potential in the northern and southern regions of Nigeria on the basis of Weibull and Rayleigh models

Abdullahi Ahmed[1], Adisa Ademola Bello[2], Dandakuta Habou[2]

[1]Department of Mechanical Engineering, Kano University of Science and Technology, Wudil, Nigeria
[2]Department of Mechanical Engineering, Abubakar Tafawa Balewa University, Bauchi, Nigeria

Email address:
abdula2k2@yahoo.com(A. Ahmed), biieeyz@yahoo.com(A. A. Bello), dandakuta@yahoo.co.uk (D. Habou)

Abstract: This paper presents an evaluation of wind energy potential in the northern and southern region of Nigeria on the basis of Weibull and Rayleigh models. The aim of this study is to know which of the locations in the regions would have more wind power density where wind energy conversion system (WECS) could be installed for electricity generation in Nigeria with excellent percentage of clean energy. From the analysis of the wind speed data collected from Nigeria meteorological station, Abuja at 10m height from years (1990 – 2006) the locations in northern region of Nigeria that were found quite viable for electricity generation are (Jos, Kano, Sokoto and Maiduguri) while for the southern region of Nigeria are (Lagos and Enugu). These locations were found to have wind power density above 100W/m². The Weibull model was found to be more applicable in estimating the power density because it returns a lower percentage error than the Rayleigh model. Probability density function in the northern region has a peak value of 1.01795 and 0.2937 in Bauchi for both Weibull and Rayleigh respectively while for the southern region the probability density function has a peak value for Weibull as 0.8347 in Calabar and Rayleigh as 0.2341 in Rivers southern region of Nigeria.

Keywords: Wind Energy Potential, Nigeria, Generation, Weibull, Rayleigh, Probability Density Function

1. Introduction

Wind energy is a free, clean and inexhaustible energy source. It has served mankind for many centuries by propelling ships and driving wind turbines to grind grains and pump water. Wind power and other forms of solar power are being strongly encouraged. Wind power may become a major source of energy in spite of slightly higher costs than other traditional sources of energy generation. Considerable progress is being made in making wind power less expensive, but even without a clear cost advantage, wind power will become important in the world energy sources. Wind energy is the world's fastest growing energy source and it can power industry, businesses and homes with clean, renewable electricity for many years to come. Wind turbines do not consume fuel due to its operation. It does not produce emissions such as carbon dioxide, sulphur dioxide, mercury, particulates or any other type of air pollution [1].

The major problem associated with extracting energy from the wind is due to the physical properties of air. The density of air is small and therefore, equipment designed to extract appreciable amounts of energy from moving air must be capable of intercepting large air areas. Also, the noise generated by wind turbines is considered and the interference with television reception, due to both rotating components and the fixed installation, may affect wind exploiting. Hazard due to failure of the moving components of the wind turbine is considered as another problem [1].

Akpinar [2] presented a work on statistical analysis of wind energy potential on the basis of the Weibull and Raleigh distribution for Agin–Elazig, Turkey. The work studied 5years measured wind speed data based on the Weibull and Rayleigh models. The Weibull distribution provides better power density estimation in all the twelve months than the Rayleigh distribution.

Shata [3], worked on evaluation of wind energy potential and electricity generation on the coast of Mediterranean sea in Egypt. Wind data for 10 coastal meteorological stations along the Mediterranean Sea in Egypt was used in the analysis. Results shows that wind energy potential along the coast of Mediterranean sea in north Egypt is quite promising.

Celik [6], presented a work on statistical analysis of wind power density based on the Weibull and Rayleigh models at the southern region of Turkey. In his study one year monthly measured wind speed data was used, in the analysis Weibull model returns smaller error values in calculating the power density when compared to the Rayleigh model.

Ojosu [8] presented a statistical analysis of wind energy potential for power generation in Nigeria, 15 years (1968–1983) monthly average wind speed data was used for six sites at four distinctive regions in Nigeria using Weibull distribution. The analysis shows that wind turbine can generate up to 97MWh/year.

In Nigeria, wind energy is still at an infant stage because is almost an untapped renewable energy source. The participation of government, private and individual to harness the utilization of this clean and inexhaustible energy source is very important for national development. Presently, the existing power plants in Nigeria are; Sayya Gidan–Gada, Sokoto with a capacity of 5kW/h, Dan–Jawa, Sokoto with a capacity of 0.75kW/h and Energy research, Benin with a capacity of 1kW/h, based on these capacity, the contribution its quite negligible to the total energy consumption in Nigeria [9].

In this study, wind energy potential for twelve locations in the northern and southern regions of Nigeria were analyzed using wind speed data at 10m height collected from Nigerian meteorological station NIMET Abuja for the period of (1990–2006). These locations are characterized by different geographical and climatoligical conditions.

This study provides useful information for developing wind sites and planning economical wind turbines capacity for the electricity production in Nigeria.

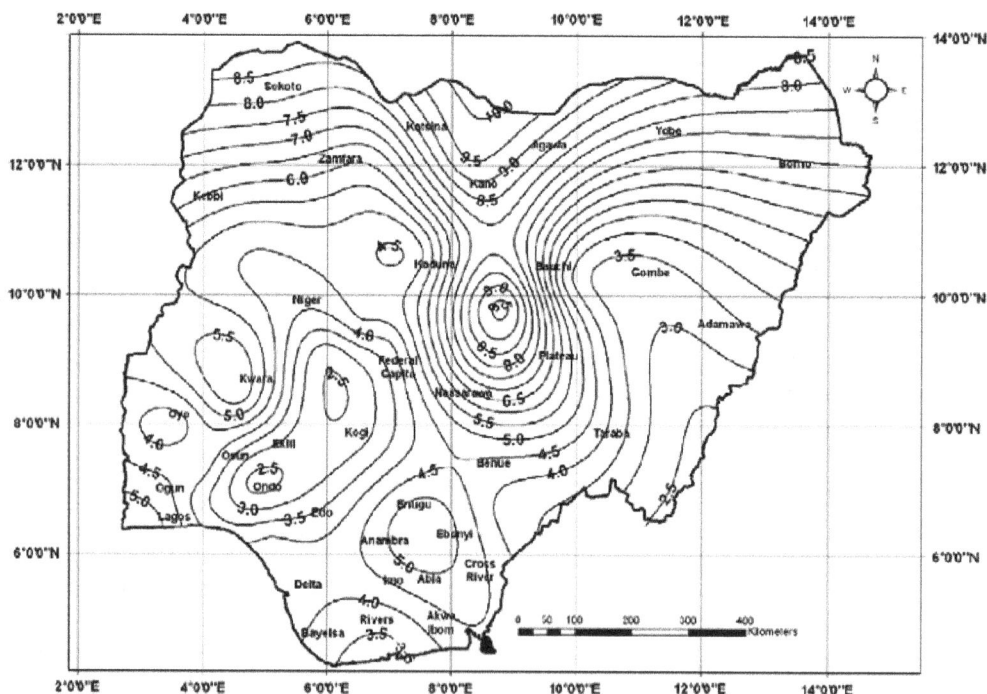

Figure 1: Predicted monthly average wind speeds (m/s) distribution at 10 m height in Nigeria (Source: Fadare, 2010).

Table 1: Geographical data for the selected locations

LOCATIONS	STATE	LAT (N)	LONG (E)	ALT(M)	ALT (FT)
KANO	KANO	$11^0$59'47	$8^0$31'0	476	1564
SOKOTO	SOKOTO	$13^0$3'5	$5^0$13'45	272	895
BAUCHI	BAUCHI	$10^0$18'57	$9^0$50'39	615	2020
MAIDUGURI	BORNO	$11^0$50'47	$13^0$9'37	299	984
ABUJA	ABUJA	$9^0$15'0	$6^0$55'60'	246	810
JOS	PLATEAU	$9^0$55'0	$8^0$54'	1217	3996
ABEOKUTA	OGUN	$7^0$9'0	$3^0$21'0	66	219
LAGOS	LAGOS	$6^0$27'11	$3^0$23'45	34	114
ENUGU	ENUGU	$6^0$26'25	$7^0$29'39	247	813
OWERRI	IMO	$5^0$28'60	$7^0$1'60	158	521
CALABAR	CROSSRIVER	$4^0$34'27'	$6^0$58'33	380	1249
PORTHARCOURT	RIVERS	$4^0$47'34	$6^0$59'50	465	1528

Source: http://www.fallingrain.com

2. Wind Speed Data and Sites Descriptions

The present study is based on a data source measured at 10m height above ground level for twelve different locations in the northern and southern regions of Nigeria.

The wind speed data were collected during the period (1990 – 2006) at 10m height using a cup anemometer at the meteorological station. Fig. 1 shows the locations and distribution of the sites and table 1 shows the geographical features of these locations. These locations were selected for the analysis based on different geographical features.

3. Wind Speed Frequency Distribution

The knowledge of wind speed frequency distribution is very important factor to evaluate the wind potential in an area. If the wind speed in the location is known, then the power potential to the site can easily be obtained. Wind data obtained with various observation methods has a wide range therefore in wind energy analysis, it is necessary to have few parameters that can explain the behaviour of a wide range of wind speed data. The simplest and most practical method for the procedure is to use a distribution function. There are several density functions which can be used to describe the wind speed frequency, the common two are the Weibull and Rayleigh models [4].

The Weibull distribution function which is a two parameter distribution can be expressed as;

$$f_w(v) = \left(\frac{k}{c}\right)\left(\frac{v}{c}\right)^{k-1} \exp\left[-\left(\frac{v}{c}\right)^k\right]$$

where f_w = Weibull probability density function $\qquad$ (1)
$\qquad k$ = shape factor
$\qquad c$ = scale factor

The corresponding cumulative probability function of the Weibull distribution is;

$$F_w(v) = 1 - \exp\left(\frac{v}{c}\right)^k$$

(2)

F_W = Weibull cumulative distribution function

The Rayleigh distribution function is a special case of the Weibull distribution in which the shape factor is 2.0. Probability density and cumulative function of Rayleigh distribution are calculated as;

$$f_R(v) = \frac{\pi}{2}\frac{v}{v_m^2}\exp\left[-\left(\frac{\pi}{4}\right)\left(\frac{v}{v_m^2}\right)^k\right]$$

(3)

$$F_R(v) = 1 - \exp\left[-\left(\frac{\pi}{4}\right)\left(\frac{v}{v_m}\right)^2\right]$$

(4)

where $f_R(v)$ = Rayleigh probability density function
$F_R(v)$ = Rayleigh cummulative distribution fuction

3.1. Wind Power Density

The power density evaluation is a fundamental importance in the assessment of wind energy in a given area. The wind power density depends on the air density, the cube of the wind speed and the wind speed distribution. Therefore, this parameter is generally considered a better indicator of wind resource than the wind speed. It can be estimated by using the equation;

$$p(V) = \frac{1}{3}\rho A V^3$$

(5)

Where P = Power in watts (W)
A = Area perpendicular to wind speed vector in (m²)
V = Wind speed in (m/s)
ρ = Average density of air in (kg/m³)

$$P_W = \frac{1}{2}\rho c^3 \Gamma\left[1 + \frac{3}{k}\right]$$

(6)

Where P_W = Weibull Power
c = scale factor (m/s)
k = shape factor (dimensionless)
Γ = Gammafunction

$$P_R = \frac{3}{\pi}\rho v_m^3$$

(7)

Where P_R = Rayleigh Power
v_m^3 = mean speed

The yearly average error in calculating power densities using both Weibull and Rayleigh functions is obtained by using equation below;

$$\text{Error (\%)} = \frac{1}{12}\sum_{i=1}^{12}\left(\frac{P_{W,R} - P_{m,R}}{P_{m,R}}\right)$$

(8)

4. Results and Discussion

From the results obtained, table 2 presents yearly average monthly variation of Weibull parameters and wind speed for the northern and southern regions of Nigeria. The Weibull shape factor (k) for northern region of Nigeria varies between 4.13 – 7.47, the scale factor (c) varies between 2.58 – 10.76 m/s and wind speed varies between 2.36 – 10.12m/s, while for the southern region of Nigeria Weibull shape (k) varies between 3.51 – 10.07, the scale factor (c) varies between 3.35 – 5.91 and the wind speed varies from 3.02 – 5.47m/s.

It can be seen from table 3 that the favourable locations for wind power generation in northern region of Nigeria are; Maiduguri, Sokoto, Kano and Jos with values for Weibull and Rayleigh varying between 162.51 – 924.45W/m^2 and 202.67 – 1256.11W/m^2 respectively, while for southern region of Nigeria the favourable locations are; Enugu and Lagos with values varying between 152.35 – 197.02W/m^2 and 172.21 – 188.73W/m^2.

Table 4, presents the percentage error for Weibull and Rayleigh models, from the results shown, it can be seen that Rayleigh models returns high percentage error in all the locations considered in this research.

Comparison of probability density function according to observed data with Weibull and Rayleigh models for the investigated sites is illustrated in the figures 2 - 5 below. It is seen from figure 2 and 3 that Bauchi shows peak value of probability density function as 1.01795 and 0.2937 for Weibull and Rayleigh, while figure 4 shows that Calabar has a peak value for Weibull probability density function as 0.8347 and Rivers has the peak value for Rayleigh probability density function as 0.2341 .

Table 3: *Weibull and Rayleigh Power density*

Locations	P_W(W/m^2)	P_R (W/m^2)	P_{mR}(W/m^2)
Northern region			
Kano	618.45	818.10	538.89
Sokoto	360.22	460.62	274.12
Bauchi	14.43	17.29	11.02
Maiduguri	162.51	202.67	120.49
Abuja	55.53	62.44	41.39
Jos	924.45	1256.11	713.95
Southern region			
Abeokuta	30.89	33.10	22.98
Lagos	172.21	188.73	119.74
Enugu	152.35	197.02	106.03
Owerri	34.74	41.78	26.92
Calabar	68.97	102.56	56.36
Rivers	35.08	41.85	26.59

Table 4: *Weibull and Rayleigh % Error*

Locations	% Error (Weibull)	% Error (Rayleigh)
Northern region		
Kano	23.33	61.29
Sokoto	30.95	65.18
Bauchi	28.04	50.74
Maiduguri	36.23	68.11
Abuja	33.48	49.84
Jos	29.28	42.87
Southern region		
Abeokuta	31.25	42.87
Lagos	40.99	58.13
Enugu	42.64	81.31
Owerri	30.77	61.46
Calabar	22.74	82.81
Rivers	32.67	58.38

Table 2: *Weibull distribution parameters at 10m height.*

Locations	k	c (m/s)	v(m/s)
Northern region			
Kano	6.71	9.35	8.74
Sokoto	6.02	7.69	7.14
Bauchi	4.71	2.58	2.36
Maiduguri	5.47	5.90	5.45
Abuja	4.13	4.12	3.74
Jos	7.47	10.76	10.12
Southern region			
Abeokuta	3.51	3.35	3.02
Lagos	3.79	5.91	5.33
Enugu	8.92	5.80	5.47
Owerri	5.62	3.55	3.27
Calabar	10.07	4.60	4.42
Rivers	5.03	3.54	3.25

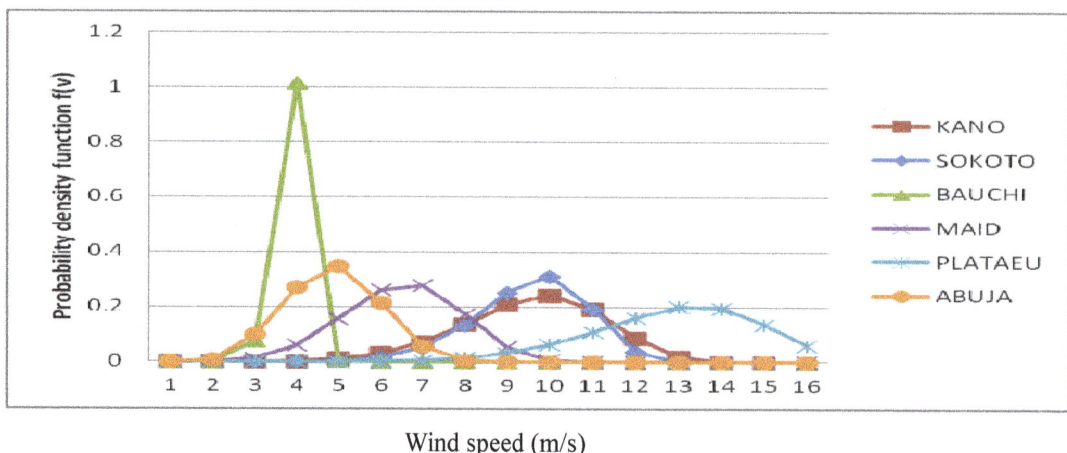

Fig 2: *Comparison of Weibull probability density functions for northern region.*

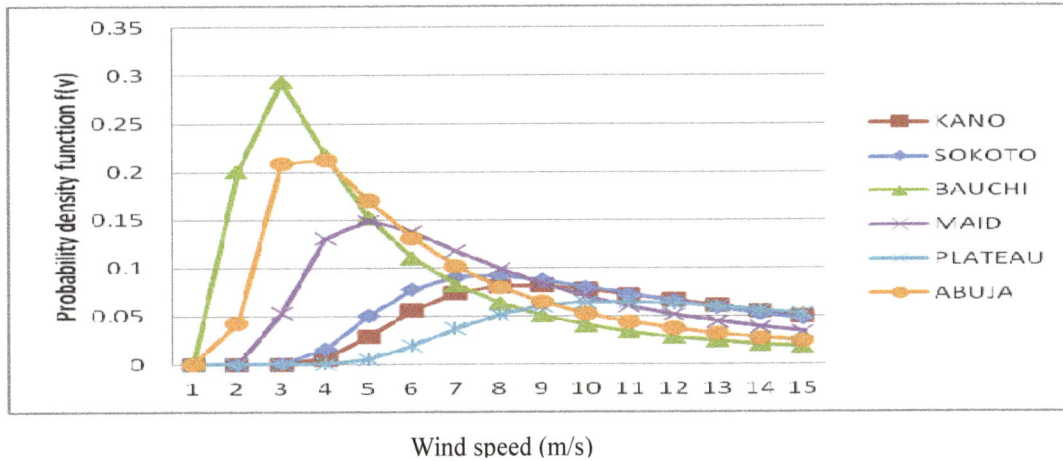

Fig 3: Comparison of Rayleigh probability density functions for northern region.

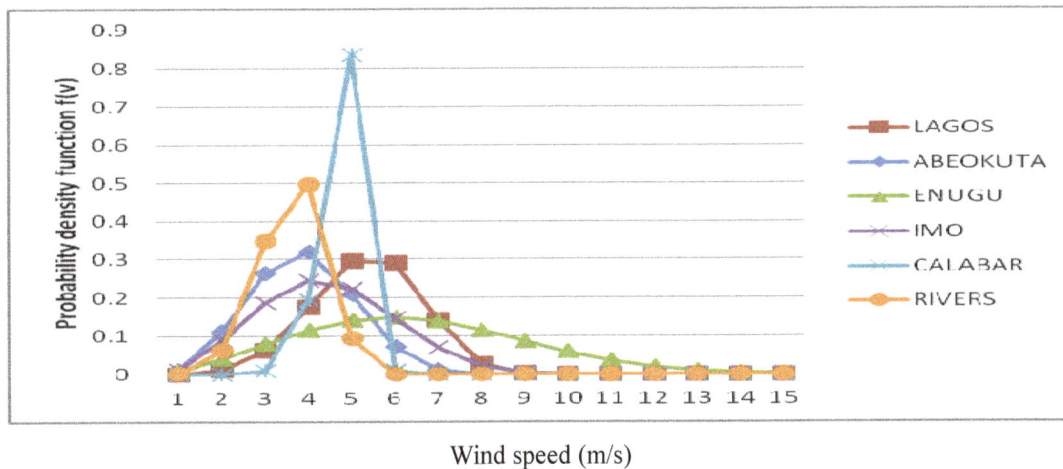

Fig 4: Comparison of Weibull probability density functions for southern region.

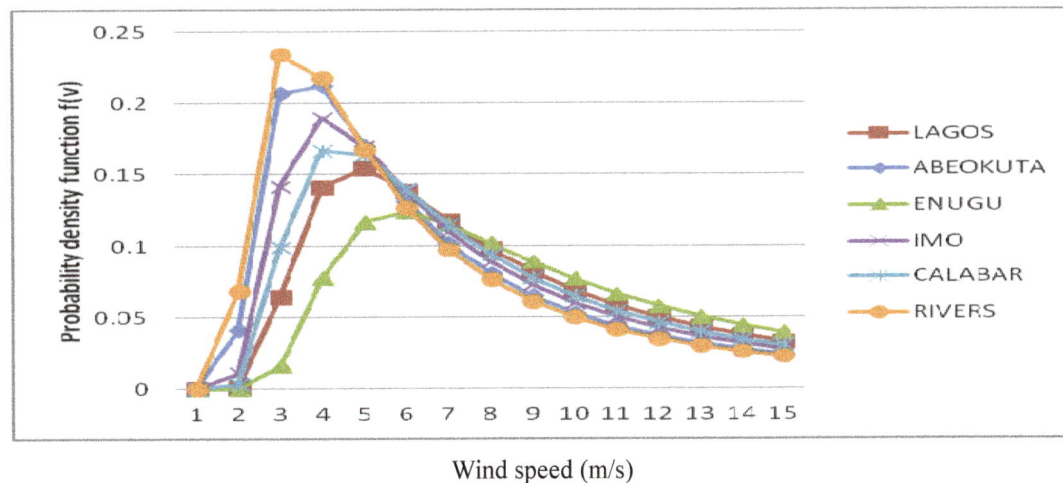

Fig 5: Comparison of Rayleigh probability density functions for southern region.

5. Conclusion

In this study, the monthly and yearly wind speed distribution and wind power density during the period (1990–2006) in the northern and southern region of Nigeria were evaluated. The wind speed frequency distribution of the locations was found by using Weibull and Rayleigh distribution functions. It can be concluded as follows;

(1). In the northern region, the highest average monthly wind speed occurred in Jos and was determined as 11.74m/s in February while the lowest wind speed occurred in Bauchi with value of 1.47m/s in December. The yearly average annual wind speeds for these locations are shown in table 2.

(2). In the southern region, the highest average monthly wind speed occurred in Enugu and was deter- mined as 6.47m/s in April while the lowest wind speed occurred in Rivers with value of 2.5m/s in November. The yearly average annual wind speeds for these locations are shown in table 2.

(3). The yearly average values for Weibull parameters (k and c) in the northern region varies between 4.71 – 7.47 and 2.58 – 10.76m/s respectively while for the southern region it varies between 3.51 – 10.07 and 3.35 – 5.91m/s respectively.

(4). In the northern region, the highest average monthly wind power occurred in Jos and was determined as 1469.81W/m^2 for Weibull and 1892.64W/m^2 for Rayleigh in February while the lowest average monthly wind power occurred in Bauchi and was determined as 3.87W/m^2 for Weibull and 3.72W/m^2 for Rayleigh in December. The yearly average annual wind powers for these locations are presented in table 3.

(5). In the southern region, the highest average monthly average wind power occurred in Lagos and was determined as 301.46W/m^2 for Weibull and 312.48W/m^2 for Rayleigh in the month of August while the lowest average monthly wind power occurred in Rivers and was determined as 15.77W/m^2 for Weibull and 18.53W/m^2 for Rayleigh in November.

(6). In calculating error in power density in both regions, the Weibull model returns smaller errors in calculating the power density compared to the Rayleigh model.

Finally, the Weibull distribution provides better power density estimation in both regions than the Rayleigh model since it returns lower error in power estimation.

References

[1]   Alsaad MK. Wind energy potential in selected areas in Jordan. Energy Conversion and Management 2013; 65:704 – 708.

[2]   Akinpar EK, Akinpar S. Statistical analysis of wind energy potential on the basis of the Weibull and Rayleigh distribution for Agin-Elazig, Turkey. Power and Energy 2004; 218: 557 – 565.

[3]   Ahmed Shata AS, Hanitsch R. Evaluation of wind energy potential and electricity generation on the coast of Mediterranean Sea Egypt. Renewable Energy 2006; 31:1183-202.

[4]   Ahmed Shata AS, Hanistsch R. The potential of electricity generation on the east coast of Red Sea in Egypt. Renewable Energy 2006; 31: 1597 – 625.

[5]   Celik AN. On the distributional parameters used in assessment of the suitability of wind speed probability density functions. Energy Conversion and Management 2004; 45: 1735 – 47.

[6]   Celik AN: A statistical analysis of wind power density based on the Weibull and Rayleigh Models at the southern region of Turkey. Renewable Energy 2004; 29:593 – 604.

[7]   Celik AN. Assessing the suitability of wind speed probability distribution functions based on the wind power density. Renewable Energy2003; 28:1563 -1574.

[8]   Ojosu JO, Salawu RI. A statistical analysis of wind energy potential for power generation in Nigeria. Journal of Solar Energy 1989; 8: 273 - 288.

[9]   Amina I, Nafi'u T, Bilyaminu A. Wind power: An untapped renewable energy resource in Nigeria. International Journal of Scientific & Engineering Research 2012; 3: ISSN 2229 – 5518.

[10]   Oztopal A, Sahin AD, Akgun N, Sen Z: On the regional wind energy potential of Turkey. Energy 2000; 25: 189 – 200.

[11]   Sambo AS.T he renewable energy for rural development. The Nigerian perspective "ISESCO" Science and Technology vision May, 2005; 1:16 -18.

[12]   Salem AL. Characteristics of surface wind speed and direction over Egypt .Solar Energy for sustainable development 2004; 4: 491 – 499.

[13]   Ulgen K, Hepbasli A. Determination of Weibull parameter for wind energy analysis of Izmir, Turkey. International Journal of Energy Research 2002; 26:495 – 506.

18

CFD modeling of devolatilization and combustion of shredded tires and pine wood in rotary cement kilns

Peter Mtui

College of Engineering and Technology, University of Dar es Salaam, ar es Salaam, Tanzania

Email address:
plmtui@yahoo.com

Abstract: Computational fluid dynamics (CFD) has been used to study the devolatilization and combustion of large particles of shredded scrap tire "rubber" and pulverized pine wood "biomass" in a rotary cement kilns. The CFD model constitutes of modeling the hydrodynamics within the rotary kiln, heat transfer, devolatilization and co-combustion of rubber and biomass blends. Equivalent particle diameters of 1 mm and 2 mm for biomass and rubber, respectively, were used to simulate the process conditions and co-combustion in cement rotary kilns. The ratios of biomass substitution were varied from 5% to 20% on mass basis. The effect of the percentage of biomass content on the temperature distribution and devolatilization rate as well as wall heat transfer rate are presented. Results indicate that up to 20% biomass blend provides improved combustion characteristics compared with combustion of scrap tire alone. Further, the devolatilization rate was found to improve remarkably when scrap tire is blended with biomass. Most importantly, the biomass blend was found to increase the flame spreading and penetration, consequently improving the wall heat transfer rate, therefore providing favorable conditions for heat transfer to the cement clinker. The present study provides additional knowledge for future investigations on the use of biomass residues and some industrial wastes as alternative fuels for rotary cement kilns. That is, the use of alternative fuels in rotary cement kilns has a potential for economic improvement and environmental sustainability.

Keywords: CFD, Rotary Kiln, Scrap Tire, Biomass, Devolatilization, Combustion

1. Introduction

It has been reported by BP Statistical Review of World Energy [1] that the cement industry consumes approximately 2% of the world's primary energy consumption. Cembureau [2] pointed out that, fuel consumption in cement production accounts for nearly 30–40% of the total production costs. Traditionally, fossil fuels, such as coal and petroleum coke have been used as primary energy sources in the cement manufacturing industry. However, due to the recent increase of fossil fuel prices, and environmental concerns, the cement industry has increased the utilization of alternative fuels as a substitute for fossil fuels. For example, the most common alternative fuels in the cement industry are tire-derived fuels (TDF) and biomass residues. The utilization of coarse large solid particles as an alternative fuel in cement rotary kilns is an attractive option. Roller [3] pointed out that the refuse-derived fuel represents a significant quantity of energy therefore recovery of part of this energy would also reduce a growing environmental problem. The larger solid fuel particles in high temperatures environment in the rotary kiln provides longer retention times which is favorable for fuel burnout. Genon et al [4] have focused on pollutant emissions from waste derived fuel combustion in cement kilns while some researchers [5] carried out a study to investigate air requirement and energy input by partly replacing primary fuel by bone meal and sewage sludge.

Combustion of large fuel particles in rotary kilns is limited in the literature. Therefore, systematic investigation of percentage blend of pulverized biomass for a rotary kiln initially fired with shredded scrap tire as alternative fuels is of interest. Understanding the phenomena in a rotary kiln by experiments is not a trivial task because of the complex heat and mass transfer which involves chemical reactions during the process. The purpose of this work is therefore, to study the devolatilization and combustion of a blends of scrap tire and biomass and on the particles transport phenomena that involves heat and mass transfer within the rotary kiln. The study was carried out using computational fluid dynamics (CFD) modeling technique.

The pyrolysis technique is an alternative for reducing scrap tires and biomass and leads into the production of oils and black carbon. Tires are known to be composed mainly of rubbers, such as natural rubber, or styrene-butadiene rubber, as well as mineral oils and black carbon [6]. Thermal degradation of tires has been studied by Yang, et al [7] who reported that devolatilization rate is strongly dependent on temperature and the particle size.

Larsen et al [8] reported the devolatilization times of cylindrical tire rubber particles with diameters ranging from 7 to 22 mm and height of 35 mm at 840 °C in an inert atmosphere to range from 75 to 300 seconds. Depending on thickness and temperature of tire, the devolatilization times were found by Chinyama [9] to be 30−100 seconds for the rubber particle size range of 6−12 mm in air at temperature from 700 to 1000°C. Generally, results reported in literature about devolatilization of large TDF particles show clear tendencies for devolatilization time to increase with increasing particle size and decrease with increasing temperature.

Previous work of Atal and Levendis [10] on the combustion of tire char was performed with a few mg of tire char particles in the range of 100−500 μm size in thermogravimetric analyzers (TGA). For tire char with particle size of 500 μm, the char conversion times are reported to be well below 1 s in air and at 1200 °C.

Devolatilization of non-spherical pine wood particles with equivalent diameters ranging 10 to 45 mm in the temperature range 650° C to 850 °C has been investigated by Diego et al [11]. The reported devolatilization times for 10 to 30 mm diameter particles were, for example, from 30 to 150 seconds at 850 °C. Di Blasi and Branca [12] studied temperature profiles in cylindrical beech wood particles with lengths of 20 mm and diameters from 2 to 10 mm. They reported the devolatilization time of 40 seconds at the bed temperature of 439 and 834 °C. Di Blasi [13] and White et al. [14] have presented exhaustive review of kinetics models for biomass pyrolysis and report general tendencies that devolatilization times increase with increasing particle size and decrease with increasing temperatures.

2. Model Description

A three-dimensional geometry of the rotary kiln was constructed using Gambit software as per Table 1. A multi-annular, multi-fuel burner was designed so that the shredded tire and pulverized biomass were independently introduced into the kiln by the fluidizing primary air. The proximate and ultimate analyses of the rubber are reported by Juma et al [15] and those of biomass suggested by Diego, et al [11] are contained in Table 2, both on dry basis. The process data and boundary conditions such as wall temperatures and wall emissivity factors were set appropriately in the model.

Table 1. *Kiln data*

Parameter	Value
Inner diameter (m)	4.5
Kiln length (m)	60
Rotational speed (rpm)	4
Tilt angle (deg)	3

Table 2. *Fuels data*

Proximate Analysis	Rubber [15]	Biomass [11]
Fixed carbon (%wt)	22.66	15.3
Volatiles (%wt)	61.61	75.1
Moisture (%wt)	1.72	8.3
Ash (%wt)	14.01	1.3
LHV (MJ/kg)	34.2	18.7
Ultimate Analysis		
C	81.24	52.9
H	7.36	6.8
O	8.92	38.9
N	0.3	0.1
S	2.18	1.3

3. Model Development

This section seeks to develop mathematical models for devolatilization and char combustion for the selected alternative fuels. The models are used under conditions similar to those in the material inlet end of cement rotary kilns.

3.1. Inert Heating

Inert heating of fuel particles takes place during which the particle temperature is less than the evaporation temperature or when all the volatiles have been consumed. The heating is mainly due to convection and radiation heat transfer.

$$m_p C_p \frac{dT_p}{dt} = h A_p (T_\infty - T_p) + \varepsilon_p A_p \sigma(\theta_R^4 - T_p^4) \quad (1)$$

where m_p is the mass of the particle, C_p is the heat capacity of the particle, A_p is the surface area of the particle, T_∞ is the local gas phase temperature, T_p is the particle temperature, h is the convective heat transfer coefficient, ε_p is the particle emmisivity, σ is the Boltzmann constant and θ_R is the radiation temperature. The heat transfer coefficient h is evaluated as:

$$Nu = \frac{h D_p}{k_\infty} = 2.0 + 0.6 \, \mathrm{Re}_D^{1/2} \, \mathrm{Pr}^{1/3} \quad (2)$$

where k_∞ and Pr are respectively the thermal conductivity and Prandtl number of the gas phase.

3.2. Devolatilization Model

Devolatilization was modeled by the two competing rates model proposed by Kobayashi, et al [16] so that:

$$R_1 = A_1 e^{-(E_1/RT_p)} \tag{3}$$

$$R_2 = A_2 e^{-(E_2/RT_p)} \tag{4}$$

where R_1 and R_2 are competing rates that may control the devolatilization over different temperature ranges so that the rate of devolatilization is given as:

$$\frac{m_v(t)}{(1-f_{w,0})(m_{p,0}-m_a)} = \int_0^t (\alpha_1 R_1 + \alpha_2 R_2) \exp\left(-\int_0^t (R_1 + R_2) dt\right) dt \tag{5}$$

where: $m_v(t)$ = volatile yield at time t

$m_{p,0}$ = initial mass of particle

m_a = ash content in the particle

α_1, α_2 = model constants

$f_{w,0}$ = mass fraction of volatiles initially present

in the particle

3.3. Char Combustion Model

The aim of this section is to derive a model for char conversion in the rotary kiln, including the main rate limiting parameters for char conversion.

Surface char combustion is accounted for by using the Kinetic/Diffusion reaction rate model, which assumes that the surface reaction rate is determined either by kinetics or a diffusion rate. In this work, the model of Braum et al [17] and Field [18] is used such that the diffusion rate,

$$R_1 = C_1 \frac{\left(\frac{(T_p + T_\infty)}{2}\right)^{3/4}}{D_p} \tag{6}$$

and the kinetic rate

$$R_2 = C_2 \exp(-E/RT_p) \tag{7}$$

so that the rate of change of particle mass, m_p is given by

$$\frac{dm_p}{dt} = \pi D_p^2 P_0 \frac{R_1 R_2}{R_1 + R_2} \tag{8}$$

where P_0 is the partial pressure of oxidant species in the gas surrounding the combusting particle and the kinetic rate R_2 incorporates the effects of chemical reaction on the internal surface of the char particle and pore diffusion. The particle size is assumed to remain constant at time Δt

while the density is allowed to decrease.

Figure 1. *Typical rotary kiln showing the flame propagation and bed temperature*

4. Transport Phenomena

CFD modeling of the thermo-chemical processes such as of reactive flows is governed by the continuity and momentum equations and includes the description of fluid flow, heat and mass transfer, and chemical reactions. The process fundamental governing equations are the conservation laws of mass, momentum, energy and species. The conservation equations can be generally written as [19]:

$$\frac{\partial(\rho\varphi)}{\partial t} + \frac{\partial(\rho u_j \varphi)}{\partial x_j} = \frac{\partial}{\partial x_j}\left(\Gamma_\varphi \frac{\partial\varphi}{\partial x_j}\right) + S_\varphi \tag{9}$$

where ϕ is a general variable such as velocity components, temperature, mass fraction, etc. For the above, ρ is the fluid density, and u is the fluid velocity vector. The variable Γ_ϕ represents the transport coefficient and S_ϕ is a source term. The conservative equations are discretized and solved iteratively so that the CFD technique enforces the conservation of mass, momentum, energy and species concentrations over the entire computational domain [20].

5. Results and Discussions

Results from this simulation are discussed for various blends of waste tire ("rubber") and wheat straw ("biomass"). Generally, the results show strong dependence on the quantity of biomass blend on the combustion characteristics in the rotary kiln.

Figure 2 shows the contour plots for the axial and radial temperature distributions in the kiln for various blends of biomass content from 5 to 20%. Results indicate that larger quantities of biomass substitution improve the radial and axial temperature distribution. This phenomenon is probably due to higher volatiles content in biomass which enhances the reactivity of the rubber. Similarly, Figure 3 depicts the axial temperature profile along the center axis of the kiln length. Up to 20% biomass substitution shows similar temperature profiles as for a kiln fired with scrap tire alone.

Figure 4 illustrates the rate of devolatilization for scrap tire blended with biomass. Results show remarkably higher devolatilization rate as the biomass is increased to 20%. As expected, the higher volatiles content in biomass enhances the overall devolatilization of the scrap tire ("rubber"). Thus, the enhanced devolatilization rate improves the char burn-rate as shown in Figure 5. Similarly, Figure 6 shows the improved wall heat transfer rate for larger quantities of biomass substitution which is consistent with increased radial temperature distribution illustrated in Figure 2.

Figure 2. *Temperature distribution along the length of the kiln with varying biomass blend*

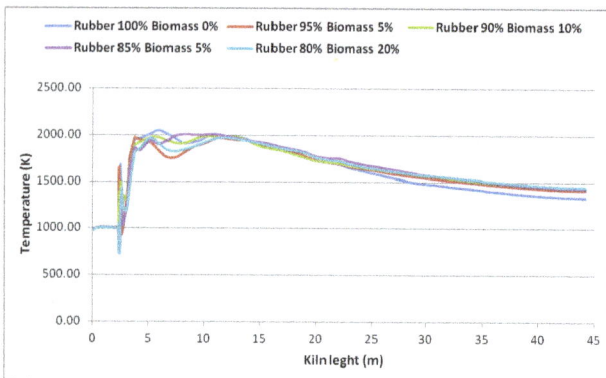

Figure 3. *Temperature profile along the center axis of kiln length with varying biomass blend*

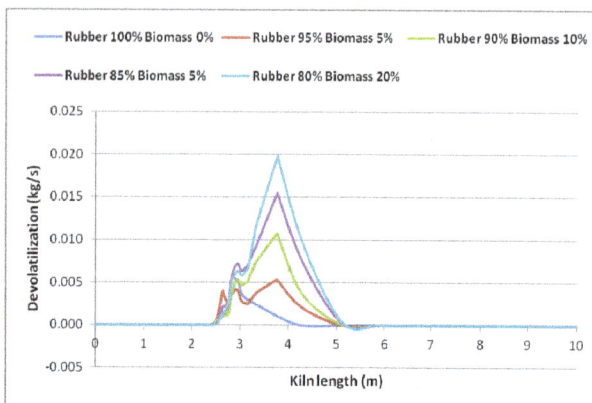

Figure 4. *Devolatilization rate along the center axis of kiln length with varying biomass blend*

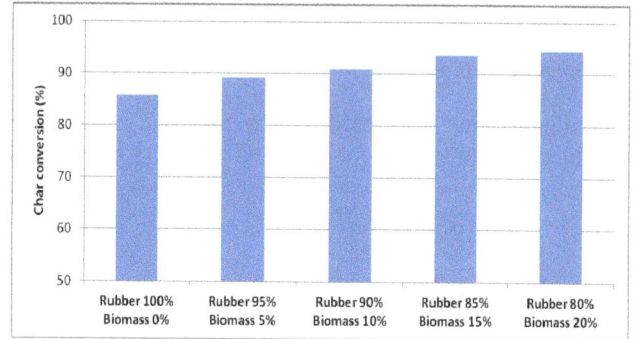

Figure 5. *Char conversion during co-combustion*

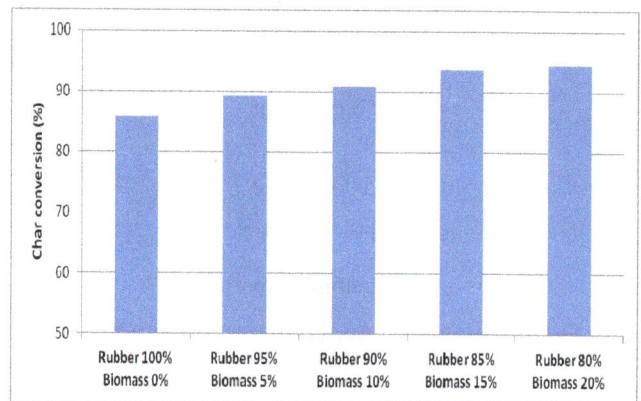

Figure 6. *Kiln wall heat transfer rate*

6. Conclusion

Computational fluid dynamics has been used to study the devolatilization and co-combustion of shredded scrap tire and pulverized pine wood in rotary kiln. The biomass blend was varied from 5% to 20% in order to investigate the influence of the quantity of the biomass blend on the combustion characteristics of rotary cement kiln originally fired with shredded tires.

Results indicate a significant increase in axial and radial flame distribution in the rotary kiln as the biomass content was increased from 5% to 20%. Consequently, the wall heat transfer rate increases as the biomass quantity was increased. The increased heat transfer rates improves the performance in the cement kiln in transferring heat to the cement clinker and improve the performance of the kiln. Further, the increased rate of devolatilization with larger quantity of biomass substitution assists the thermal degradation of the waste tire therefore accelerates the char burnout. The co-combustion of biomass in rotary cement kilns fired with scrap tire may be thought of as an efficiently way to improve the reactivity of scrap tire fuel.

Overall conclusion is that co-combustion of waste tire with biomass substitution nearly up to 20% provides improved combustion characteristics, therefore, offering an opportunity of using alternative fuel blends in rotary cement kilns.

References

[1]   BP Statistical Review of World Energy; June 2009, http://www.bp.com.

[2]   Cembureau,: Alternative Fuels in Cement Manufacture; 1997, http://www.cembureau.be

[3]   R.C. Poller, Reclamation of waste plastics and rubber recovery of materials and energy, J. Chem. Tech. Biotechnol. 30_1980.152–160.

[4]   G. Genon and E. Brizio, "Perspectives and limits for cement kilns as a destination for RDF," Waste Management, Vol. 28, no. 11, pp. 2375–2385, Nov. 2008.

[5]   U. Kaantee, R. Zevenhoven, R. Backman and M. Hupa, "Cement manufacturing using alternative fuels and the advantages of process   modelling," Fuel Processing Technology, Vol. 85, no. 4, pp. 293-301, Mar. 2004.

[6]   Conesa J.A., Font R., Marcilla R. Mass spectrometry validation of a kinetic model for the thermal decomposition of tyre wastes. Journal of Analytical and Applied Pyrolysis 43; pp. 83-96; 1997.

[7]   Yang J., Tanguy P. A., Roy C. Heat transfer, mass transfer and kinetics study of the vacuum pyrolysis of a large used tire particle. Chemical Engineering Science 50; pp. 1909-1922; 1995.

[8]   Larsen, M. B.; Schultz, L.; Glarborg, P.; Skaarup-Jensen, L.; Dam-Johansen, K.; Frandsen, F.; Henriksen, U. Fuel 2006, 85, 1335–1345.

[9]   Chinyama, M. P. M; Lockwood, F. C. J. Energy Inst. 2007, 80 (3), 162–167

[10] Atal, A.; Levendis, Y. A. Fuel 1995, 74 (11), 1570–1581

[11] de Diego L. F., F. García-Labiano, A. Abad, P. Gayán, J. Adánez Coupled drying and devolatilization of non-spherical wet pine wood particles in fluidized beds. Journal of Analytical and Applied Pyrolysis 65 (2002) 173–184

[12] Di Blasi, C.; Branca, C. Energy Fuels 2003, 17, 247–254

[13] C. Di Blasi, Modeling chemical and physical processes of wood and biomass pyrolysis, Progress in Energy and Combustion Science 34 (2008) 47–90

[14] J.E. White, W.J. Catallo, B.L. Legendre, Biomass pyrolysis kinetics: a comparative critical review with relevant agricultural residue case studies, Journal of Analytical and Applied Pyrolysis 91 (2011) 1–33.

[15] M. Juma, Z. Koreňová, J. Marks, J. Annus, L. Jelemensky. Pyrolysis and Combustion of Scrap Tires Petroleum & Coal 48 (1), 15-26, 2006

[16] H. Kobayashi, J. B. Howard, and A. F. Saro Coal Devolatilization at High Temperatures. In 16th Symp. (Int'l.) on Combustion. The Combustion Institute, 1976

[17] Braum M. M. and P. J. Street. Predicting the Combustion Behaviour of Coal Particles. Combustion Science and Technology, 3(5), 231-243. (1971)

[18] Field M. A. Rate of Combustion of Size-Graded Fractions of Char from a Low Rank Coal between 1200K and 2000K. Combustion and Flame, 13, 237-252. (1969)

[19] Elaine S. Oran and Jay P. Boris Numerical Simulation of Reactive Flow, Cambridge University Press, Second Edition, Second edition (2001)

[20] Fluent 12 User Guide Copyright @2009 by ANSYS, Inc.

19

Study on physical and chemical properties of crop residues briquettes for gasification

Khardiwar Mahadeo[1, *], Anil Kumar Dubey[2], Dilip Mahalle[3], Shailendra Kumar[4]

[1]Senior Research Associate, Anand Agricultural University, Muvaliya Farm, Dahod-389 151, Gujarat, India
[2]CIAE, Principal Scientist, Central Institute of Agricultural Engineering, Bhopal, India
[3]Assistant Professor, College of Agricultural Engineering and Technology, Dr. PDKV Akola-44104, Maharastra, India
[4]Senior Research Associate, Anand Agricultural University, Muvaliya Farm, Dahod-389151

Email address:

msk1987res@gmail.com (K. Mahadeo), anilkumardubey@gmail.com (A. K. Dubey), dmmahalle@gmail.com (D. Mahalle),
shailendrakgangwar@gmail.com (S. Kumar)

Abstract: The selection of material was done on the basis of availability and need of the gasifier design. Moreover, agricultural production in the country is increasing day by day with the agricultural mechanization, providing tremendous volume of agricultural residue every year. The volatile matter of soybean briquette, pigeon pea briquette and mix briquette was found to be in the range of 77.07 - 79.14%. Highest volatile mater found in briquette of mix biomass. Regarding volatile matter content, soybean briquette, pigeon pea briquette and mix briquette materials include about 70% volatile probably due to their high moisture content. Taking into account both ash and volatile matter contents, soybean briquette, pigeon pea briquette and mix briquette seem to be the best material for pyrolysis and gasification. Bulk densities were in the range of 598-675 kg/m^3. The bulk density was found more with pigeon pea briquette. Highest ash fusion temperature was found to be 1210 °C, with pigeon pea briquette. The ash fusion temperatures indicate clinker formation. The Calorific values were found to be in the range of 4107-4520, cal/kg. The soybean briquette has higher calorific value than other briquette. Crop residues briquette can be used effectively as energy fuel for gasifier.

Keywords: Agricultural Crop Residues, Biomass, Biomass Briquette, Gasification, Physical Properties, Proximate Analysis, Pyrolysis

1. Introduction

A recent study showed that agricultural residues are the most potential considering their quantitative availability. In order to characterize the physical and chemical properties of crop residues briquette used as feedstock for energy conversion process. The selection of material was done on the basis of availability and need of the gasifier design. Moreover, agricultural production in the country is increasing day by day with the agricultural mechanization, providing tremendous volume of agricultural residue every year. Biomass fuels continue to representing the primary source of energy for more than 50% of the world population and amount to about 14% of the total energy global consumption [8]. On the other hand, biomass corresponds to the most common form of renewable energy, which becomes of increasingly interest since it is considered as playing an important role in mitigating global warming and securing

fuel supply. During the past decade, interest in biomass came back on international stage since it is the most promising alternative to fossil fuels. Moreover, biomass is readily available in many countries worldwide, especially those of the developing world. This renewable energy resource could be sustainable when appropriately managed and assure locally fuel supply at competitive cost. Using biomass as fuel offers also environmental benefits. When correctly operated and controlled, the combustion of biomass produces significantly less nitrogen oxides and sulphur dioxide than that of solid fossil fuels such as coal or lignite, and is CO_2-neutral since the emitted carbon dioxide will serve for the biomass regrowth, and thus not to enhance global warming. Biomass is produced by green plants through photosynthesis using sunlight. Biomass contains organic matter which can be converted to energy. This energy can be replenished by human effort. Biomass today accounts for over one-third of all energy used in the developing countries. The estimated power generation potential from biomass in India is about 19000 MW. There are two methods to produce energy from biomass; gasification and combustion route. In combustion route, biomass is burnt to produce steam. The steam is used for power generation through turbines. In gasification process, biomass is converted into producer gas and the producer gas is used for thermal or electrical application. Biomass can be burnt in a boiler for production of high pressure superheated steam using conventional combustion technology that would generate power through steam.

In India, well-recognized as one of the first rank exporters of agricultural and food products, biomass has been the traditional energy source, especially in rural areas for decades. Various types of biomass are available in India mostly in the form of non plantation resources: (i) agricultural residues, (ii) residues from wood and furniture industry, (iii) animal manure, (iii) municipal solid wastes and landfill gas, and, (iv) wastewater. Policy Research for Renewable Energy Promotion and Energy Efficiency Improvement indicated that agricultural residues are the most potential considering their quantitative availability. Actually, agricultural residues, including residues from paddy (rice husk, rice straw), sugarcane (bagasse, leaves and trashiers), maize (corncob, stalk), cassava (stalk, sludge cake), and palm (empty fruit bunch, shell, fiber), annually amount a total of 98 million tons, and 41 million tons can be used as energy resource, but only 50% of the available quantity are currently used for energy purpose. For comparison, in China, available agricultural residues represent a total of 939 million tons annually, and 551 million tons can be used as an energy resource, but only 266 million tons are effectively used [3].

However, to use biomass efficiently for energy production a detailed knowledge of its physical and chemical properties is required. These properties, more specifically average and variation in elemental compositions, are also essential for modeling and analysis of energy conversion processes [9]. Actually, information on concentration and speciation of some elements is useful both for energy and environmental issues, e.g. concentration and speciation of alkali will help to better design biomass power generation system or of heavy metal to assess the potential environmental impacts. Therefore, the investigation of chemical elemental characteristics of biomass fuels would help finding for them suitable and appropriate energy conversion technologies, but also for different existing conversion technologies to effectively use biomass feedstock. Extensive research to determine the physical and chemical properties of the indigenous available biomass resources has been conducted in several countries and international networks [14].

Gasification as a process for energy conversion has been used extensively for charcoal and woody biomass, but very little work has been reported for loose agro wastes except rice husk, at small power plant levels (about 1 MW). Current estimation of the net annual bio-residues availability for power generation in India stands at 100 million tones (t) per year amounting to about 15,000 MW capacities [2, 11]. Typical residues generated from agro industries are rice husk, coconut shell, corncobs, coir pith, tapioca waste, groundnut shells, coffee husk, etc. Bagasse from the sugar industry has a captive use for both heat and electricity. There are other wastes generated from industries where wood or woody like material is used as raw material; as in industries manufacturing paper, plywood, furniture, pencils, etc., where sawdust is available in abundance. Typically, 5–20% of the feedstock remains as waste depending upon the industry. Some of these residues are used as fuel in combustion systems either for heat or power generation or a combination of both. The power generation is packaged with steam turbines in the capacity range of 4 MW and above. The concept of captive power generation using wastes generated in-house is common in industries such as sugar, paper, and rice mills. These industries require heat in addition to electricity for process application. In recent times, cogeneration has been promoted in several countries, leading to improvement in the overall energy efficiency. For instance in India, the sugar sector is adopting cogeneration packages where exporting electricity to the state grid is financially attractive under the rules of the state electricity regulatory commissions. Even though there are several case studies where captive power generation systems have been successfully implemented, there is also enough evidence that a large amount of these raw materials is being inefficiently utilized, thus contributing to pollution of the environment.

Among several kinds of biomass, agricultural residues have become one of most promising choices. They are easily available and environmentally friendly. Nevertheless, the majority of them are not suitable to be utilized as fuel without an appropriate process since they are bulky, uneven and have low energy density. These characteristics make this kind of waste for difficult in handling, storage transport and utilization. One of the promising solutions to overcome these problems is the briquetting technology. The technology may be defined as densification process for improving the handling characteristics of raw, material and enhancing the volumetric calorific value of the biomass. The biomass are available in different forms like rice husk, coffee husk, coir pith, jute sticks, bagasse, groundnut shells, mustard stalks, cotton stalks, bamboo dust, caster seed, palm husk, soybean husk pigeon crop residues has traditionally been a handy and valuable source of heat energy all over the world in rural as well as the sub urban areas. In spite of rapid increase in the supply of, access to and use of fossil fuels, agri residue is likely to play an important role in developing countries in general and India in particular, in the foreseeable future. Thus, developing and promoting techno-economically viable technologies to utilize crop residue for power generation, remains a pursuit of high priority. The effective agro-residues utilization for gasification will impart the boost for utilization of the briquetting and gasification technology in our country. The outcome of the project will certainly enhance the spectrum of use of agro residues for gasification to generate the electricity on decentralized mode. The project will lead to provide the technology for generation of fuel, prepared from different agro residues for their application in a gasifier. Thus, the outcome will be highly useful for the gasifier users to attain the better gasification by using wide variety of agro residues. pea stalk and wheat straw etc. The use of biomass as a source of energy is of interest worldwide because of its environmental advantages. During recent decades, biomass use for energy production has been proposed increasingly as a substitute for fossil fuels. There is large variability in crop residues generation and their uses in different regions of country depending on the cropping intensity, productivity and crops grown [14]. They estimated total available crop residues in India as 523.4 Mt/year and surplus as 127.3 Mt/year. The annual surplus crop residues of cotton stalk, pigeon pea stalk, jute & mesta, groundnut shell, rapeseed & mustard, sunflower were 11.8, 9.0, 1.5, 5.0, 4.5, and 1.0 Mt/year, respectively. The residues of most of the cereal crops and 50% of pulses are used for fodder. Coconut shell, stalks of rapeseed and mustard, pigeon pea and jute & mesta, and sun flower are used as domestic fuel. Biomass as energy is gaining importance as a renewable source to strengthen the country's agriculture as a prime player in Indian economy. The use of biomass for thermal energy is ancient but the biomass as a renewable energy source implying clean combustion process is more recent. In the last three decades the Ministry of New and Renewable Energy (MNRE) of Govt. of India, has encouraged R&D in developing gasification systems in India. Therefore the objectives of this paper is used to find a clean and more efficient source of renewable energy from crop residues for this reason we must know their chemical and physical characteristics, in order to choose the best energetic conversion process.

2. Material and Methods

Raw material of crop residues i.e. soybean and pigeon pea stalk available at energy and power division CIAE Bhopal, Briquettes were prepared using a briquetting machine based on piston-press technology in which soybean residues and pigeon pea residues are punched or pushed into a die of 60 mm by a reciprocating ram by high pressure. These briquettes are broken in the length of 60-100 mm manually. Open core downdraft gasifier used for its efficient operation. Following observations were recorded viz. Physical and Chemical properties of crop residues briquette are obviously the most vital parameter which decides the consistent and efficient operation of the gasifier. Following properties will be determined.

2.1. Physical Properties of Crop Residues Briquette

The physical properties such as moisture content, overall length and diameter, bulk density, tumbling resistance, and resistance to water penetration of crop residue briquettes were determined. The briquettes selected for determine physical and chemical properties of three type of briquette

i.e. soybean briquette, pigeon pea briquette, Mix briquette of (soybean + pigeon pea). To measure the overall length and diameter of briquettes, scale and Vernier caliper was used.

2.1.1. Moisture Content

The moisture content of a solid is defined as the quantity of water per unit mass of the wet solid. The moisture content plays an important role in the formation of briquette and subsequently its combustion. Moisture content of biomass at the time of harvesting varies drastically. The moisture content of biomass was measured by oven dry method. Initially the sample with the known weight was kept in oven at 105°C for 24 hours. The oven dry sample is then weighed. The moisture content of sample was calculated by following formula. [1]

$$\text{M.C.} = \frac{W_1 - W_2}{W_1} \times 100 \qquad (1)$$

Where,

W_1 = Weight of sample before drying, g
W_2 = Weight of sample after drying, g

2.1.2. Bulk Density

Water displacement method was used to measure the volume of individual briquette. The briquettes were coated with wax, in order to prevent any water absorption during merging process. Each briquette was weighed and then coated with wax. The wax-coated briquettes were weighed and then submerged into water in suspension position and weight of displaced water was measured and recorded as the volume of the wax briquettes. The volume of each briquette was calculated by subtracting the volume of coating wax from the volume of wax briquettes. The volume of coating wax was obtained by dividing its weight of the wax obtained by subtracting original weight of briquette from the weight of wax briquette by its volume. [13].

Volume of sample = Volume of waxed sample − Volume of wax

$$\text{Volume of sample} = V - \frac{W_3 - W_2}{\text{Density of wax}} \qquad (2)$$

Where,

W_1 = Initial weight of sample, g
W_2 = Weight of sample + string, g
W_3 = Weight of waxed sample + string, g

2.1.3. Tumbling Resistance

The Tumbling test was conducted to find out percentage loss of weight of single briquette subjected to tumbling action. Each briquette was weighed and placed in the cuboids formed by angle iron frame having dimensions of 30×30×45 cm and fixed over hollow shaft diagonally was used to conduct the tumbling test. The sample of briquettes was put inside and cuboids is rotated for15 minute. After 15 minutes of tumbling action the briquette was taken out weighed and percent loss was calculated by using formula. [13]

$$\text{Weight loss (\%)} = \frac{W_1 - W_2}{W_1} \quad (3)$$

Where,

W_1 = Weight of briquette before tumbling, g

W_2 = Weight of briquette after tumbling, g

$$\text{Tumbling resistance (\%)} = 100 - \text{percent weight loss} \quad (4)$$

2.1.4. Resistance to Water Penetration

It is measure of percentage water absorbed by a briquette when immersed in water. Each briquette was immersed in 150 mm of water column at 27°C for 30 s. The percent water gain was calculated and recorded by using following formula. [13].

$$\text{Water gain by briquettes (\%)} = \frac{(W_2 - W_1)}{W_1} x100 \quad (5)$$

Where,

W_1 = Initial weight of briquette, g

W_2 = Weight of wet briquette, g

$$\text{Resistance to water penetration (\%)} = 100 - \text{Water gain (\%)} \quad (6)$$

2.2. Chemical Parameter of Crop Residues Briquette

Chemical properties are very important to determine the fuel quality. Study of proximate analysis of biomass was carried out for determination of volatile matter, fixed carbon, ash content, Ash fusion temperature and Calorific value, in the biomass. The ASTM D 3172, ASTM D 3177, ASTM D 3175, ASTM D 1875, ASTM D 3286 respectively was used for the study [1].

2.2.1. Volatile Matter

The same sample from previous determination of moisture content is used to determine the percentage of volatile matter. The sample in the covered crucible is then heated in the muffle furnace at temperature of 950 ± 20°C for 7 minutes. crucible was taken out the and first brought down its temperature to room temperature rapidly (to avoid oxidation of its contents) by placing in a cold iron plate and then transferred warm crucible to desiccators to bring it to room temperature. Take the final weight of crucible and contents. Percentage of volatile matter of the sample is determined by using the following formula. [1]

$$\text{volatile matter, \%} = \frac{b - c}{a} \times 100 \quad (7)$$

Where, a = initial weight of sample, 1g.

b = final weight of sample after cooling
(Heating temperature 107± 3 °C for 1 hour).

c = final weight of sample after cooling
(Heating temperature 950 ± 20 °C).

2.2.2. Fixed Carbon

The residue remaining after volatile matter release has been expelled, contains the mineral matter originally present

and non volatile or fixed carbon. The fixed carbon was thus calculated as follows. [1]

$$\text{Fixed carbon (\%)} = 100 - (\% \text{ moisture } + \% \text{ Ash } + \% \text{ V.M}) \quad (8)$$

2.2.3. Ash Content

The same sample from previous determination of volatile matter content is used to determine the percentage of Ash content. The sample from in the crucible was then heated without lid in a muffle furnace at 700 ± 50°C for an hour. The crucible was then taken out, cooled first in air, then in desiccators and weighed. Heating, cooling and weighing is repeated till a constant weight is obtained. The residue was reported as ash on per cent basis. Percentage of ash is to be determined by using the following formula. [1]

$$Ash(\%) = \frac{\text{Weight of ash left}}{\text{Weight of sample taken}} \times 100 \quad (9)$$

2.2.4. Ash Fusion Temperature

The standard tests for fusibility of coal and coke ash are based on ASTM D-1875.[1] The biomass is dried, ground, and placed in the muffle furnace at 750°C in the atmosphere of air till constant weight is obtained. The residual ash is then finely ground a solution containing 10% dextrin, 0.1% salicylic acid and 89.9% H_2O by weight, is added to ash. To make it a plastic mass, the mass is moulded to a cone shape by pressing it into a suitable mould. These cones are taken out and allowed to dry. The dry cones placed on a refectory base are then inserted in a high temperature furnace and heated to 800 °C. After about 15 minutes interval the temperature of the sample is raised at an increment of 50°C during each interval the shape of the cone is observed. The temperature range at which the initial rounding off or bending of the apex of the cone is observed can be termed as 'ash deformation temperature'. As the temperature is further increased, the same sample has a tendency to fuse into a hemispherical lump. The temperature range during which the phenomenon is observed can be taken as 'ash fusion temperature.'

2.2.5. Calorific Value

A known mass of the given sample was taken in clean crucible. The crucible was then supported over the ring. A fine magnesium wire, touching the fuel sample, was then stretched across the electrodes. The bomb lid was tightly screwed and bomb filled with Oxygen 25 atmospheric pressure. The bomb was then lowered into copper calorimeter, containing a known mass of water. The stirrer was worked and initial temperature of the water was noted. The electrodes are then connected to 6-volt battery and circuit completed. The sample burns and heat was liberated. Uniform stirring of water was continued and the maximum temperature attained was recorded, the experimental setup for determination of calorific value using Bomb calorimeter. The calorific value of the different crop residues briquette was determined by using Bomb Calorimeter. The calorific value of the briquette was determined by using the following

formula [6]

$$\text{Calorific value (kcal/kg)} = \frac{(W+w) \times (T_1 - T_2)}{X} \qquad (10)$$

Where,

W = weight of water in calorimeter (kg),

w = water equivalent of apparatus,

T_1 = initial temperature of water ($^{\circ}$C),

T_2 = final temperature of water ($^{\circ}$C),

X = weight of fuel sample taken (kg)

Fig. 2.1. Different crop residue briquette produces at energy enclave CIAE, Bhopal

3. Results and Discussion

Properties of crop residues briquette suitable for gasification properties of soybean briquette, pigeon pea briquette, mixed biomass briquette were studied prior to use this fuel in the gasifier for power generation. The results such as overall length and diameter, moisture content, bulk density, volatile matter, ash content, fixed carbon, tumbling resistance, resistance to water penetration, ash fusion temperature and calorific value of crop residues briquette. (Fig.2.1)

3.1. Physical Properties of Briquette

3.1.1. Overall Length and Diameter

Analyzed for the briquette were produced from 60 mm and 30 mm diameter. The length of Soybean briquette, Pigeon pea briquette, mixed briquette were found to be 60-85, 65-90, 65-80, respectively. The diameter of briquette of soybean, pigeon pea and mix were varied from 59.9-60 mm. where as its diameter was found to be 60 mm, 30 mm and 60 mm. respectively. The gasification system is designed for selected sizes of briquette fuel were within the acceptable limit.

3.1.2. Moisture Content

Moisture content is of important interest since it corresponds to one of the main criteria for the selection of energy conversion process technology. Thermal conversion technology requires biomass fuels with low moisture content,

while those with high moisture content are more appropriate moisture content of lesser than 15%, and hence more suitable to serve as feedstock for thermal conversion technologies. The ground biomass used to produce briquette were measured different properties shown in table 3.1. Moisture content of Soybean briquette, Pigeon pea briquette, mixed briquette were found to be 8.75, 9.12 and 8.9 per cent, respectively. The acceptable limit of moisture content in the fuel for efficient gasification should in the range of 6 – 10 per cent. The values of moisture content of fuel were found within the limit and as per the requirement of gasification process.

Table 3.1. Physical properties of different crop residues briquette

Type of briquette	Diameter (mm)	Length (mm)	Bulk density (kg/m³)	True Density (kg/m³)
Soybean	60	60-85	618	1130
Pigeon pea	60	70-95	625	1150
Pigeon pea	30	65-90	675	1125
Mix biomass	60	65-80	598	1120

3.1.3. Bulk Density

It has been stated that bulk density of biomass briquettes depends on density of the original biomass [5]. Bulk density play vital role in transportation and storage efficiency. bulk and true densities of the briquette are depicted in fig.3.1 where as bulk density varies between 598 to 675 kg/m³.several researcher have reported that densification would result in bulk densities in the range of 450 to 700 kg/m³ depending upon feedstock and densification condition [12]. True density of the briquette varies between 1120 to 1150 kg/m³. Obviously, for all crop residues briquettes, Comparison between each briquette indicated that Pigeon pea briquette had the highest density followed by Soybean briquette and mixed briquette respectively. The bulk density of pigeon pea briquette was found more shown in (table 3.1). High densities of fuel represent high-energy value. Bulk density of all three fuels showed their suitability as a fuel in gasifier.

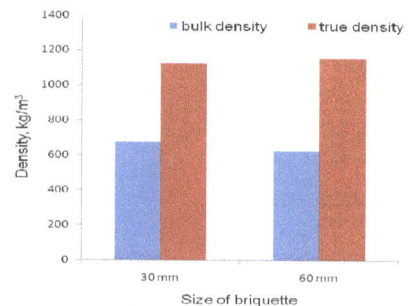

Fig 3.1. Effect of briquette size on bulk density and true density

3.1.4. Tumbling Resistance

Tumbling resistance test was performed for Soybean briquette, Pigeon pea briquette and Mix briquette. The average material loss from briquette was observed less than 4 percent during tumbling test for Soybean briquette, Pigeon pea briquette, and mixed briquette. The tumbling resistances of briquette were found to be 94.7 to 97.1 percent. The variation in tumbling resistance are given in (fig 3.2) above values indicated that the briquettes can be handled easily during transportation without any damage.

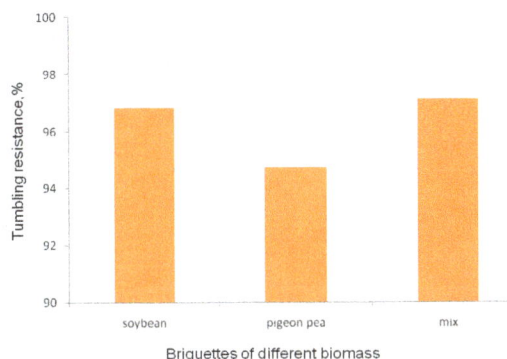

Fig 3.2. *Variation in tumbling resistance of different briquette*

3.1.5. Resistance to Water Penetration

This test was performed for Soybean briquette, Pigeon pea briquette and Mix briquette. It is observed that resistance to water penetration for given briquette was 58.4 to 71.4 and percent. Resistance to water penetration depends on the true density of material. As true density of Soybean briquette and Pigeon pea briquette was more shown in (fig 3.3) therefore, these briquettes are more resistant to water penetration. These briquettes can be stored for long duration without any adverse effect.

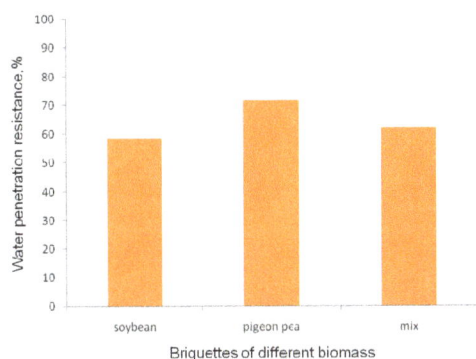

Fig 3.3. *Variation in resistance of water penetration of different briquette*

3.2. Proximate Analysis of Briquette

The proximate analysis of selected briquette were result obtained are given in (table 3.2) The volatile matter of soybean briquette, pigeon pea briquette and mix briquette was found to be 76.96, 77.07and 79.14 percent, respectively. Regarding volatile matter content, soybean briquette, pigeon pea briquette and mix briquette materials include about 70% volatile probably due to their high moisture content. Taking into account both ash and volatile matter contents, soybean briquette, pigeon pea briquette and mix briquette seem to be the best candidates for pyrolysis and gasification. The above analyses revealed that considerable amount of volatiles are available in the material of gasification.

Table 3.2. Proximate analysis of briquette of different biomass

Briquette	Volatile matter (%)	Fixed carbon (%)	Ash content (%)	Calorific value (kcal/kg)
Soybean	76.96	16.46	6.58	4520
Pigeon pea	77.07	15.88	7.05	4107
Mix biomass	79.14	13.52	7.34	4180

The ash content of Soybean briquette, Pigeon pea briquette and Mix briquette, were found to be 6.58, 7.05 and 7.34 percent, respectively. The ash content of biomass influences the expenses related to handling and processing to be included in the overall conversion cost. On the other hand, the chemical composition of the ash is a determinant parameter to consider for the operation of a thermal conversion unit, since it gives rise to problems of slagging, fouling, sintering and corrosion. It is desirable to use lower ash content fuel for gasification. This fuel was found suitable for gasification. For efficient gasification ash content of the fuel should be below 10 per cent [10]. High ash content fuel seriously interferes with the operation of gasifier and increase pressure drop.

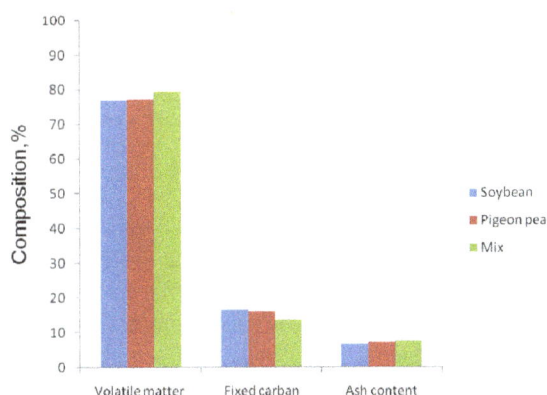

Fig 3.4. *Proximate analysis of different crop residues briquette*

Whereas fixed carbon content in Soybean briquette, Pigeon pea briquette and Mix briquette were found to be 16.46, 15.88 and 13.52 percent, respectively. It was found maximum in soybean briquette. Proximate analysis of different biomass briquette were volatile matter76.9-79.2, fixed carbon 13.5-16.5, which is comparable with woody biomass, (Prosopis juliflora) and (Leucaena leucocephala)in given (table 3.3) whose performance was already known. Meanwhile biomass briquette is suitable for gasification [7].

Table 3.3. Proximate analysis of different woody biomass

Woody biomass	Volatile matter (%)	Fixed carbon (%)	Ash content (%)
Prosopis juliflora	83.05	15.94	1.7
Leucaena leucocephala	82.17	16.94	1.47

3.3. Ash Fusion Temperature

The parameters that decide ash fusion are temperature and the residence time the particles are subjected to. The temperature in the oxidation zone can vary between 1,473 and 1,673 K and hence most of the agro residue ash can fuse in this zone if the char reaches such temperatures. The problem becomes serious if there are any traces of foreign matter like sand and mud. In order to determine the ash fusion conditions that are related to the operating flux conditions inside the reactor simple experiments were performed. The experimental setup consists of an inverted downdraft gasification stove with air being supplied in a controlled manner with the help of a blower and flow measuring device [4] Ash fusion temperature of Soybean briquette, Pigeon pea briquette and Mix briquette was found to be 1147 °C, 1210 °C and 1183 °C respectively(fig.3.5)

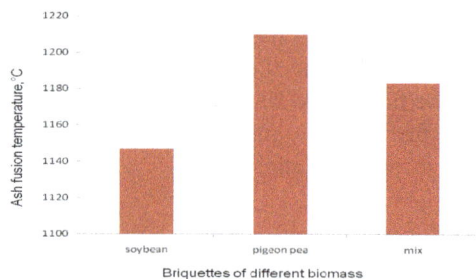

Fig 3.5. Variation in ash fusion temperature of different briquette

3.4. Calorific Value

Calorific value of different crop residues briquette are given in (table 3.2).Lower Heating Values indicate energy effectively released by the biomass fuels, and so the necessary quantity to feed in the energy conversion unit. Our results showed that among all analyzed samples, Soybean briquette has the highest calorific value, followed by Pigeon pea briquette and Mix biomass briquette.

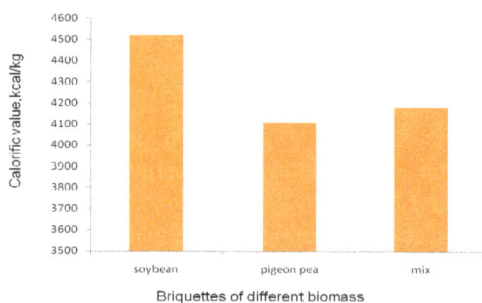

Fig 3.6. Variation in calorific value of different briquette

4. Conclusion

In order to characterize the physical and chemical properties of crop residues briquette used as feedstock for energy conversion process. The volatile matter of soybean briquette, pigeon pea briquette and mix briquette was found to be in the range of 77.07 - 79.14%. highest volatile found in briquette of mix biomass, Regarding volatile matter

content, soybean briquette, pigeon pea briquette and mix briquette materials include about 70% volatile probably due to their high moisture content. Taking into account both ash and volatile matter contents, soybean briquette, pigeon pea briquette and mix briquette seem to be the best candidates for pyrolysis and gasification. Bulk densities were in the range of 598-675 kg/m³. The bulk density was found more with pigeon pea briquette. The highest ash fusion temperature was found to be 1210°C, with pigeon pea briquette. The ash fusion temperature indicates clinker formation. The Calorific values were found to be in range of 4107-4520, kcal/kg. The soybean briquette has higher calorific value than other briquette. Crop residues briquette i.e. Soybean briquette, pigeon pea briquette and mix briquette can be used effectively as energy fuel for gasifier.

Acknowledgement

The authors are grateful to Dr. Pitam Chandra, Director, Central Institute of Agriculture Engineering, Bhopal (MP) for providing facilities and valuable guidance to carry out the study.

References

[1] ASTM. Annual book of ASTM standards. Philadelphia: American society for testing of materials, 1983; p. 19103.

[2] Akshay Urja (2008) Volume 2, Issue 4, 2008, an MNRE publication

[3] Cuiping, L., Changzhi, W., Yanyongijie. and Haitao, H. (2004). Chemical elemental characteristics of biomass fuels in China, Biomass and Bioenergy, 27, pp. 119-130.

[4] Dasappa S., H.S. Mukunda, P.J. Paul, and N.K.S. Rajan (2003) Biomass to Energy: The Science and Technology of the IISc Bioenergy systems Dept of Aerospace Engg, Indian Institute of Science, (155pp.), ABETS 2003.

[5] Demirbas A. and Sahin A. 1998. "Evaluation of Biomass Residue 1. Briquetting Waste Paper and Wheat Straw Mixtures." Fuel Processing Technology. 55: 175-183.

[6] Dara S.S., (1999) A practical handbook of Engineering Chemistry, 60-62.

[7] Kalbande, S.R., M.M.Deshmukh, H.M. Wakudkar and G.wasu. 2010. Evaluation of gasifier based power generation system using different woody biomass. ARPN Journal of Engineering and Applied Sciences, 5(11), 82-88.

[8] Mc Kendry, P. (2002). Energy production from biomass (part I): Overview of biomass. Bioresource Technology, 83, pp. 37-46.

[9] Nordin A. (1994) Chemical elemental characteristics of biomass fuels. Biomass and Bioenergy, 6(5), pp. 339-347.

[10] Savitri G ,Ubonwan C, Suthum P and Jittawadee D (2006) Physical and Chemical Properties of Thai Biomass Fuels from Agricultural Residues The 2nd Joint International Conference on "Sustainable Energy and Environment" C-048 (P)1- 23November 2006, Bangkok, Thailand.

[11] Sheshagiri GS, Rajan NKS, Dasappa S, Paul PJ (2009) Agro residue mapping of India, Proceedings of the 17th European Biomass Conference and Exhibition.

[12] S. Sokhansanj and A.F.Turhollow, "biomass densification: cubing operation and cost for corn stover," applied engineering in agriculture, vol.20, no.4, pp 495-499, 2004.

[13] Tayade, S.R. 2009. Evaluation of differentb briquette making for gasifier. Unpublished M.Tech. Thesis, PGI,Dr. PDKV, Akola.

[14] Thailand Research Fund (2006). "Policy Research on Renewable Energy Promotion and Energy Efficiency Improvement in Thailand" Project – First Progress Report.

[15] Pathak, B.S. 2006. Crop residues to energy.Enviroment and agriculture 2006: 854-869

20

Euler-Lagrange Modeling of entrained flow gasification of coke-biomass slurry mixture

Peter Mtui

College of Engineering and Technology, University of Dar es Salaam, P. O. Box 35131, Dar es Salaam, Tanzania

Email address:

plmtui@yahoo.com

Abstract: An Euler-Lagrange CFD method for co-gasification simulation of slurry mixture of pulverized petroleum coke and up to 20% biomass (wheat straw) in a pressurized entrained flow gasifier is proposed to increase the biomass contribution to green electricity generation. The gas phase is modeled as a continuum and the solid phase is modeled by a Discrete Phase Modeling (DPM) using a soft sphere approach for the particle collision dynamic. The model takes into account detailed gas phase chemistry, modeling of the pyrolysis and gasification of each individual particle, particle shrinkage, and heat and mass-transfer between the gas phase and the dispersed phase. The coke was blended with 5–20% wheat straw on mass basis. The effect of the percentage of biomass blended with coke on the flow field, gas and temperature distribution, syngas composition and particles trajectories are presented. Most important result is the quality and quantity of syngas produced when blended up to 20% biomass is similar to that of coke gasification. Additional observation is that the reactivity of coke was greatly improved by the presence of biomass. The overall conclusion of this his study is that co-gasification is possible provided that operation is properly adapted.

Keywords: CFD, Co-Gasification, Syngas, Coke, Biomass, Slurry

1. Introduction

Biomass fuels an important source considered renewable and environmentally friendly which has recently spurred interest to mitigate global warming and the limited fossil resources such as coke and coal. Co-gasification of biomass in entrained coke gasification is one of options currently undertaken.

Despite the long tradition of utilizing the combustible fuel gas from gasification of solid fuels, there still is a lack of detailed scientific knowledge about the complex interactions between the gasification reactions and the fluid dynamics of entrained flow gasifiers.

In order to improve the gasification efficiency and syngas composition, numerous mathematical models for coal and biomass gasification in fluidized beds have been developed. For example, one-dimensional models have been reported by Radmanesh [1] and the more complex Euler-Lagrange models Zhou [2] and Rong [3]. Simpler models such as equilibrium models are suitable for design optimization of gasifiers. Lathouwers [4] considered more comprehensive Euler-Euler modeling approach that considers particulate phase with interpenetrating and interacting with the gaseous phase in reacting fluidized beds. Agrawal [5] reported that the Euler-Euler models allow a relatively realistic description of the time dependent processes in non-reacting.

Euler-Lagrange models are more targeted towards fundamental investigations of the chemical and fluidized bed flows (Zhou, [2]). In Euler-Lagrange models, the trajectory and the state of each individual particle is tracked in space and time by integrating the equations of motion, energy, and mass for each particle in the system. For dense particle system with multiple contacts between particles the soft-sphere the Discrete Phase Model is usually applied. Therefore, the Euler-Lagrange approach potentially offers the most accurate description not only of the particle motion (translational and rotational, particle-particle collisions) but also of chemical reactions and heat and mass transfer between the dispersed phase and the gaseous.

1.1. Model Description

Co-gasification of coke and biomass is modeled in a two-stage industrial entrained flow gasifier. The shape of the two stage gasifier is based on the work reported by Bockelie, et al [6] for an industrial gasifier consuming 3000 tons per day. Figure 1a illustrates a similar two-stage gasifier used in this work that contains three levels of symmetrically placed injectors. The upper level injectors are oriented opposed each other and only water slurry of coke-biomass is introduced. The two bottom levels of injectors are oriented tangentially to create strong swirl flow where water slurry of coke-biomass introduced to the gasifier. Oxygen is also introduced at the two bottom levels. Figure 1b illustrates the computed particle flow path in the gasifier.

As for fuel properties, the proximate and ultimate analyses of the coke are taken from the work of Furimsky [7] and for wheat straw biomass from Chunggen [8] which are contained in Table 1. Except for biomass co-firing, the gasifier process conditions and other necessary data such as wall boundary conditions were those reported by Bockelie, et al. [6]. A 1.6 m diameter 10 meter high pressurized entrained flow gasifier shown in Figure 1 was modeled using the CFD software Fluent [9]. The model results presented in this work were validated against the work of Bockelie, et al [6] who performed coal gasification in a similar pressured entrained flow gasifier.

2. The Governing Conservative Equations

The mathematical modeling of the fluid flow is based on a set of coupled conservation governing equations of mass, momentum, energy, and chemical species, similar to Ferziger, et al. [10] and Chung [11]. Properties of fluid (density, the viscosity, the specific heats, the molecular diffusivity, the thermal conductivity, the radiation properties etc.) are given as a function of the state variables.

2.1. Mass Conservation

The continuity equation is a mass balance states that the overall mass of the gaseous phase system is conserved. The gas phase conservation equation of mass can be written as:

$$\frac{\partial \rho}{\partial t}+\frac{\partial}{\partial x_i}(\rho u_i)=S_m \quad (1)$$

where S_m is the mass source term which accounts for the mass transfer from solids (or liquid) phase to the gas phase.

2.2. Momentum Conservation

The Navier-Stokes equation for the conservation of momentum can be written as:

$$\frac{\partial(\rho u)}{\partial t}+\frac{\partial}{\partial x_i}(\rho u_i u_j)=-\frac{\partial p}{\partial x_i}+\frac{\partial \tau_{ij}}{\partial x_j}+\rho F_i \quad (2)$$

where F_i is the sum of all external forces (in our case it is

only gravity) and τ_{ij} is the viscous stress tensor is given by as:

$$\pi_{ij}=\mu\left(\frac{\partial u_i}{\partial x_j}+\frac{\partial u_j}{\partial x_i}\right)-\frac{2}{3}\mu\delta_{ij}\frac{\partial u_l}{\partial x_l} \quad (3)$$

where the molecular viscosity μ is introduced, depending on the fluid properties and δ_{ij} is the Kronecker symbol.

2.3. Species Conservation

The conservation equation of chemical species can be written as follows:

$$\frac{\partial(\rho Y_a)}{\partial t}+\frac{\partial}{\partial x_i}(\rho u_i Y_a)=\frac{\partial}{\partial x_i}\left(\frac{\mu}{Sc_a}\frac{\partial Y_a}{\partial x_j}\right)=S_a \quad (4)$$

for n number of species, $a=1,.....n$. The Schmidt number of the species a, defined as:

$$Sc_a=\frac{\mu}{\rho D_a} \quad (5)$$

2.4. Energy Conservation

The conservation equation of energy can be written as:

$$\frac{\partial(\rho h)}{\partial t}+\frac{\partial}{\partial x_i}(\rho u_i h)=\frac{\partial \tau_{ij}}{\partial x_j}+\frac{\partial q_i}{\partial x_i}+\rho u_i F_i+S_h \quad (6)$$

where h is the total specific enthalpy and for a multi-component medium it takes the following form:

$$h=\sum Y_i h_i \quad (7)$$

where Y_i is the mass fraction of species i in the mixture and h_i is the total enthalpy defined as:

$$h_i=h_{ref,i}^0+\int_{T_{ref}}^{T}Cp_i(T)\,dT \quad (8)$$

where $h_{ref,i}^0$ and $Cp_i(T)$ is the enthalpy of formation and the specific heat at a constant pressure, for species i. The reference temperature is given by T_{ref}.

3. Turbulent Modeling

Turbulence gaseous phase is expressed with k-ε two-equation model in which the solution of two separate transport equations allows the turbulent velocity and length scales to be determined independently. The general form of the governing equations for the gas phase is given as follows:

$$\frac{\partial}{\partial x_i}(\rho u_i k) = \frac{\partial}{\partial x_i}\left(\frac{\mu_e}{\sigma_k}\frac{\partial k}{\partial x_i}\right) + G - \rho\varepsilon) \qquad (9)$$

$$\frac{\partial}{\partial x_i}(\rho u_i \varepsilon) = \frac{\partial}{\partial x_i}\left(\frac{\mu_e}{\sigma_\varepsilon}\frac{\partial k}{\partial x_i}\right) + \frac{\varepsilon}{k}(C_1 G - C_2\rho\varepsilon) \qquad (10)$$

where ρ is the fluid density, G represents the generation of turbulence kinetic energy, C_1 and C_2 are constants, σ_k and σ_ε are the turbulent Prandtl numbers for k and ε respectively.

4. Particle Trajectory

In addition to solving transport equations for continuity, momentum, energy, turbulence kinetic energy and dispersion of species, the dispersed second phase is simulated in a Lagrangian frame of reference where the trajectory of the particles is calculated by integrating the force balance of particles.

$$\frac{du_p}{dt} = F_D(u - u_p) + \frac{g(\rho_s - \rho)}{\rho_s} + F_x \qquad (11)$$

where $F_D(u - u_p)$ is the drag force per unit particle mass and is defined as

$$F_D = \frac{18\mu}{\rho_s D_p^2}\frac{C_D \text{Re}_p}{24} \qquad (12)$$

where u and u_p is the velocity for gaseous phase and particle velocity, respectively.

5. Chemistry

The chemistry of gasification and combustion solid carboneous fuel is modeled in three stages, namely: volatilization, solid phase and gaseous phase reactions.

5.1. Devolatilization

The evolution of volatile gases from solid carboneous fuel is accounted for using the single rate devolatilization model. The single rate model similar to Badzioch and Hawksley [12] assumes that the rate of devolatilization is first-order and dependent on the amount of volatiles remaining in the particle.

$$\frac{dm_p}{dt} = -k\left[m_p - (1 - f_{v,0} - f_{w,0})m_{p,0}\right] \qquad (13)$$

where $m_{p,0}$ is the initial particle mass (kg), m_p is the instantaneous particle mass (kg), $f_{v,0}$ is the fraction of volatiles initially present in the particle, and $f_{w,0}$ is the mass fraction of devolatilizing material. The Arrhenius rate constant k (s⁻¹) is given as:

$$k = A\exp(-E/RT) \qquad (14)$$

There exists extensive literature on the kinetics if devolatilization and gasification. A simple case during volatilization, one assumes a single step process where the volatile matter in the coke and biomass undergo thermal decomposition shown in equation (5.3) where α_i is the stoichiometric coefficients for species i. The solid specie (char) undergoes heterogeneous reactions as discussed in Section 5.2.

$$FUEL = \alpha_1 CHAR + \alpha_2 CO + \alpha_3 CO_2 + \alpha_4 CH_4 + \alpha_5 H_2 + \alpha_6 H_2O \qquad (15)$$

5.2. Char Combustion

Surface char combustion is accounted for using the kinetic/diffusion reaction rate model, which assumes that the surface reaction rate is determined either by kinetics or a diffusion rate. The model of Baum & Street [13] is used in which the diffusion rate as follows:

$$R_1 = C_1\frac{[(T_p + T_\infty)/2]^{0.75}}{D_p} \qquad (16)$$

and the kinetic rate

$$R_2 = C_2\exp(-E/R * T_p) \qquad (17)$$

the above are weighted to yield a char combustion rate:

$$\frac{dm_p}{dt} = -\pi D_p^2 P_0\frac{R_1 R_2}{R_1 + R_2} \qquad (18)$$

where P_0 is the partial pressure of oxidant species in the gas surrounding the combusting particle and the kinetic rate R_2 incorporates the effects of chemical reaction on the internal surface of the char particle and pore diffusion. In the model, the particle size is assumed to remain constant at time Δt while the density is allowed to decrease.

5.3. Gaseous Phase Combustion

The combustion chemistry of gaseous phase is modeled using the Mixture Fraction/PDF approach, which simplifies the combustion process into a mixing problem. This model involves the solution of transport equations for one or two conserved scalars (the mixture fractions). Therefore, instead of solving individual species transport equations, the thermo-chemical properties of the fluid are derived from the predicted mixture fraction distribution. Since the Mixture Fraction/PDF model does not require the solution of multiple species transport equations it is more computationally efficient than the species transport model.

The present work assumes the coke-biomass volatiles and the char are treated as a two fuel streams. The equilibrium chemistry model is used, which assumes that the chemistry is fast enough for chemical equilibrium to exist. The mixture fraction is written in terms of the atomic mass fraction as

$$f = \frac{Z_i - Z_{i,ox}}{Z_{i,fuel} - Z_{i,ox}} \qquad (19)$$

where Z_i is the elemental mass fraction for species i. The subscripts ox and $fuel$ denote the values at the oxidizer and fuel stream inlets respectively. The mixture fraction is a conserved scalar and its value at each of the computational control volume is calculated through the solution of the transport equation for the mean (time-averaged) value of f.

$$\frac{\partial}{\partial t}(\rho \overline{f}) + \frac{\partial}{\partial x_i}(\rho u_i \overline{f}) = \frac{\partial}{\partial x_i}\left(\frac{\mu_t}{\sigma_t} \frac{\partial \overline{f}}{\partial x_i}\right) + S_m \qquad (20)$$

The source term, S_m, is due to mass transfer of reacting coal particles into the gas phase. In addition to solving for the Favre mean mixture fractions, Fluent [9] solves a conservation equation for the mixture fraction variance, $\overline{f'^2}$ which is calculated by the following transport equation:

$$\frac{\partial}{\partial t}(\rho \overline{f'^2}) + \frac{\partial}{\partial x_i}(\rho u_i \overline{f'^2}) = \frac{\partial}{\partial x_i}\left(\frac{\mu_t}{\sigma_t} \frac{\partial \overline{f'^2}}{\partial x_i}\right) + C_g \mu_t \left(\frac{\partial \overline{f}}{\partial x_i}\right)^2 - C_d \rho \frac{\varepsilon}{k} \overline{f'^2} \qquad (21)$$

where the constants σ_t, C_g and C_d are 0.85, 2.86 and 2.0, respectively [Fluent User Guide, 2009] [9]. Under the assumption of chemical equilibrium all thermo-chemical scalars (species fractions, temperature and density) are uniquely related to the instantaneous fuel mixture fraction.

$$\phi_i = \phi(f) \qquad (22)$$

where ϕ_i represents the instantaneous species concentration, temperature or density. In a non-adiabatic system, equation (22) is generalized as

$$\phi_i = \phi(f, H) \qquad (23)$$

where H is the instantaneous enthalpy.

The effects of turbulence on combustion chemistry are accounted for by using an assumed shape probability density function approach. The probability density function, $p(f)$, which describes the temporal fluctuations of the mixture fraction, f, in the turbulent flow is used to compute the averaged values of variables that depend on f.

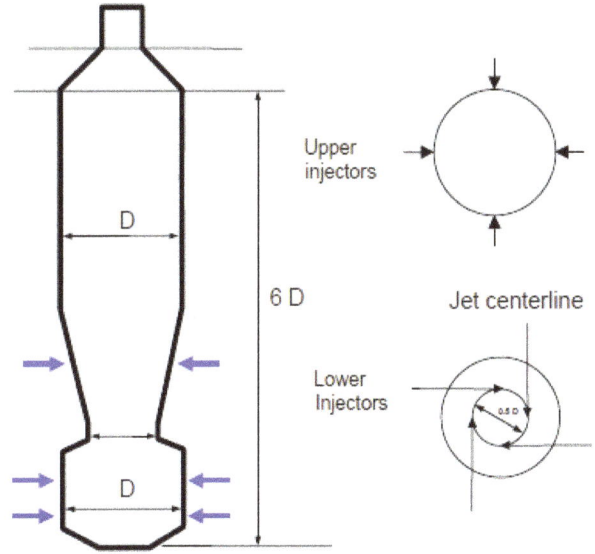

Figure 1a: *Two-stage entrained gasifier showing the three levels slurry and oxygen injection.*

Figure 1b: *Computed particle swirling flow path for tangential firing (this work)*

Table 1: *Fuels Data [8]*

Proximate analysis (dry basis)	Coke	Wheat Straw
Fixed carbon (%wt)	89.3	15.6
Volatiles (%wt)	4.85	79.5
Moisture (%wt)	0.44	7.7
Ash (%wt)	5.4	4.91
LHV (MJ/kg)	32.4	18.5
Proximate analysis (dry basis)		
C	82.7	47.3
H	1.72	5.68
O	1.81	41.6
N	1.75	0.54
S	6.78	< 0.01
Particle Size Distribution		
Minimum diameter (mm)	25	50
Maximum diameter (mm)	200	1000
Mean diameter (mm)	110	451
Spread parameter [-]	4.4	2.31

Table 2: *Reactor geometry and process data [9]*

Reactor Data	Diameter (m)	1.6
	Height (m)	10
	Operating pressure (bar)	18
Process Data	Coke & biomass feed rate (kg/s)	32.274
	Water flow rate (kg/s)	11.188
	Oxidant flow rate (kg/s)	23.128
Oxidant	Oxygen (%)	95
	Nitrogen (%)	5

6. Results and Discussions

Illustrated in Figure 2 are the contours of the concentration of CO (mole fraction) along the height of the two-stage entrained flow gasifier. Results indicate higher concentrations of CO for the biomass substitution in the range of 5% and 10%. This phenomenon may be explained by the increased reactivity of coke due to the presence of biomass which has higher volatiles content. Figure 3 shows similar trend of increased rate of volatilization as the quantity of biomass substitution is increased to about 15%. Note a second peak in Figure 3 corresponds to the location of slurry injection of at the upper level about 4m high as shown in Figure 1a.

Shown in Figure 4 is the rate of char burnout with varying quantities of biomass content. Similar to devolatilization, the char burnout is enhanced by the presence of biomass which is more reactive than coke. Similarly, the second peak corresponds to the location of the upper level injection of slurry. Figure 5 concentration of syngas (CO and H2) at gasifier outlet as the biomass content is varied. Results indicate that the CO and H2 concentrations are similar for biomass substitution up to 20%. Further, the rate of production of syngas is nearly independent of biomass substitution for the cases studied up to 20% as shown in Figure 6.

7. Conclusion

The gasification of a mixture of petroleum coke and biomass in a 3000 ton/day entrained-flow gasifier has been performed. The characteristics of syngas flow rate, gas composition, heating value, and carbon conversion were determined and compared. The average concentrations of syngas (H2, CO) produced from coke gasification were 43% and 17%, respectively. In the case of co-gasification, the maximum concentrations of H2 and CO were 46% and 22%, respectively at 10% biomass blend. These results indicate that the co-gasification of coke-biomass slurry in an entrained-flow gasifier may be an excellent method to efficiently use petroleum coke which has low reactivity.

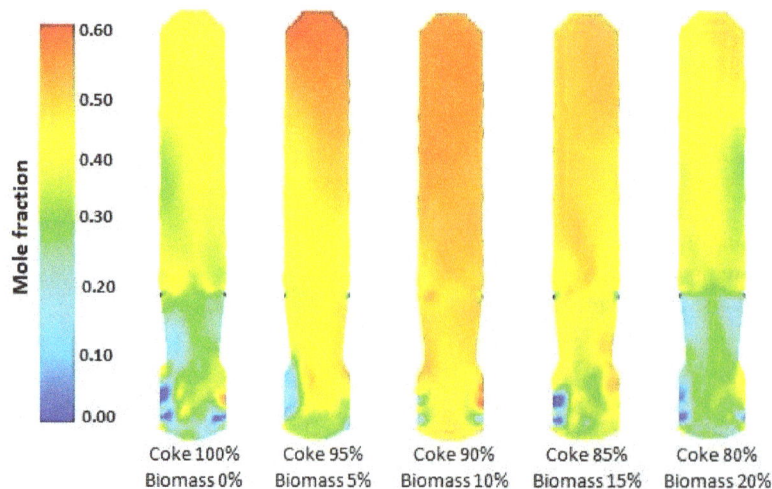

Figure 2: *Concentration of CO along the gasifier height with varying biomass blends*

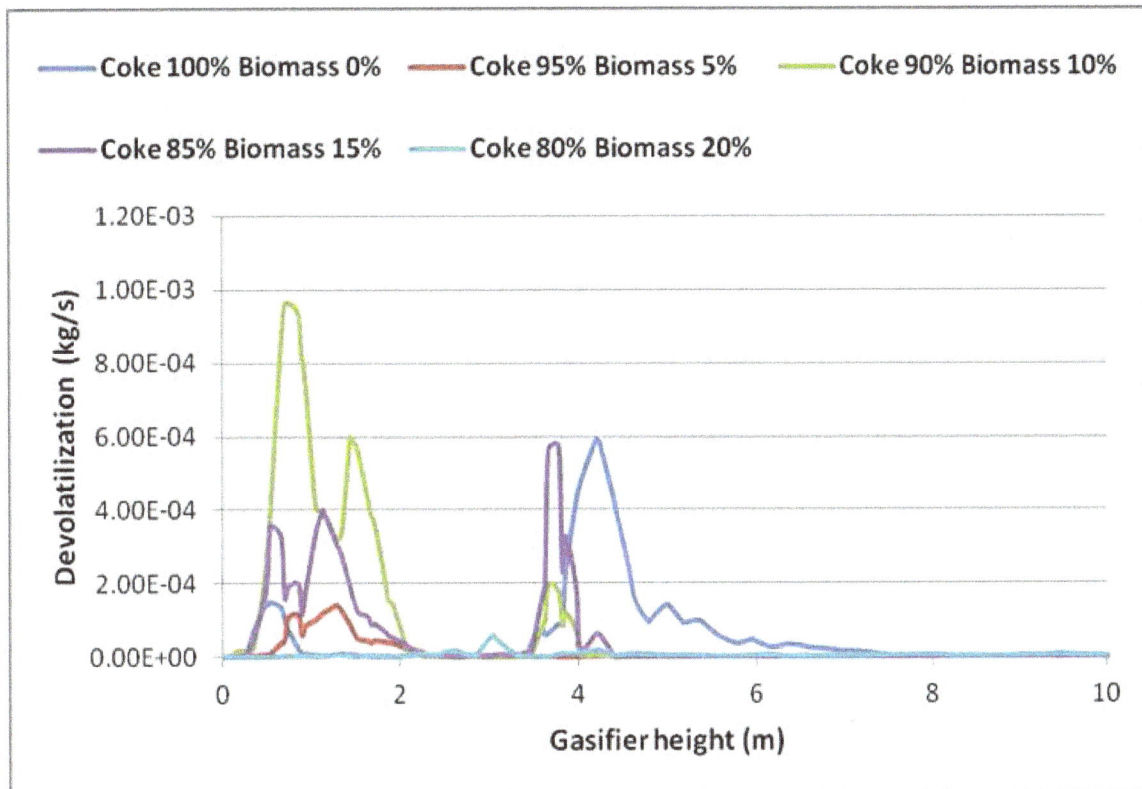

Figure 3: Devolatilization rate along the central axis of the gasifier height

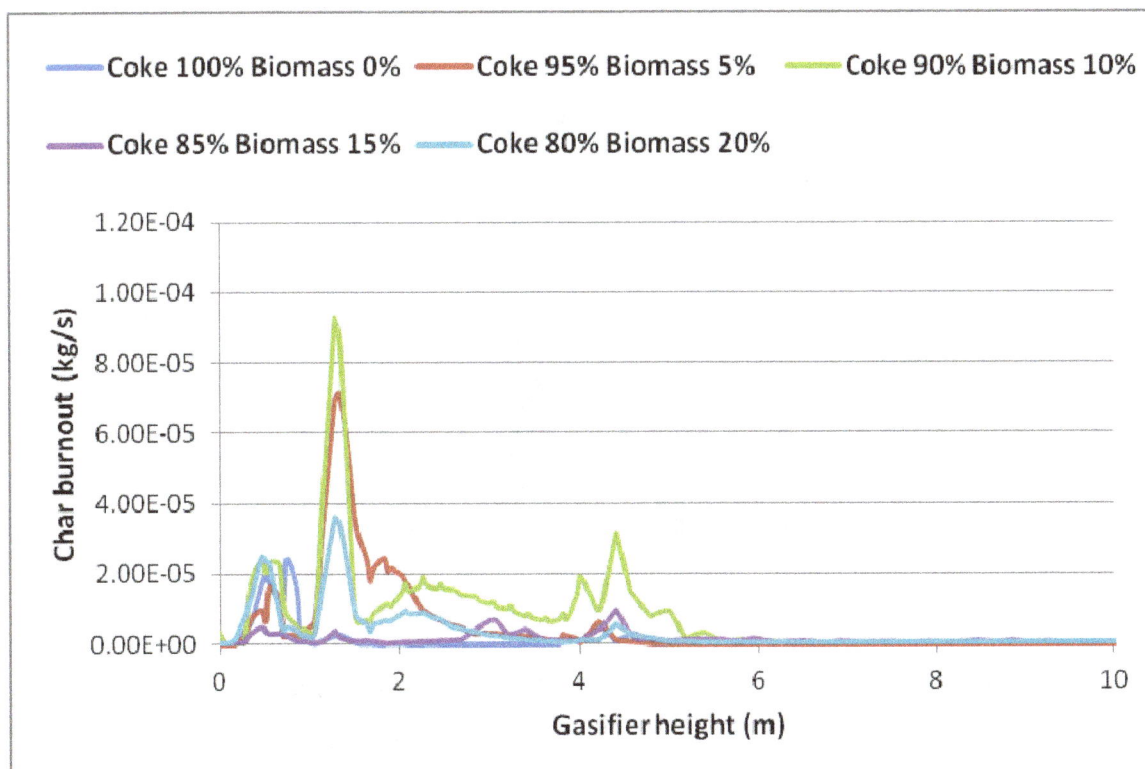

Figure 4: Char burnout rate along the central axis of the gasifier height

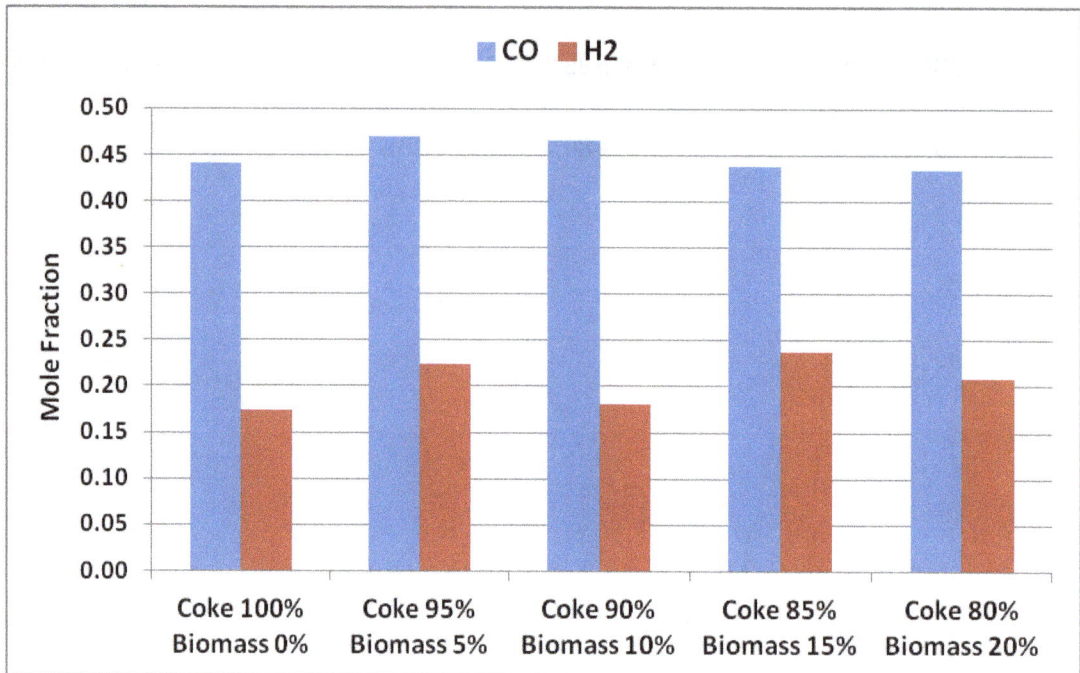

Figure 5: Concentration of syngas (CO and H2) at gasifier outlet

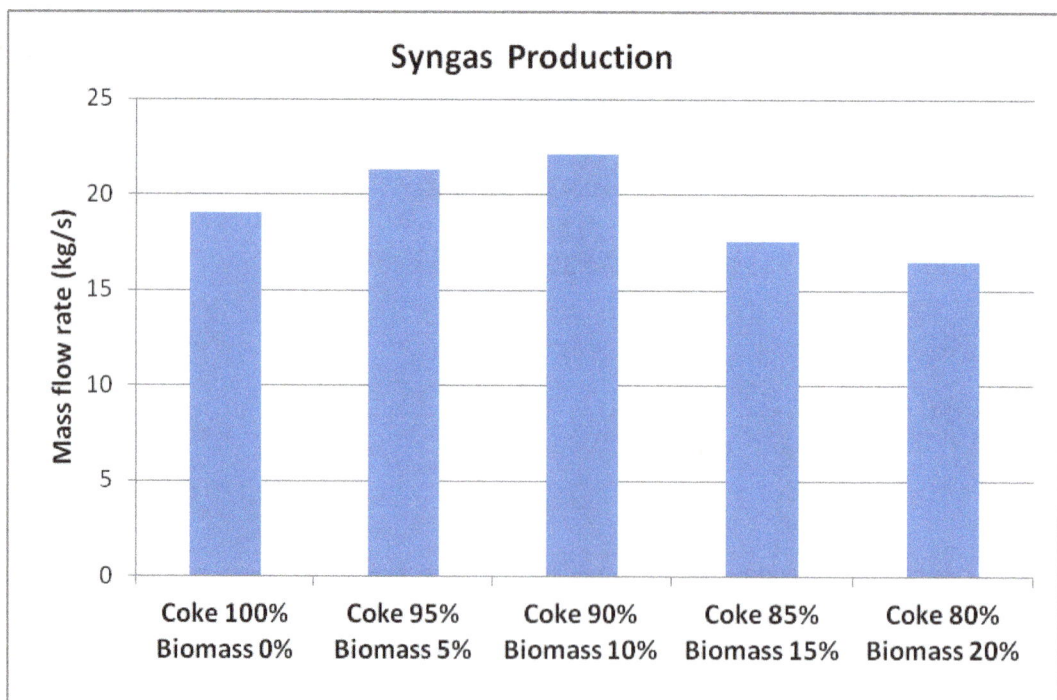

Figure 6: Syngas proaduction rate at the gasifier outlet

References

[1]   Radmanesh, R, J. Chaouki, and C. Guy. Biomass gasification in a bubbling fluidized bed reactor: Experiments and modeling. AIChE Journal, 52(12):4258–4272, 2006.

[2]   Zhou H, G. Flamant, and D. Gauthier. DEM-LES of coal combustion in a bubbling fluidized bed. Part I: Gas-particle turbulent flow structure. Chemical Engineering Science, 59:4193–4203, 2004.

[3]   Rong D and M. Horio. DEM simulation of char combustion in a fluidized bed. In 3rd Int. Conf. on CFD in the Minerals and Process Industries, pages 469–474. CSIRO, Melbourne, Australia, 1999.

[4]   Lathouwers D. and J. Bellan. Modelling of dense gas-solid reactive mixtures applied to biomass pyrolysis in a fluidized bed. Int. J. of Multiphase Flow, 27:2155–2187, 2001.

[5]   Agrawal K, Loezos P. N. , Syamlal, M and S. Sundaresan: The role of meso-scale structures in rapid gas-solid flows. J. of Fluid Mechanics, 445:151–185, 2001.

[6]   Bockelie M, Denison M, Chen Z. Linjewile T, Senior C and A. Sarofim: Modeling of Entrained Flow Gasifiers, Gasification Technology Conference 2002, October 27-29, 2002, San Francisco, CA USA

[7]   Furimsky, E., *Gasification of oil sand coke: Review*. Fuel Processing Technology, 1998. **56**(3): p. 263-290.

[8]   Chunggen Yin, Soren Knudsen, Lasse Rosendahl and Soren Lovmand: Modeling of Pulverized Coal Biomass Co-firing Swirling-Stabilized Burner and Experimental Validation Proceedinds of the International Conference on Power Engineering-09 (ICOPE-9) 16-20, 2009, Kobe, Japan

[9]   Fluent 12 User Guide Copyright @2009 by ANSYS, Inc.

[10] Ferziger, J. and M. Peric. Computational Methods for Fluids Dynamics. Springer Verlag, 1999.

[11] Chung T. J., Computational Fluid Dynamics. Cambridge University Press, 2002.

[12] Badzioch, S., and Hawksley, P. G. W. (1970). Kinetics of Thermal Decomposition of Pulverized Coal Particles, *Ind. Eng. Chem. Process Des. Develop. , 9*(4), 521-530.

[13] Baum, M. M., and Street, P. J. (1971). Predicting the Combustion Behaviour of Coal Particles. *Combustion Science and Technology, 3*(5), 231-243.

21

Traditional uses of ethnobotanical plants for construction of the Hut and hamlets in the Sitamata Wildlife Sanctuary of Rajasthan, India

Kanhaiya Lal Meena, Vimala Dhaka, Prakash Chandra Ahir

Department of Botany, M.L.V. Government College, Bhilwara - 311001, Rajasthan, India

Email address:

kanhaiyameena211@yahoo.com (K. Lal Meena)

Abstract: An extensive survey of the Sitamata Wildlife Sanctuaryof Rajasthan has been made to document the information about ethnobotanical plants being used by them to construct Hut and hamlets. 31 plant species of angiosperms have been recorded along withtheirplant parts used toconstruct of various types of Hut and hamlets in the sanctuary.

Keywords: Ethnobotanical Plants, Huts And Hamlets, Sitamata Wildlife Sanctuary, Rajasthan, India

1. Introduction and Review

The Sitamata wildlife sanctuary, is one of the pride owner of most unique ecosystem with first richest biodiversity in Flora & fauna in Southern Rajasthan. It is one and only of the important natural habitats for flying squirrel in India. As the name itself explains the sanctuary is also associated with mythological events, it is believed that Devi Sita (wife of Lord Ram) stayed here during the period of her exile in the ashram of Rishi Valmiki. The Valmiki ashram was situated in the sanctuary, thus sanctuary bears the name of Davy Sitamata and her famous temple is situated in the heart in the forest area. It is spread over the Aravalli and Vindhyanchal mountain ranges and in this forest teak trees of timber value are abundantly present.

The Sitamata wildlife sanctuary is situated in between 74°25' - 74°40' E longitudes and 24^004' - 24^023' N latitude. It is situated in the south region of the Rajasthan in ChittorGarh, PratapGarh and Udaipur districts and is extended in tehsils Barisadari, Chhoti sadari, Dhariya wad and PratapGarh. The sanctuary covers an area about 422.95 Sq. Km in which the total reserved forest area is 345. 37 sq. The km and the protected forest area are 77.57 sq. Km (Fig. 1).

The Sitamata wildlife sanctuary of Rajasthan, located at the trijunction of Aravalli & Vindhyan Hill Ranges as well as Malwa Plateau, which harbors its unique and diverse biodiversity. It is important mainly because it forms the northwestern limit of Teak-bamboo forests and the fauna occurring therein. It is exceptional for diversity and the interspersion of habitats, which includes an area of teak stands, wetlands, perennial streams, gentle undulating mountains, natural deep gorges and fine grooves of mixed woodlands.

The network of rivers (Jakham, the Karmoi and the Sitamata) and accompanied riparian vegetation is main characteristic of this sanctuary. All this has resulted in diverse micro and macro habitats that are home to quite a few conservation significant floral species like *Acacia catechu* (L. f.) Willd., *Anogeissus latifolia* (Roxb. ex DC.) Wall. ex Guill. & Perr., *Boswellia serrata* Roxb. ex Cocls., *Buchanania lanzan* Spreng., *Celastrus paniculata* Willd., *Chlorophytum tuberosum* (Roxb.) Baker, *Dendrocalamus strictus* (Roxb.) Nees, *Ficus benghalensis* L., *Lagascea mollis* Cav., *Lannea coromandelica* (Houtt.) Merrill., *Madhuca indica* J. F. Gmelin, *Sterculia urnes* Roxb. and *Tectonia grandis* L. f. are major tree species *viz.* Starred Tortoise, Marsh crocodile or Mugger, Long-bill Vulture, White-rump Vulture, Scavenger Vulture, Pangolin, Ratel, Four horned antelope and Leopard. The forest is interspersed with about 30 villages and their agriculture field that creates a typical mosaic. The agricultural activities coupled with the heavy biotic pressure on domestic livestock, illicit cutting of wood, timber and bamboo and other Minor Forest Produces collection including encroachments, both inside and the periphery exerts enormous pressure on the vegetation are reported.

Further, though it is endowed with rich natural resources, it is affected by natural calamities and hazards like drought, fire, flood and storm, with drought being a common phenomenon. Various aspects of the study area were mostly extracted from the management plan.

Prior to the publication of Hooker's flora (1872-97), local flora and lists were available for several regions of India but nothing was known in Rajasthan. The lacuna was, however, very amply filled during the middle part of the century and large number of technical and semi technical

papers were published.

The work on the flora and present day Botany has been pioneered by Linnaeus with his classic publication Genera plantarum (1737). In India the work on floral exploration had been initiated by Roxburgh (1820-1824) who published "Flora Indica". Later eminent botanists like Hooker and Thompson (1855) published an introductory essay to the flora India and the publication of "Flora of British India" (1872-1897).

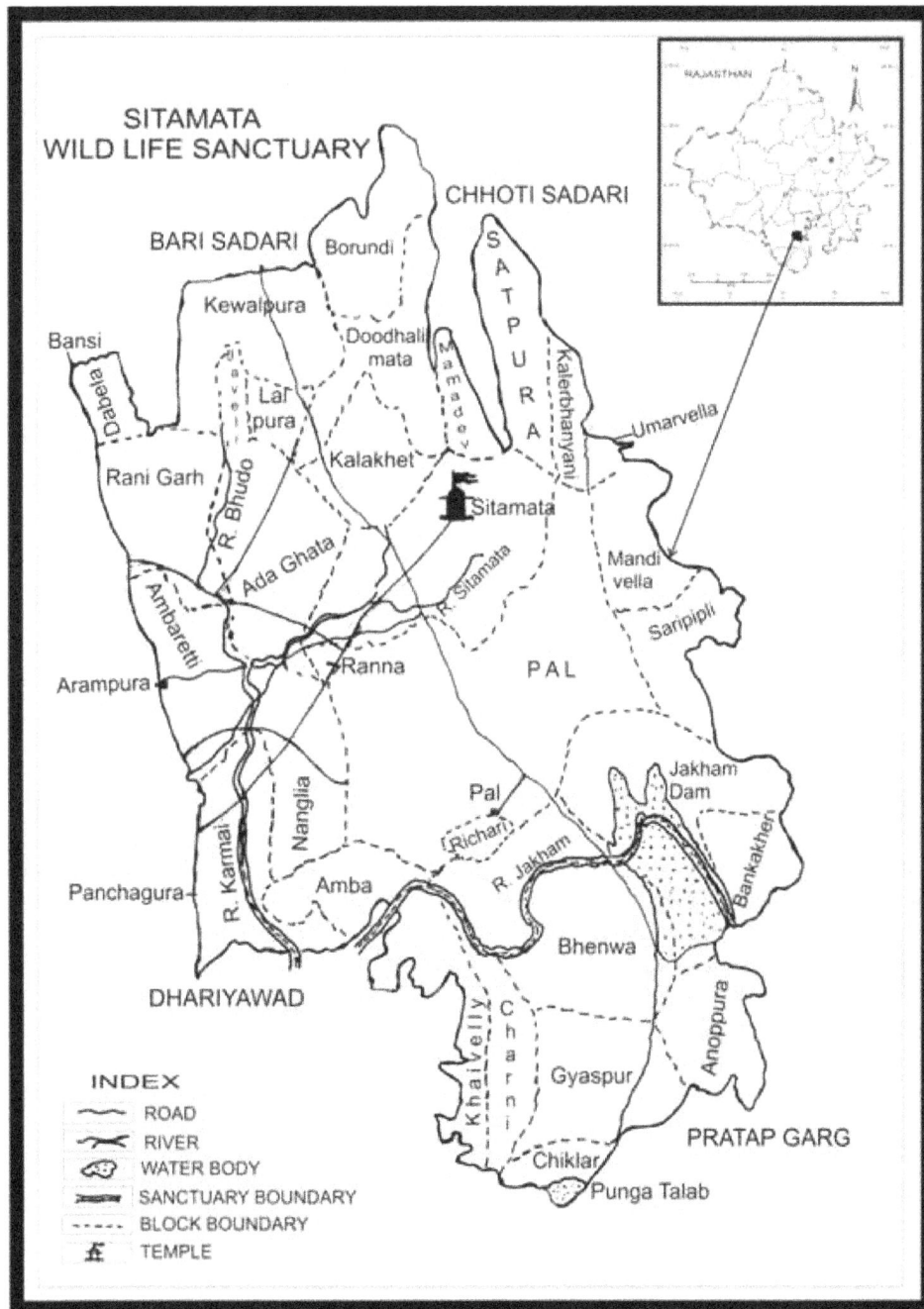

Fig. 1. The Sitamata Wildlife Sanctuary.

In as many as 60 papers, the nomenclature, notes on additions to the flora of Rajasthan or new records for India from this State and notes on extended distribution of

various taxa have been reported. Out of 42 publications on floristics, the grasses were dealt in 7, sedges in 3 and both the grasses and sedges in 2 papers, contribution on halophytes in one, while the hydrophytes or marshland plants in 5 and weed flora of various places have been reported in 2 papers. The aspects of economic and harmful plants and the plant introduction, afforestation etc. have been covered in more than 50 papers (Jain, 1970). Shringi (1981) has enumerated the grasses of Jhalawar district. Account of medicinal plants from tribal area of Rajasthan has been provided by Sebastian & Bhandari (1984a, 84b, 1988), Vyas (1987), Rajawat (1990), Sharma (1991), Singh & Pandey (1998), Gupta & Jiyalal (1997), Katewa & Arora (1997), Katewa & Guria (1997), Katewa & Sharma (1998), Sharma & Asawa (1999), Sharma (2002), Deora *et al.* (2002), Rathore (2002), Trivedi (2002), Jain *et al.* (2004, 2005b), Joshi, 1995, Katewa *et al.* (2003), Katewa & Galav (2005), Katewa & Jain (2006), Meena & Yadav (2006, 2007, 2008, 2009, 2010a, 2010b, 2010c, 2010d), Meena(2011, 2012), Meena *et al.* (2013), Ahir *et al.* (2012), Meena 2013 added further the ethnobotanical work from Rajasthan.

2. Material and Methods

Ethnobotanical survey was conducted in the Sitamata Wildlife Sanctuary, Rajasthan, India. Field trips conducted with by local people. Generally tribals, who know about the Sanctuary villages, tribes generally do not want to give all the information about plants because they believe that when the plant is disclosed its properties will be lost. For this reason the information collected from the tribes is an important aspect of ethnobotanical study. The peoples who can provide information about plants were consulted and includes experience men and women, elders, birth attendants, bhopa, woodcutters, shepherds and headmen of the community. For authenticity about medicinal properties of plants the information collected during fieldwork were verified at different places through different informants and in different seasons.

Each of the plant species recorded have been collected with the help of the informants and photographs were also taken. The species were identified with the help of reputed flora of India Series- 2 (Flora of Rajasthan volume 1-3,Shettey & Singh (1987-1993) Botanical Survey of India) and recently published Flora of South Central Rajasthan by Yadav & Meena (2011). The voucher specimen was deposited in the Herbarium of Department of Botany, MLV Government College, Bhilwara.

3. Observations

Cultural exuberance of the tribes of the Sitamata wildlife sanctuary is rightly being depicted in several of its elements. The tribes believe in their own god "*Bheru bawji*" as well as "*Mataji*" and they prey with them for all purposes. They construct small houses or huts, with one room and veranda made of mud and wood like *Anogeissus latifolia* (Roxb. ex DC.) Wall. ex Guill. Perr., *Terminalia catappa* L., *Nyctanthes arbor-tristis,Bambusa* sp. etc. and are built on

the slopes of the hills with their agriculture fields. Maize is the staple food grown by all tribalfamilies. The tribal family honors their guest by a dish prepared from goat or hen. Different duties are allotted for both male and female. The men do the physical labor such as ploughing, harvesting, building the houses as well as hunting. The female carries domestic duties including cooking, nursing and milking the animals and also taking care of the children.

Marriage too like any other tribal communities is held in high position. The tribal communities have permitted freedom in selecting their partners. Young males between the ages of 10 -24 generally marry females who are between 14 - 18 years.

Dresses of tribes are quite exquisite. Both the male and females of tribal communities developed an individual style of dressing. Several silver ornaments are in fashion. The woman usually wears black, blue or red blouses with huge petticoats. Men are noted for their white or red turbans.

4. Huts and Hamlets

Livings in hamlets that comprise huts scattered sparsely constitute a village. The number of huts in a village may range from a small one like ten-twelve spread over two or more square kilometers. When large the village is called a "*Kheri*", Dhani is usually made of several huts and tribals of different casts inhabit in different streets. In same localities, however, the tribes live in clusters of several shelters, on plain ground at the foot of the hills. Such shelters denote the accepted authority of one i.e. the father over the sons or the eldest brother over the younger.

Rare still are clustered double storage hutments called "*medi*". Compact shelter-partners are also seen in villages with a mixed population of tribal and non tribal or in the outskirts of small towns. The typical tribal village is situated between the forest or adjacent to it, near source of natural water supply, whether a river or where water can be obtained through wells or baoris (Step wells)

A tribal generally constructs his hut, close to his agricultural field and often in the field (Plate 1 A). Usually, a tribal's field lies on the slopes below his hut on a hillock. Rarely, affluent tribes, may own two huts, one in the field if it is far away. Besides the huts and fields each village has sacred spots with deities houses in constructing*devras* or in constructing sites sheltered by trees like *Acacia leucophloea, Ficus religiosa* or *Ficus benghalensis.* The *devras* are also called as "thanak" Sacred spots may be in the center of the village or on the outskirts (Plate 1 B).

A large *Azadirachta indica* or *F. Religious* or *F. benghalensis* tree or a grove forms the community gathering place. Meetings are also held at the *Devras* or in the headman's hut. In front of the headman's hut there is an another open house which serves the purpose of entertaining and lodging guests of the village. Such place is called as *"Pol"*. The pol is an open structure with a rectangular roof of timber supported by 3-5 long pillars lenters and covered with a thick mat of *Butea monosperma* leaves & straw of *Triticum aestivum* or with sticks of *Acacia nilotica* and strips of *Bambusa,* sometimes it is covered with Kelu (kelu is made by soils). It is at the pol that a guest is lodged or the folks assemble for their night palaver or discussion.

The tribal houses are essentially rectangular in construction with the roof sloping down from a common point beyond the upper ends of the two long walls of the huts. These two longer walls are called "Chanda". The roof style prevents overheating by direct scorching sunrays in summers and facilitates the torrential water drops to speedily flow down in rains. A house usually has only one room (sometimes more) separated in 3-4 portions, a corner forms the kitchen, an adjacent one the *"Dormitory"*. While the cattle & the pet animals are kept on the outer side of the hut (Plate 1 C). The single entrance is usually on one side of a long wall, though an entrance through the smaller wall is also not uncommon. The house almost never has windows. The walls and the ground are plastered with clay mixed with cow dung or straw of *Triticum* or *Hordium* spp.

The cattle may also be tied at the side of the hut inside a fenced enclosure or a semi open construction. Within the hut compound there may also be a raised platform called Dhariya or some times dagla.

A typical tribal house consists of :

i.   The front wall (L.N. Barnewali Bheetri)
ii.   The back wall (L.N. Pachhewali Bheetri)
iii. The side wall (Chanda)
iv.   The roof (Tapri)
v.   The gate

The four walls of a rectangular house are made up of mud and leaves of *Saccharum officinarum.* Now a days due to low rainfall, the *Saccharum* is not available and therefore, straw of *Triticum aestivum* is commonly used. On the walls, the roof is made using *Phoenix sylvestris* or *Acacia catechu* stem as girder and leaves of *phoenix sylvestris* to cover it. On two small front and back walls a lenter is applied which is known as Myar, on the Myar three short pillars are used known as *Mulvari*, the *mulvari* are supporting polls for three lenters, girders applied on side walls. These long lenters are called *Khankwari*. On khankwari transverse sticks of wood of *Acacia nilotica* or *Anogeissus pendula* are used but recently the bamboo is commonly used. After this the horizontal longitudinal strips of bamboo are applied and then this is covered by kelu.

The gate is made up of wood of *Acacia nilotica.* Sometimes a rectangular screen made up of frame of bamboo filled with *Phoenix sylvestrics* leaves. This screen is tied along one end to a pole to keep it moving for opening and closing.

Mats prepared from *Phoenix sylvestris* leaflets are used to sitting in gatherings. A symbol of God Ganesha locally known as Mandpo prepared from wood of *Magifera indica* is used at the gate of the hut during marriage. A basket made of Bamboo is generally used as a trap for chickens as well as for fishing.

Slight variations in selection of timber observed in the neighboring villages are enumerated.

Beams : (Myar, Adia) *Phoenix sylvestris, Acacia nilotica.*
Pillars : (Khankwari) *Phoenix sylvestris, Acacia nilotica.*
Poles : *Anogeissus pendula*
Gates : *Acacia nilotica, Azadirachta indica.*

Uses of plants with their different parts in the construction of huts and hutments are given in table 1.

***Table 1.** Plants and plant materials used for making huts.*

Sr. No.	Name of species	Plant parts used	Uses
1.	*Acacia nilotica* (L.) Willd. ex Del. ssp. *indica* (Benth.) Brenan	Stem & branches	Framework of roof, windows and doors
2.	*Acacia catechu* (L. f.) Willd.	Stem	Beams
3.	*Acacia leucophloea* (Roxb.) Willd.	Stem	Beams
4.	*Acacia senegal* (L.) Willd.	Stem	Supporting pillars
6.	*Anogeissus latifolia* (Roxb. ex DC.)Wall. ex Guill. & Perr.	Stem and tender branches	Framework of roof, windows and doors
8.	*Azadirachata indica* A. Juss.	Stem and branches	Framework of roof, windows and doors
9.	*Bambusa arundinacea* (Retz.) Roxb.	Culms	Framework of roof, Thatching of hut walls.
10.	*Boswellia serrata* Roxb. ex Coleb.	Stem	Pillars and doors
11.	*Butea monosperma* (Lam.) Taub.	Stem, branches and leaves	Doors, pillars and framework of roof
12.	*Cajanus cajan* (L.) Mill.	Stem	Roof
13.	*Calotropis procera* (Ait.) Ait. f. ssp. *hamiltonii* (Wight) Ali	Stem	Framework of roof
14.	*Capparis decidua* (Forssk.)Edgew.	Stem	Framework of roof

Sr. No.	Name of species	Plant parts used	Uses
16.	*Diospyros melanoxylon* Roxb.	Stem	Framework of roof, windows and doors
17.	*Gossypium herbaceum* L.	Stem	Doors and walls
18.	*Madhuca indica* J. F. Gmelin	Stem	Doors and beams
20.	*Phoenix sylvestris* (L.) Roxb.	Stem and leaves	Beams, walls, gates and roof
23.	*Saccharum bengalense* Retz.	Culms and leaves	Thatching of walls and roof.
24.	*Tectona grandis* L.	Stem & Leaves	Beams & roof
25.	*Terminalia bellirica* (Gaertn.) Roxb.	Stem	Beams
26.	*Terminalia crenulata* Roth	Stem	Beams
27.	*Triticum aestivum* L.	Culms and leaves	Thatching of walls and roof.
28.	*Typha angustata* Bory & Chaub.	Culms and leaves	Walls and roof
29.	*Zey mays* L.	Culms and leaves	Walls and roof
30.	*Ziziphus mauritiana* Lam.	Stems and branches	Roof
31.	*Ziziphus nummularia* (Burm. f.) Wight & Arn.	Stems and branches	Framework of walls.

5. Local Terms Related to Huts & Shelters

Shelter : Asro
Hut : Jhunpari, tapri
Raw materials for making a hut : Jugar.
Rising walls on both sides of huts : Chando
Wall behind hut : Paseet
Wall : Bheet
Foundation : Anchhot
Small room : Orri
Roof : Chhappar
Plastered ground of hut : Lippen

Balcony : Mundari
Crosswise pole : Danda
Pillars : Thambo
Main vertical pole : Thambo
Gate constructed for entry into hut compound as well as field : Tati
The knob for closing door : Kari
Latch : Hakri, sankli
Window : Kiwari
Plastering (by clay / cow dung) : Lippen
A place meant for cattle resting : Dharyo, Chhapari
Sites for tying cattle (Thann) : Barro
Watch place (Machaan): Daglo, dagro (Plate 1D)

Plate 1A: A typical Hut of tribes surrounded by Agricultural field.

Plate 1B: Thanak of Tribes in Sitamata Wildlife Sanctuary.

Plate C: Huts for cattle and pet animals.

Plate 1D: Watch place (Machaan): Daglo, dagro

6. Discussion

The observations recorded during the present investigations of the flora of the Sitamata wildlife sanctuary may now be discussed in the light of researches carried out by earlier workers in this context.

The remarkable feature of Rajasthan is the Aravalli range, perhaps the oldest folded mountain range in the world. It intersects Rajasthan from end to end, diagonally running from Delhi to the plains of Gujarat for a distance of about 692 km. Within Rajasthan the range runs from Khetri in the northeast to Khed Brahma in the south-west for a length of about 550 km. The elevation of the Aravalli range gradually rises in south-west direction, as it is 335 m at Delhi and in Rajasthan 792 m at Khetri, 913 m at Harshnath, 920 m at Kho, 1055 m at Raghunathgarh, 1100 m at Bijapur and 1727 m at Mt. Abu. Further, south-west wards, the elevation gradually decreases to the plains of Gujarat.

The loftiest and the most clearly defined section of the Aravalli is in between Mt. Abu and Ajmer where the range stands like a barricade. Beyond Ajmer to the northeast, there are gaps in the Aravalli range near Sambhar, east of Sikar etc. Structurally, it is composed of rocks belonging originally to the Delhi system, folded in a Synclinorium occupying the site of geosyncline which has been deeply eroded. Aravalli range divides the whole of Rajasthan into two natural divisions i.e. three fifth lying on north-west and two fifth on the east and southeast.

The present status of forest everywhere is a matter of deep concern as they are gradually declining and disappearing from the countryside. Their presence in agricultural lands, grazing, fragmentation of the grove-owning families, erosion of cultural & religious beliefs, introduction of *Lantana camara* L., *Parthenium hysteriophorus* L., *Spigelia anthelmia* and taboos are the major reasons. In view of this, and due to failure of pure legal protective measures in guaranteeing conservation, it has become imperative to search for alternative solutions based on indigenous knowledge of the people. The tribes of this region conserve medicinal plants and the forest patches rich in biodiversity and play an important role in their sustainable manner with their socioeconomic and religious practice with the belief in nature worship inherited from generation after generations. Tribe believes that if the habitat of the medicinal plant species is protected theses species will be multiplied without any conservation practices.

The forest represents a long tradition of environmental conservation by the tribes of this Santuary. Therefore, there is an urgent need not only to protect forest, but also to revive and reinvent such traditional practices of nature conservation and environmental management. Initially only two species namely *Commiphora wightii* (Arn.) Bhandari and *Rosa involucrata* Roxb. were included in the threatened species from state by BSI but now this list has been increased and including the number of species in this category. The species like *Anogeissus sericea* Brandis var *nummularia* King ex Duthie *Ceropegia vincaefolia* Hook. *emend.* Ansari and *Chlorophytum borivilianum* Sant. *et* Fernand is rare and included in the red data book of Indian plants and some species like *Citrullus colocynthis* (L.) Schrad., *C. wightii* (Arn.) Bhandari and *Tecomella undullata* (Sm.) Seem, are reported as threatened species(Meena & Yadav, 2006, Meena, 2012). The gum of *C. wightii* (Arn.) Bhandari hasan importance in international trade and it appears to be being extracted at unsustainable rates, causing declines, so presently it is included in IUCN Red List of Threatened species. In Rajasthan these species are widely distributed in the western part of the state, but here it is also reported as rare and threatened species in this district. Among the present species many are facing various threats in this region.

The danger of extinction on such species are ahead, therefore it is necessary to ensure the survival of germplasm by their protection, conservation, multiplication and maximum afforestation of such medicinal and economically important plant species. Because of limited resources of this tribe it is essential for a biotechnologist to come forward for *in-situ* conservation through tissue culture, establishment of botanical gardens or *ex-situ* conservation by way of protecting the forest region of the state. The political as well as involvement of NGOs may play an important role in the protection of this valuable area of the state.

Acknowledgement

Authors are highly thankful to all the tribal informants for their cooperation and help during ethnobotanical study. Help rendered by Shree Jetha Ram Ji, Laxman Ram Ji, Limba Ram Ji, Modan Ji, Nana Ram Ji, Ram Chandra Ji, Salira Ram Ji and Vikram Ji by way of providing traditional knowledge is thankfully acknowledged. Thanks are also due to Shri Bhanwar Singh Ji Meena, Additional Superintendent of Police, district Sirohi for his cooperation during these studies. Authors are thankful Principal, Vice Principals for providing the facilities. Financial assistance provided by UGC Bhopal is gratefully acknowledged.

References

[1]　P.C. Ahir, S. Hussain& Dhaka, V. 2012.Some Traditional Ethnoveterinary Medicinalplants of Kumbhalgarh wildlife sanctuary, Rajasthan, India. In: Meena, K. L. (Ed.) *Proc. Nat. Conf. Biod. Cons: Caus. Cons. andSol.* MLV GC Bhlwara pp - 207-211.

[2]　Deora, G.S., Singh, G.P. & Jhala, G.P. 2002. Ethno-medico-botanical diversity of Kotra, Udaipur. In (Eds.) Biodiversity : Strategies for conservation. *A.P.H. Publishing Corporation,* New Delhi. 317-329.

[3]　Duthie, J.F. 1903-1929. *Flora of the Upper Gangetic Plains and of the adjascettt Siwalik and Sub - Himalayan tract.* Calcutta.

[4]　Gupta, R.S. & Jiyalal. 1997. Traditional herbal remedies. In Devendra Sharma (Ed.) *Compendium on phytomedicine,* Council for Development of Rural Area. New Delhi. 375-379.

[5]　Jain, A., Katewa, S.S., Choudhary, B.L. & Galav, P.K. 2004. Folk Herbal medicines used in birth control and Sexual diseases by tribals of Southern Rajasthan, India. *J. Ethnopharmacology.* 90(1):171-177.

[6]　Jain, A., Katewa, S.S., Galav, P.K. & Sharma, P. 2005b. Medicinal plant diversity from the Sitamata wild life sanctuary. Chittorgarh district India. *J. Ethnopharmacology* 102 (3): 543-557.

[7]　Katewa, S. S. & Arora, A. 1997. Some plants of folk medicine of Udaipur district, Rajasthan. Ethnobotany. 9 : 48 - 51.

[8]　Katewa, S.S. & Galav, P.K. 2005. Traditional Folk Herbal Medicines from Shekhawati Region of Rajasthan. *Indian Journal of Traditional Knowledge* 4 (3): 237-245.

[9]　Katewa, S.S. & Guria, B.D. 1997. Ethnomedicinal observations on certain wild plants from Southern Aravalli hills of Rajasthan : *Vasundhara.* 2 : 85-88.

[10]　Katewa, S.S. & Jain, A. 2006. *Traditional Folk Herbal Medicines.* Apex Publishing House, Udaipur.

[11]　Katewa, S.S. & Sharma, R. 1998. Ethnomedicinal observation from certain watershed areas of Rajasthan. *Ethnobotany* 10 : 46-49.

[12]　Katewa, S.S., Chaudhary, B.L., Jain, A. & Galav, P.K. 2003. Traditional uses of plant biodiversity from Aravalli hills of Rajasthan. *Indian Journal of Traditional Knowledge* 2 (I) : 27-39.

[13]　Meena K. L. 2010. *Morinda coreia* Buch.-Ham.: A new record to the flora of Rajasthan. *J. Indian. Bot. Soc.* Vol. 89. (1-2): 210 - 212.

[14]　Meena, K. L. & Yadav, B. L. 2006. Some important medicinal plants of Bhilwara district Rajasthan, India. Proceedings National Conference on Biodiversity Conservation. MLV Govt. College, Bhilwara. pp. 84 - 89.

[15]　Meena, K. L. & Yadav, B. L. 2007. Some ethnomedicinal plants of Rajasthan. In: P. C. Trivedi (Ed.) Ethnomedicinal Plants of India. pp. 33-44. Aavishkar publishes Distributors, Jaipur.

[16]　Meena, K. L. & Yadav, B. L. 2008. Floral resources of Rajasthan with special reference to Sitamata wildlife sanctuary. Geographical aspects. Proceedings of the 35th National conference of Rajasthan geography Association. MLV Government College, Bhilwara. Vol. IX : 56 - 65.

[17]　Meena, K.L. & Yadav, B.L. 2006. Some important medicinal plants of Bhilwara District Rajathan, India. *Proc. Nat. Conf. Biod. cons.* MLV GC Bhlwara PP - 84-89.

[18]　Meena, K.L. & Yadav, B.L. 2007. Some ethnomedicinal plants of Rajasthan. In P. C. Trivedi (Edt.) *Ethnomedicinal plants of India.* pp.33-44. Aavishkar publishers distributors, Jaipur (India).

[19]　Meena, K.L. & Yadav, B.L. 2010a. Some Ethnomedicinal Plants of Southern Rajasthan. *Indian Journal of traditional knowledge.* Vol. 9 (1): 169 - 172.

[20]　Meena, K.L. & Yadav, B.L. 2010b. Some Traditional Ethnomedicinal Plants of Southern Rajasthan. *Indian Journal of traditional knowledge.* Vol. 9 (3): 471 - 474.

[21]　Meena, K.L. & Yadav, B.L. 2010c. *Spigelia anthelmia* L. (Spigeliaceae) : A New Generic Record to the Flora of Rajasthan. *J. Indian. Bot. Soc.* Vol. 89. (3&4): 258 - 261.

[22]　Meena, K.L. 2011. Ethnobotany Of Garasia Tribe, Rajasthan, India. LAP LAMBERT Academic Publishing, 66123 Saarbrücken Germany.

[23]　Meena, K. L. 2013.Flora of Wildlife Sanctuary. Discovery Publishing House Pvt. Ltd. New Delhi.

[24]　Meena, K. L., Ahir, P.C. and Dhaka, V. 2013.Ethnomedicinal survey of medicinal plants for sexual debility and birth control by tribals of Southern Rajasthan, India. The Journal of Ethnobiology and Traditional Medicine. Photon 118: 238-244.

[25]　Mena, K.L. 2012.Angiospermic Diversity of District Bhilwara from Rajasthan, India. *The Journal of Biodiversity.* Photon 112: 193-204

[26]　Rajawat, K.S. 1990. *Traditional system of tribal medicine and medicinal herbs.* Project Report, M.L.V. Tribal Research Institute. Udaipur.

[27] Rathore, M.S. 2002. Studies on traditional uses of wild plants by tribes in Kotra region, Udaipur, Rajasthan, Ph.D. Thesis, M.L. Sukhadia University, Udaipur.

[28] Roxburgh, W. 1820-1824. *Flora Indica* (Edited by W. Carey & N. Wallich). Serampore. 1(1820); 2(1824).

[29] Sebastian, M. K. & Bhandari, M.M. 1984a. Magico-religious beliefs about plants among the Bhils of Udaiour district of Rajasthan. *Folklore* 4:77-80.

[30] Sebastian, M. K. & Bhandari, M. M. 1984b. Medicoethnobotany of Mt. Abu, Rajasthan. J. Ethnopharmacol. 12(2) : 223 - 230.

[31] Sebastian, M. K. & Bhandari, M. M. 1988. Medicinal plant lore of Udaipur district, Rajasthan. Bull. Med. Ethnobot. Res. 5(3-4): 133 - 134.

[32] Shetty, B. V. & Singh, V. (Edits.) 1987. *Flora of Rajasthan.* Vol. I. BSI, Howrah.

[33] Shetty, B. V. & Singh, V. (Edits.) 1991. *Flora of Rajasthan.* Vol. II. BSI, Howrah.

[34] Shetty, B.V. & Singh, V. (Edits.) 1993. *Flora of Rajasthan.* Vol. III. BSI, Howrah.

[35] Shringi, O.P. 1981. Botany of Jhalawar District, Rajasthan-1. Grasses. *J. Econ. Tax. Bot.*2 : 85-105.

[36] Singh, V. & Pandey, R.P. 1998. Ethnobotany of Rajasthan, India. *Scientific Publishers* (India), Jodhpur.

[37] Trivedi, P.C. 2002. *Ethnobotany*. Aavishkar Publishers, Distributors, Jaipur India.

[38] Vyas, M.S. 1987. *Ferns and Fern allies of Rajasthan : Ethnobotany and Biochemical Analysis.* Ph.D. Thesis. University of Jodhpur, Jodhpur (India).

[39] Yadav, B. L. & Meena, K. L. 2011. Flora of South Central Rajasthan. Scientific Publishers Jodhpur.

22

Natural resources as a factor socioeconomic development

Goran Rajović[1], Jelisavka Bulatović[2]

[1]Street Vojvode Stepe 252, Belgrade, Serbia
[2]College of Textile Design, Technology and Management Street Starine Novaka 20, Belgrade, Serbia

Email address:

dkgoran.rajovic@gmail.com(G. Rajović), jelisavka.bulatovic@gmail.com(J. Bulatović)

Abstract: The paper discusses the natural resources of the northeastern of Montenegro as a factor of socio-economic development. Based on the properties of relief, we have selected three relatively homogeneous regions for economic development. Area sub region alluvial plains of rivers, river terraces, lake sediments Berane, Andrijevica and Polimlja Valley has the most favorable conditions for intensive agricultural production, summer tourism, construction and transport development. Diversity and complementarily of water resources is the main characteristic of the considered region, which is of particular importance for the future economic development. In the water resources, we looked at the possibilities of exploiting hydropower, water supply of population and industry, agriculture water supply. Thanks to the geological structure of the valley Berane there are significant reserves of brown coal and lignite (total reserves are 176.231.197 tons). Program development and production of coal in the valley Berane would cause intense regrouping and integration of industrial enterprises and caused the need for capacity expansion (Beran Village, Dolac). In the region, appear to ore metals: lead, zinc, copper, iron and pyrite. From non-metallic mineral resources, occupy their presence and reservoir construction materials: gravel, sand and decorative stone. Agricultural land is an important part of the natural wealth of the region. In the period 1964-2005 in the agricultural land there was a change in the way of exploitation to reduce the area under fields and gardens and pastures and increased the area under orchards and meadows.

Keywords: Northeastern Montenegro, Morphological Units, Water Resources, Mineral Resources, Agricultural Lands, Forests

1. Introduction

Northeastern Montenegro is a geographical unit, which includes the basin Lima in Montenegro, whose total area is 2.557 km². Subject region with around 1.304.1 km² or about 51% of the total surface of the catchment in are Republic. Northeastern territory of Montenegro includes three municipalities: Berane, Andrijevica and Plav, with an area of 1.486 km², which is on the 2003 census, 54.658 people lived or 36, 8 in / km². Natural resources as a factor of socio-economic development of the region were not investigating. So that our research records based on similar research Pavlović et al (2009), pointed to the forefront several obvious socio-economic problems:

1. First, because of geopolitical and economic crisis, the region is up to the Second World War was one of the most economically underdeveloped regions of the former Yugoslavia,

2. Second, only since the sixties of the twentieth century, there was a greater degree of valorization of the natural resources of the region, which led to a rapid economic development,

3. Third, the disintegration of the Federal Republic of Yugoslavia or the State Union of Serbia and Montenegro, and the border region receives a peripheral position in Montenegro, which in even more negative impact on economic trends.

Results of the survey were using to determine the author of those natural resources can cause faster economic trends in the region and to provide uniform spatial-functional development. Northeastern Montenegro has minerals, forests, waterpower and arable land, but they are not properly used. That was causing by unfavorable geographical and historical conditions in which the economy of the region has evolved. Given the natural resources, the holder of the economic development in the region, should be agriculture and tourism. Because not only agriculture and tourism to economic development through are its function with this development, but also their transformation into modern forms of economic development itself represents.

2. Research Methodology

Geographical study of natural resources has so far devoted little attention. Activities in this area are describing as partial consideration of this issue at some conferences and publications in the field of economy, tourism, biology, urban planning.... This research aims to meet the professional and the public with the natural resources of northeastern Montenegro, in the geographical context of their exploitation and use. Objective of this study it was possible to realize the combined use of different research methods. The core of the methodological procedure used in this paper makes geographic (spatial) method. Specifically, in terms of administrative and territorial affiliation, northeastern Montenegro includes three municipalities: Berane, Andrijevica and Plav. Application of statistical methods was necessary to define the quality characteristics of quantitative research. Permeated through the entire text of the analytical method and, thanks to which we are able to recognize, define and quantify the potential economic and social constraints of natural resources in terms of socio-economic development. Since work has essentially synthetic character, used the results published in the international literature. Among them this time emphasize this: Gylfason (2001), Ploeg (2001), Albrecht (2004), Hopwood and Mellor and O'Brien (2005), Rajović (2007a), Barbier (2007), Arezki and Ploeg (2011).

3. Analysis and Discussion

3.1. Morphology and Opportunities and Economic Valorization

The relief of the region is an important factor in economic development. According to the economic importance of the northeastern Montenegro, distinguish the two major morphological groups: Lim valley and mountainous area. For ease of determining morph metric relief benefits, primarily the impact of morph metric traits on agriculture, tourism and construction in the region hypsometric emphasize three areas: lower, middle and upper. In are lower hypsometric areas in terms of economic development. Spotlight area sub region alluvial plains of rivers, river terraces, lake sediments Berane, Andrijevica and Polimlja basin (A), which includes the area sub region of Plav-Gusinje Basin (B) and area sub region, which includes high-low regions mountainous areas of low relief and high-mountain relief of 948 m el to 1.100 m el (C).

I Area (A) sub region alluvial plains of rivers, river terraces, lake sediments Berane, Andrijevica and Polimlja basin (Figure 1) has the most favorable conditions for intensive agricultural production, summer tourism,

construction and transport development. These are areas with a slope of up to 3° and underexposed exposures. Length of the growing season with Td $\geq$ 10 ° C is greater than 150 days and the sum of active temperatures of Td $\geq$ 10 ° C higher than 2100 ° C, allow the variety of plant vegetables. However, low values of relative humidity during April (62%) increases the risk of spring frost and dew, and make these areas less favorable for fruit production. Adverse climatic characteristics are associated with a small amount of rainfall during July and August. In summer, (July and August) monthly mean relative humidity in the afternoon (14 h) is below 45%. This low value of saturation of air is with water vapor, a very negative impact on agricultural crops. Large amplitude fluctuations of groundwater in alluvial deposits and the growing use of these waters makes it difficult for irrigation during the summer period. Therefore, the further back from the riverbed increases the depth of underground aquifers and less irrigation. Summer low flow, lack of access to coast, distance from the riverbed, are reducing the possibility of using river water for irrigation. For the alluvial flat alluvial are rivers connected with the land, which was the most important aspect of the production for possible cultivation of most crops. The river terraces as the dominant soil types of various production values, there are eutric camisoles, vertisols, pseudogley and amphigley. According to the natural advantages river alluvial plains and river terraces are suitable for intensive agriculture, particularly crop production (Rajović, 2009). This sub area is very important for the class trip, a certain summer tourist season and has characteristics of a distinct seasonal occurrence due to climate, or rather the air temperature. Average air temperature in the area during July and August is around 18°C, and mean air temperature over 20° C, it cannot taken as an absolute rule. First, the local population acclimated to river water temperature are conditions corresponding to an average value equal to or greater than 15°C. Season bathing tourism and recreation at appropriate points may last from 30 to 90 days. This fact cannot ignored no matter what it is that the temperature conditions of a relatively modest measures conducive to the development of swimming, and therefore dismissed the coastal population and recreational functions. In relation to the recreational use of available resources of the area value assessment can do in terms of benefits of rowing sports, especially kayaking and canoe. The development of these activities strengthened by almost guaranteed a sufficient amount of water flowing in Lima, but the average decline (Murino - Andrijevica overall fall 75 feet, Andrijevica - Berane 85 m). At the same profile is registered and mean annual discharge of water, which meets the kayaking as one of the aspects of sports and recreational activities. Mountain water flows in what one of Lim ((except in the sector through Berane Basin) can use for kayaking and canoe. Rowing, sailing, kayaking, walking and hiking tourism are possible on the rivers of this district (Rajović, 2010).

I area

■ I area - A
■ I area - B
■ I area - C
▨ II area
■ III area

Figure 1. Morphological conditions of northeastern Montenegro according to the degree benefits of natural conditions development and collocation economy

Despite favorable conditions for development of agriculture and tourism are this spatial entity characterized by the favorable conditions for development of construction and transport. Any form of development (settlement, infrastructure, industrial facilities...) Indicates the specific is requirements in relation to certain morph metric characteristics. Morph metric requirements for construction, we have defined over from the construction of settlements and roads. The construction of the village is very small gradients (up to 1°) are not optimal, because the removal of atmospheric and water channel requires the formation of slope. However, since this region dominated by gradients of 1 - 3°, unexposed surface, good structure height (height ratio as an indicator of energy efficiency infrastructure, and express transport accessibility in relation to overcoming height differences) and the construction season lasts about 260 days, we have very good conditions for the construction of settlements and roads. Compared to the corresponding properties of climate, mean annual air temperature around 80-10°C, relative humidity below 75% make this area suitable for habitation and livelihood of the population. Eating on development traffic characterized

primarily for the winter half year. With regard to mean maximum thickness of snow along the route of the main roads in the valley of Lima does not exceed 50 cm in January, when the largest amount of snow and the number of days with snowfall lasting from October to May (when the snow melts already in contact with the ground), this area has good conditions for the flow of road traffic. Considered a whole has the capacity for industrial development because of lake sediments Berane lowland reservoirs are lignite and brown coal, and the river are the locality river Trebačka amount of building stone in the bed of the river Lim deposits of gravel and sand (Rajović and Bulatović, 2012 a).

I Area sub region that includes Plav-Gusinje Basin (B) (Figure 1) have similar features as the previous agro climatic area sub region. It is characterizing by an inclination of 0-3°, and the unexposed southern exposure, altitudes up to 948 m evolution. The dominant soil type is fluvisol, locally present district cambisoles, eutric camisoles, podzol, planohistol (Lake Plav). Different shades of brown forest allow fruit production. This area belongs to the second class of so-called very favorable land for agricultural production. Length of growing season with Td ≥ 10° C over 140 days and the sum of active temperature Td ≥ 10° C is about 1200°C, allow the cultivation of vegetable crops. In summer, (July and August) monthly mean relative humidity in the afternoon (14 h) was 46% below the already low value of saturation of air with water vapor, a very negative impact on agricultural crops. This allows the area and the development of summer tourism. Mean air temperatures during July and August is around 17°C, relative humidity is about 66%, and the water temperature about 16°C. Swimming season lasts about 45 days. A sufficient amount of water flowing in Lima and the average decline in Plav - Murino (87 m), mean annual water discharge (Lim at the Plav is 17,8 m³/s) meet the needs of canoeing. Rowing, sailing, walking, underwater sports and fishing are possible in the Plav Lake. In addition, there are solid opportunities for the development of some sports and recreational activities during the winter season. Us then be influenced by negative values of temperature (which takes about 30-60 days), which would surface, except for some parts of Plav Lake, should be used as a natural ice rink with a previously detailed observation and studious appearance, thickness, quality and capacity of ice as hydrological phenomena present during the winter tourist season. As regards the construction and transport the region from the standpoint of urban building belongs to the class III suitable terrain (due no exposure and vertical belt (948 m) in length no period with frost (124 days)) and class II the point the construction of transport infrastructure (for vertical belt). This sub area can be difficult to characterize performance of road transport; the mean maximum thickness of snow along the route of the main roads can reach 131 cm. The thickness of snow cover can be very difficult obstacle to traffic, especially on windy passes and villages, where the wind forms a high snowdrifts that existing machinery can break a long time.

Area sub region high-mountain regions of low relief and low-medium landscape mountainous terrain up to 1.100 m above sea level (C) (Figure 1) characterized by mild forms of relief and side slopes of 6° to 9°, greater depth of land covers (luvisols, vertisols, eutric camisols, districts cambisol, sometimes represented rendzina), relief forms are relatively favorable for agricultural production. The land is suitable for the production of various agricultural crops, orchards, and above 1000 m as is mainly woodland (beech-fir forests, oak woods and forests of black and white pine), pastures and meadows. Bases Balja, the area around the rural settlements in the valley of the river Kralje Kraštica, Trešnjevik (relief and slope of the form (3° - 9°)), are favorable for the production of certain fruits and vegetables. They represent the following area districts cambisol, iatric camisoles, rankers, colluvial soil et al. Dulipolje the settlements around Zlorecica, which flows into rivers Perućica and Kutska (slope 3° - 6°), good production potential of land (marsh gluey soil, eutric camisols, rendzina, districts camisoles, land, meadow), suitable for growing various crops, plants such as alder, field ash, oak, birch, and various types of forests (beech, oak, pine...). Kutski river valley of the river can seen as favorable for the production of certain crops (barley, oats, and corn) and fruit production, from the mouth to the settlement Zlorečica and Cecuni. Further to the source of the geomorphologic features, which make up the system on a particular area are not favorable for agricultural production but mostly there are pastures, meadows and forests (Rajović and Bulatović, 2012 b). Areas on the left side of Lima from the expansion of Luge until Pepić (inclination of 3°- 6°) with the land (eutric camisols, colluvial soil, vertisols, amphigley land) are suitable for the production of vegetable crops, cultivation of meadows and forests. Areas on the right side of Lima, which include base Rasojevića head, Javorišta, Grahova, Koradzinog hill, Prijedola part between the mouth of the river Piščevske (slope 6°- 9° and 12°) with the dominant land and districts and eutric cambisols, are relatively favorable for agricultural production. Rural areas of the territory above the valley of Plav-Gusinje of 948 m evolution to 1.100 m above the sea level mainly characterized by slopes of 12° to 20°, with the dominant land: calcocambisol, podzol, brunipodzol, rankers, a sporadically occurring eutric and distric cambisoles. Valley of grasshoppers and Bilećko streams, rivers and Jasenička of the Novšić and rivers Velika, as well as part of the left and right side Vrulje are mostly inclined to 9° - 12° can be used for agricultural production, while other parts are mostly forest land (forests of spruce, fir, and pine, oak). Mean air temperature in the course of the growing the region of the period about 12°C, relative humidity of 68%, and the length of the growing season with Td ≥ 10° C for 130 days and the sum of active temperatures 1800°C allows growing of certain vegetable crops. From the standpoint of tourism, this is not the region attractive because of summer and winter holidays are no conditions. An interesting detail can be sulfuric water in the village of Kralje (see Rajović and Bulatović, 2013 a; Rajović and Bulatović, 2013 b). It can be developed or transitional excursion tourism. From the point of building roads and settlements, this region a Class III benefits (due to adverse vertical belt and the northern exposure). The region has opportunities for industry development; in Zagorje coal deposits are located, and the stand Zabrdje deposits of lead and zinc deposits Đulići marble.

II Area - is relating to the belt of 1100-1700 m above sea level, locally cut by deep river valleys cut into. This spatial unit characterized mainly with severe forms of the relief angle 12° - 20°. This region has rolled land cover, with the dominant land: rendzina, podsol, calc-camisoles calkocambisol, calkomenasol, rankers, and in places and districts camisoles suggesting that the predominantly grassland and forest vegetation (forest pine, spruce, beech, oak, fir). This relief unit is suitable for cattle breeding. Length of growing season with Td ≥ 10 ° C from 91 to 130 days, with the sum of active temperature Td ≥ 10° C to 1600° C to 2300° C, the mean air temperature in the vegetation period is 9,5° C - 12° C (Rajović, 2012). Given that, for each food crop biologically determined minimum, the area near the river valleys (for example, Kutski River or at the foot of the Vlahova and Javorišta) it is possible to grow certain crops (wheat, barley, oats, peas, beans, and rye), orchards and grasslands (dominated eutric camisols and rendzina). From the standpoint of tourism, this area provides opportunities for the development of health and sports and recreational tourism. Moderately and slightly favorable for the development of winter tourism, which provides spaces: Bjelasice, Komova, Cmiljevice, Kofiljače. Average amount of precipitation (snow) from 101 cm to 130 cm. Slopes are mostly from 12° - 20° and the altitude belt above 1.300 m above sea level considered relatively favorable in terms of alpine lake disciplines. Great Šiško and Bukumirsko Lake provide opportunities for the development of picnic and summer tourism, as the summer air temperature of 14° -15° C and the water temperature to 20°C. In winter the lake under the ice and can be used as a natural ice rink with the previous detailed inspection. Given the presence of mineral deposits (lead, zinc, iron, pyrite), provides the foundation for industrial development. Construction season is from 230 to 250 days, but for certain work (e.g. work with concrete), this period coincides with the length of the free period without freezing temperatures, ranging from 67 to 117 days. The absolute amount of snow in this region may be greater than 200 cm, which is a serious obstacle to the flow of traffic. In this area, there are numerous mountain pastures. During the winter and summer but here come a large number of climbers from various countries. They are as attractive characteristic of this region and provide a basis for further development of sports, mountain and hunting tourism, as well as the development of ecological tourism.

Figure 2. *Mineral resource in Northeastern Montenegro*

III Area - include the high mountain belt above 1.700 m as in this region exacerbated by the relief, thermal and land conditions. The slope of the spatial structure of the completely dominant slopes is over 18°, and slopes over 20°. Most land is represented as calkomenasol, litisoli, rendzina and podsol, so this area under forest vegetation and mountain pastures with blueberry and juniper (except where the parent material litisoli). Length of growing season with Td ≥ 10° C is less than 90 days, the sum of active temperature Td ≥ 10° C for 1.100° C and the mean daily temperature is less than 4.9°C, a maximum height of snow is greater than 240 cm in the winter months. That is to say that this area suitable for tourism development, there are six lakes (Ridsko, Visitorsko, Pešića, Little Šiško, Big and Small Ursulovačko), which may be used for the development of sports and recreation, sports and various events, sports and hunting , excursion tourism. In addition, in this region there are numerous mountain pastures. (Do Kobila, Lisa, Štavna...). During the winter, but summer here comes a large number of climbers from various countries. They like the attractive character of this region and provide a basis for further development of sports, hiking and hunting tourism to develop eco-tourism. It is known to stay at this height suitable for athletes, healthy people, but many patients and normalize are situation by improving defense power of the organism. However, this area is most suitable for the development of sports, skiing, mountaineering and rock climbing (Rajović, 2006).

3.2. Water Resources

In northeastern Montenegro by hydropower are most important Lim River and its tributaries. This was the most affected rivers water pattern, morphology and geological structure of the terrain through which the flow. The morphology of the terrain (riverbed slope) in the catchment provides excellent opportunities for hydro-energy utilization. In 1956/57 Group of experts "The energy project" from Belgrade drafted a report "Study of accumulation Plav". It provides hydro-energy use of Basin Water Lake with three variants of the rise of water (elevation 945 m, 933 m and 917 m) and with a dam Gradac hill, not far from the exit from Lima Lake. All three variants provide for fully immersion not only Plav Lake already and valley River Ljuča.The projected hydropower's "Plav" is a 24 MW, and would mean are first of 11 planned hydroelectric Lim (www.beranetown.net). According to data from the Regional Business Center Berane (www.nasme.me), which refers to "the study of energy use and its tributaries Lima" only in Plav, it is possible to build 15-hydropower formidable strength of 42.595 MW. Several power plants would have a very good performance in terms of power, potential, and the annual cost per KW / h. For example, hydropower "Jara" on Komarača River is most interesting because it has the following characteristics: an installed capacity of 6.8 MW, the installed flow 3, 85 m3/s, waterfall The 220 m, and annual output of energy 21, 57 GWh. The project envisages the construction time of one year, with the length of the inlet pipe was 4.1 km, length 570 m steel pipe with a diameter of 0,8 mi final prices KW/h. So hydro "Jara" according to the study mentioned above, constitute by far the cheapest hydroelectric power plant in the former Yugoslavia. In conclusion, in this part of northeastern Montenegro to the present day, there is not a power plant, even though there are more on Lim suitable place for their construction. "Assembly of Andrijevica 2012 adopted the" Decision on credit obligations of the municipality in the amount of 1.200.000.00 Euros "for the realization of the construction of mini-hydro power plant on the cities water "Krkori." Project involves the reconstruction of water intake (catchment); replace asbestos-concrete pipes with a length of 4 km and construction of plants for the production of electricity at the breaking chamber city water. Designed mini-hydro power up to 1 MW are size of the facility 12 h 8 m, height 5, 5 m. Private company from Andrijevica "Igma Grand" in 2012 received a concession for the construction of small hydropower installed capacity of 950 kilowatts, and the planned annual production of 3.383 GWh on the river Bradavec for a period of 30 years and with the appropriate concession fee. In this way, with the production of electricity in the region, create the conditions for job creation. In addition to energy use, the accumulated water could be using for industry, water supply, irrigation of agricultural land.... The problem of water supply of population and industry are usually solving together. Water supply Berane Andrijevica and Plav in a modern way began in the early sixties. In doing so, they generally stronger capped karts springs. Water supply Berane devoted the greatest attention, because of population growth and industrial development. The construction of urban water supply in Berane began in 1962 and this regulation "the Monastery springs" whose capacity was 85 l/s. Today it serves as a backup source of water supply for the urban part of Berane. Since 1989, the water supply is part of the urban and suburban areas Berane in made from water Lubnice - Berane, which is supplied with water from the catchment "Merić" fountainhead. "Length of distribution network (primary and secondary) is 160 km long and covers about 70% of the municipal territory, and uses about 65% of Berane (www.nasme.me). Rural part of Berane Polica, Upper Budimlja, Dapsića, and Petnjik supplied with water from "Dapsićkog hot" with a capacity of 49 l/s. According to data from the Municipal Secretariat for Economy Berane the Municipal Assembly are per capita water consumption in 1975 year 0, 31/106/m3, 1985 1,21/106/m3, 1991 1,91/106/m3. If the calculated average in m³/day then it ware 0,003 in 1975year 0,006 in 1985 and 0, 01 in 1991. On the other hand, the water consumption in the industry ranged from $1,07/106/m^3$ in 1975year $1,16/106/m3$ in 1985 and $1,80/106/m^3$ in 1991. In the mentioned period, total water use in Berane for population and industry, was $1,38/106/m^3$ u 1975year, $2,37/106/m^3$ in1985 and $3,75/106/m^3$ in 1991. According to data from the Municipal Secretariat for Economy Berane Municipal Assembly in 2000, the total need for water Berane amounted to 5,62 million m^3, of which the needs of the population 2,96 million m^3 and 2.66 million m^3 industry. Plav has a dense network, but not abundant karts springs and wells, which reduces the possibility of its permanent supply of drinking water. For now, supplied with water from springs and aquifers ("Đurički alluvion of the River"), with an average yield (15 l/s), less than the total needs. The Plav feels chronic water shortages in the summer period, particularly in rural locations Gusinje and Murino. Insufficient water supply of the population followed by many other rural towns: Prnjavor, Brezojevice, Kruševo, Martinoviće ... It is important to point out that the water pipes in the Plav obsolete and made of asbestos (main lines - profile of 200 mm and 110 mm) and plastic - other lines. This circumstance causes huge losses in the network, which is estimating at more than half the amount of water transported in the pipeline. Length of the water is about 5, 2 km, profiles 300 and 280 mm, the primary distribution line is a distance of about 5 km, and the total length of the secondary network is about 10 km. In Gusinje is also plumbing, plastic pipes overall length of 6.5 km. Primary local area network made of asbestos with a length of about 4 km and profiles 150 and 100 mm, and a secondary plastic pipe length of about 6 km. In Murino in identical supply pipeline built of plastic, with a total length of 5, 5 km in length and a local network length is about 2 km and derived from plastic pipes (www.nasme.me). According to data

from the Municipal Secretariat for Economy Municipal Assembly Plav, blue water consumption in 1991 was 3.05 million m3, of which the industry consumed 1.18 million m3, and the population of 1, 87 million m³, i.e. 61, 31% of the total. Total needs for Plav Water in 2000 amounted to 3, 97 million m³, of which the needs of are the population is 1/2 of the overall needs, i.e. 1, 99 million m^3 and 1, 98 million m^3 and industry. The first water system for organized water supply in built in 1931 Andrijevica a capacity of 0, 3 l/s. At the end of the eighties, "Water of economic organization for the development and utilization of water of Montenegro" - Podgorica did a "major water project for Andrijevica", taking into account the then current situation and future social and economic development of urban settlements in the immediate environment which includes rural areas: Andželati, Božići, Bojovići, Đulići, Kralje, Prisoja, Seoce and Slatina. Water supply began 1982 are the arrangement of the springs "Krkori". Projected water system in built from cast-iron pipe diameter of 300 mm (www.nasme.me). After the road Andrijevica - Cecuni - Kuti in 1984, made amendments to the "Main Project" which is performed by the current water supply system, which relates to revise previously constructed road route over the pass "Pear". Finally, in the "Project of reconstruction of water supply system "source "Krkori" - Andrijevica is been designed, the second time, by making the correction of pipeline related in order to improve hydraulic conditions and that is part water supply system displace on her more accessible part it is along the road to by source "Krkori." So today, besides urban settlements Andrijevica from this water supply system supply and rural: Đuliće, Bojoviće, Seoce, Božiće, Prisoja, Slatina, Zabrđe and Trešnjevo (www.andrijevica.me). According to data from the Municipal Secretariat of Economy of the Municipality Andrijevica, total water consumption in 1991 amounted to 2, 06 million m3, of which the industry consumed 0, 26 million m3, and the population of 1, 8 million m3. Benefits for irrigation of farmland in the region are different. Flies are primarily available water, quality land types and morph metric predisposition. Therefore, the use of river water for irrigation in the northeastern part of Montenegro, in time and space is limited. The striking discrepancy between the amount of available river water and the required amount of water for irrigation is related to the July and august. So Plav at the station, Lim has a mean monthly flow of 13, 7 m³/s (July), 6, 8 m³/s (August), and the station Berane 20, 8 m³/s (July) and 11, 4 m³/s (August). During the summer months (July, August, September), during the lapse of the river 9 -12% of the annual flow, which is near Plav 17,8 m³/s, and near Berane 37,4 m³/s (Rajović, 2009). In addition, many of the rivers in this period are below the biological minimum flow, and a deficit of irrigation water more pronounced. In relation to morph metric analysis relief benefits, the most favorable conditions for irrigation have alluvial Berane, Andrijevica and Polimlja valleys. Special benefit the construction of reservoirs on tributaries of Lima is contained in their hypsometric favorable position in relation to arable land at the bottom of Berane, Andrijevica and Polimlja Valley and Plav-Gusinje basin. Optimal opportunities to build reservoirs valleys provide aggressive and mosquitoes. Channel network of reservoirs in these flows could easily usher along Plav-Gusinje basin. For downstream basins should be considering in are formation of small reservoirs Zlorečica, Bistrica and Lješnica. Therefore, water management solutions in should be sought within the Spatial Plan of Montenegro, which refers to that part of, water management and water law, where it is necessary to set the focus on the construction of the regional water supply, irrigation and drainage (Rajović,2010). Accurately establishing the required amount of water for irrigation is not been performed. In the region of 29.787 ha of arable land is irrigating only about 1.000 ha. To irrigate small plots of vegetables and water to wells or water from are riverbeds. Most irrigated by surface is in Berane valley 550 ha. In the region, there are a small number of mineral springs but unexplored so neither valorized. Among them the attention it deserves, the thermal mineral springs in the village of kings. Taking into account classification (Leko et al, 1922) as well as minimum and gases in groundwater in establishment group mineral, thermal mineral springs in are village Kralje belongs to the sulfur waters. These are water in which the boundary between fresh and mineral water is around 0,001 and the minimum for inclusion in the thermal waters of 0,010, which is very fitting parameters of this fracture sources. Certainly this and other thermo-mineral springs that requires special attention and work on their tourist valorization. Experience and research have shown that the thermal waters are suitable for the treatment and rehabilitation of cardiovascular diseases, respiratory diseases, diseases of billiard tract and pancreas in diseases of the gastrointestinal tract, metabolism, kidney and urinary tract, gynecological diseases, neuropsychiatric diseases, skin diseases, children's diseases. The combination of this with each other complementary (positive) tourism means progress in the evaluation of mineral water (Marković, 1979). It should be borne in mind that these resources can be using for other purposes, primarily in sports - and excursion manifestation.

3.3. Deposits of Cool, Metal Ores and Non-Metal Ores

Coal is an important natural resource in the region and provides a solid base for the development of the energy sector. The sediments of the upper Oligocene and lower Miocene in Berane and basin Polica there are deposits of lignite-brown coal. Berane basin is representing by gravel, sand, clay and marl. Developed three coal seams are the main coal, the first and second footwall. The main coal seam thickness ranges from 1-10 m, locally to separating, which means that the thickness inlays consisting of marls, ranging up to 1 m. Tertiary main coal seam are carbonaceous and sandy clay. The thickness of are first coal seam varies from 1,2 m to 3,8 m and was developing

in the district Petnjik - Dapsića. Tertiary other footwall carbon layer is gray and gray-green clay, and the thickness ranges from 2, 0 m to 4, 5 m. Thus, coal Berane coal basin type is lignite-brown coal (Nikolić and Dimitrijević, 1990). Basin Polica is representing by sandstones, sands, sandy clays, clays and marls, and it was concluding six coal seams. Their number and distribution in different parts of the basin is variable. The carbon layers are often separating, and their thickness ranges from 0, 2 m to 7, 2 m. Coal Basin shelf also is among the lignite-brown coal. Based on petro graphic and chemical composition and macroscopic properties, coal Berane - basin Polica in largely built of detritus - Textile. The mean content of the petro graphic components of coal provides 86, 6% detritus - Textile, detritus 3% - 10, 3% gel and textiles - the gel. From the data of technical analysis can been seeing that coal contains 16 - 20% moisture, 10 - 17% ash, and 2% sulfur, 45-50% of coke, about 30% fish-s, about 35% of volatile and over 52% of combustible substances. In addition, the effect of thermal coal amounts to 16.700 KJ (GTE), and 13.400 KJ (DTE) (Nikolić and Dimitrijević, 1990).

Total reserves of coal in Berane-basin are Polica 176.231 197 tons. It is widespread in all parts of Berane lowland (Budimlja, Petnjik, Zagorje, Polica, Beran Selo, and Dolac). Excavation of coal in this basin started in the sixties in the district "Budimlja" and was completing in the late seventies, when I started building a new mine investment district "Petnjik" that began production in 1981 in the eponymous pit, where and today is exploitation of coal. Production of coal ranged from 10.000 in 1960 to 107.000 tons in 1989. In the same period, coal production has increased from 276.000 to 2.159 million tons, or at a rate 8,25%.Brown Coal Mine "Ivangrad" not escaped the fate of the collapse of the economic giants in our country. At the beginning of the nineties was sinking more and more in an uncertain economic future. In 2004 found to be insolvent. On the ninth bid for 1, 51 million Euros, the Greek company "Balkan energy" purchased the mine in 2008 and received its exploitation concession for 20 years. With the obligation to build a thermo-block of 110 megawatts over the next four years, invest another 120 million (www.mans.co.me). However, the extraction and processing of coal are still waiting, and mining for now is only considering a potential resource periodically. Program Development in Berane coal basin (lignite and brown coal ≈ 180 million), would cause the intense regrouping and integration of industrial enterprises in this part of northeastern

Montenegro, which could cause are the need for capacity expansion (Beran Village, Dolac). Communication between them is relatively inexpensive and applicable to the road network, which mainly goes through the river Lim. The roads are second and third rows and are oriented in three directions and the direction north to Bijelo Polje, east to the south to in Rožaje and Andrijevica. Through these routes was connectivity with other parts of Montenegro, and Serbia. The nearest railway station on the Belgrade-Bar, located in Bijelo Polje, at a distance about 35 km from the mine. If we add to all this in Berane basins, there are immense reserves of marl. Marl especially is on the right side of Lima, mostly on the shelf, Jasikovac, but in the hamlet Đurake. These marls same qualities match the requirements of the cement industry. According Lutovac (1957) only on Jasikovac reserves could be providing for the production of two hundred years should be annually produced 80.000 tons of cement (Boričić et al, 1967). In the considered geographic space are detecting and the appearance and bearing the following metals: lead, zinc, copper, iron and pyrite. They occur in sedimentary and volcanic rocks: Paleozoic, Lower and Middle Triassic. Occurrence of lead and zinc were discovered east of Konjuhe on the right side of the river Perućica stand on the site. Demonstrate an area of about 2 km. They occur in the Permian sediments, Lower and Middle Triassic. Mineralization occurs in the form of wires and impregnation, and their thickness ranges from 0, 2 m to 1, 0 m. The content of zinc in the wires is very variable, in some trials reaches 5%, while the middle is 0, 3%, while the content of lead is far smaller and does not exceed 0, 1%. Besides lead and zinc in Konjusi occurs and copper, whose average content is about 0, 15%. Association Konjusi minerals are: pyrite, sphalerite, chalcopyrite, galena and others. The mineralization is genetically related to diorite and quartz diorite, and was created at higher temperatures and lower levels (Group of authors, 1982). On the right side of the river, Lima in the area between the axes Kostreš, Omarska, and head were discovered also the appearance of lead and zinc. They occur in quartz-keratophyre, tuffs, volcanic breccias and limestones. Mineralization is manifesting in the form of wires. The content of lead and zinc ore FACING pieces ranging up to 6% of lead and zinc. In this area of land in some trials of zinc content ranges up to 1, 1%, and lead to 0, 8%. Mineral association of the hatchet is pyrite, sphalerite, galena and chalcopyrite. On the eastern slopes of the wider area of Bjelasica in Zabrđe and Šestaverca and wire thickness, not exceeding 1 m and the provision is rarely following for longer than 20 m. The content of zinc in the ore wired King's Brook is 2, 5% or 3%, in Border Creek to 0, 5% in Vaćevinama to 0, and 2%. Occurrence of copper in found in the creek beside Konjuh stand. They occur with lead-zinc mineralization. Copper content of this locality ranges from 0, 1% to 0, 2% levels (Group of authors, 1982). Occurrence of iron was discovering in the mountains Bjelasica in a number of localities: Kurikuće, Lubnice, Zekova Glava, Crna Glava, Strmi Pad, Konjusi (site of Klina). Occurrence of iron in Bjelasica occurs in the form of hematite, which is to say that the hematite occurs along the plate with red charts, which lie over keratophyre and below the slope of layered limestone. Iron content in these localities ranges from 17% - 33%. In the Clinical Konjuh iron ore presented hematite, and manifests as a contact lens on the volcanic rocks and limestone. Occurrence of pyrite have found in a number of localities on the mountain Bjelasica: Zekova Glava, Kurikuće, Lubnica, Crni Vrh. Pyrite occurs in the form of

wires in the form of impregnation, as well as volcanic-sedimentary rock formation. Wire pyrite in the Paleozoic sediments ranges from 0, 1 to 1 m (www.andrijevica.me). From nonmetallic mineral resources on the observed geographic, there are deposits of building materials: gravel, sand and decorative stones. Numerous deposits of gravel and sand in found in the bed of the river Lima (Plav, Andrijevica, and Berane). Only in Bandović Most, the amount of gravel and sand, available for an annual extraction is estimating at approximately 100-120.000 m³. In the northeastern part of Montenegro there are limestone quarries in the Triassic. There is a certain amount of building stone, which can be exploiting locally, but the conditions are unfavorable for continuing exploitation. When it comes to the exploitation and processing of marble and ornamental stone, it should be noted that there are multiple sites of different architectural building stone and marble, the most significant: the site Trebačka River, Seoce, Piševska River, Babov Stream, Pčelinjak, Žoljevica. Outcrops of volcanic rocks in the upper reaches of the river Trebačka appear on the left and right, from an altitude of about 850 m, while only the riverbed is covering with blocks of the same rocks. For natural cane, the rock is pale green, and the fresh green and gray-green layout. On the surface are also observed cracks some of which are open and wide and ten centimeters fall allow larger blocks rocks. In the cutting path that leads from the hamlet Gunjaje, Steppe to summer pasture at an altitude of 1.170 m to 1.185, were discovered brecciate limestone's. According to the geologic map of Montenegro 1:10.000, these rocks are represented and northwest of the hamlet Gunjaje. At about twenty meters in a south east direction, there is a slit in which the traces of mine holes, which means that the stone used probably as a quarry for the construction of residential buildings (www.andrijevica.me). Outlet Piševo rocks in the river are an integral part of the volcanogenic massive axes, which covers an area about 25 km². From rock to been discharged and are represented by andesitic keratophyre, volcanic breccias and tuffs, with mutual crossings. The presence of andesitic is most pronouncing on the western slopes Piševa and in the middle and lower reaches of the river Piševske. These rocks are particularly revealing in the notches of the forest road that goes along with Piševsku River with her right hand. The rocks are gray-green to blue-green color. Scorn resulting changes of carbonate rocks, mainly Lower Triassic bio-turbine formation, build terrain in high stream flows Bradevec, Babov and Malinovac. The rocks are best discovering in the bed of the stream Babov, which occur at an altitude of 1.170 m to 1.600 m. Scorn, are very compact and sail rock, usually striped texture. Their color is mostly gray and grayish-green, are being observing and one yellow-green and jonquil, which of course depends on the mineral composition. Brecciate limestone and dolomite limestone were discovered on the ridge of the bee, at an altitude of 1.150 m to 1.450 m, followed by delivery to the west, southwest and nameless streams, gullies actually. At the ridge, the terrain is covered and covered, so that the boundary between the reddish limestone and gray crystalline limestone, which are below them, masked and unclear (www.andrijevica.me). Žoljevica on the hill is the cradle of architectural - building stone. This deposit build medium - gray and white Triassic massive limestone mesmerist. Mesmerist white limestone on the surface, covering an area of about 3.000 m², and its thickness is about 30 m, while mesmerist gray limestone, covers an area of 30.000 m² and has a thickness of about 50 m. Resource estimates of gray marble B + C1 category, amount to 2.223 million m³. Reserves gray-white marble and white marble belong to the C1 category and amount to about 60.000 m³. Decorative stones of this deposit is very decorative, perfectly polished, and attains a high sheen. Due to their physical - mechanical properties of rock, 'Žoljevice" can be used to produce plates for covering horizontal and vertical surfaces, objects in the construction industry (www.andrijevica.me).Geological studies performed during the 1955 and 1963, defined the following characteristics of stone: the size and shape of the bearing, physical and mechanical properties of white and gray varieties, so that the cutting test, the white variety have broken down completely, while the gray variety, obtained plates of excellent quality and a beautiful shine . Based on the above data it can in say that the bay "Žoljevica" fully defined in terms of quantity and quality. However, it is not marble survey received sufficient attention, despite the fact that the site be considered after the Arandjelovac and Prilep, can be considered as one of the most significant in the former Yugoslavia (Lutovac, 1973).

Wonderful marble with "Žoljevice" we should "valorize" the art-tourism event, "Marble and Sounds." Far lead us to emphasize, what are the riches and what possibilities the marbles and marble breccias, provided the northeastern part of Montenegro, for its economic development. There are various estimates of mineral reserves in the northeastern part of Montenegro. However, research that is smaller or greater intensity exercise, lead to the discovery of new ore deposits, and are not prone to such estimates. However, we believe that mining in limited geographic space, given the presence of mineral resources, i.e. their diversity and reserves, only part used the opportunities that it provides the raw material base (Rajović and Bulatović, 2013 a).

3.4. Agricultural Areas and Forests

Agricultural land is an important part of the natural wealth of the region. The structure of land use in certain categories is of special importance because it is the result of development and intensity of agriculture and it expresses the degree depending territorial conditions for the development of certain types of agricultural production (Todorović, 1985). In the period 1964-2005 in the agricultural land of the region there has been a change in the manner of utilization to reduce the area under fields and gardens and pastures (Table 1). In contrast increased area the under orchards and meadows. Arable land in the reporting period was reducing from 8.440 ha in 1964 to

7.368 ha in 2005 up to 1.072 ha. Area under orchards increased in are same period from 1.826 ha to 2.334 ha or 508 ha. During the reporting period of are meadows slightly increased from 19.926 ha to 20.502 ha or 576 ha. Land pastures in the period 1964 - 2005 decreased from 40.286 ha to 37.821 ha or 2.465 ha (Table 2). Given the state of mind of the livestock and the degree of degradation of pastures, it is logical to expect a further decline in this category of land. Point out only the following information, which fully reflects the condition of livestock in the region. In the period of 1994-2005 the total number of cattle has decreased from 41.506 to 27.593 head or 13.913 throat, pigs from 4.264 to 2.480 or 1.784 head cattle, poultry from 92.261 to 58.770 a piece, or a piece for 33.491. Only the sheep remained almost same. In 1994 the region has grown 68.534 sheep and 68.660 in 2005. On livestock development tendencies in the region was affecting by several factors: the extensive character, throat low productivity management retarded way.... (Rajović, 2012).

Table 1. Agricultural area by categories use in northeastern Montenegro 1964 and 2005

Year	1964		2005	
Category land	u ha	%	u ha	%
Agricultural areas	70.478	100	67.379	100
Fields and gardens	8.440	11,98	6.722	9,98
Orchards	1.826	2,59	2.334	3,46
Meadows	19.926	28,27	20.502	30,43
Grasslands	40.286	57,16	37.821	56,13

Source: Statistical Office of Montenegro, Census of Agriculture (the relevant year), the calculation of data by the author

Table 2. Sowing structure arable area northeastern of Montenegro in 1964 and 2005

Year	1964		2005	
Category land	u ha	%	u ha	%
Fields and gardens	8.840	100	6.722	100
Grains	6.350	75,24	1.127	16,77
Industrial Crops	26	0,31	-	-
Vegetables	1.305	15,46	2.680	39,87
Fodder crops	759	8,89	2.036	30,29
Uncultivated arable land	-	-	879	13,08

Source: Statistical Office of Montenegro, Census of Agriculture (the relevant year), the calculation of data by the author

According to the data from Table 1 in agricultural land in 2005 there was the arable land 9,98%, 3,46% orchards, meadows and pastures 30,43% 56,13% (Table 3). Such a large percentage of meadows and pastures in are overall structure of agricultural land, indicating the mountainous character of the region. Arable land is the most important category land. However, statistics show that spontaneously abandoned arable land or planning translated into other categories of land, or alienating for non-agricultural purposes. Threaded with a reduction in the area under arable land, there comes a change in the structure of its use. The use of the structure of the arable land in the period 1964-2005, noted the positive changes in the direction of increasing the area under vegetable crops (1.305 ha- 2.680 ha), cattle fodder (759 ha -2.036 ha). Adverse changes in the structure of use are contained in the fact that the area under cereals decreased (6.350 ha- 1.127ha), abolished under industrial plants (26 ha - 0), and increase the area under fallow land (for the 879 ha). The increase has been causing by the phenomenon of elderly households that are not able to cultivate their property. However, the raw surfaces are result of negligence of state and local governments by the surplus of agricultural products. According to the data from Table 2 in the structure of the arable area in 2005, there was the corn 16,72%, 39,87% Vegetables, animal fodder 30,29% and 13,8% uncultivated arable land. Thus, the formation of such a structure using arable land, in addition to natural conditions, demographic trends are affected, the inability to use modern agricultural mechanization, irrigation, tradition …(Rajović and Bulatović, 2013 c)

Table 3. Utilization of agricultural land in northeastern Montenegro 2005

Categories of land and culture	ha	%		
		Participation in group	Fields and gardens	Farmer surface
I. Fields and gardens	6.722		100	9,97
A. Grains	1.127	100	16,76	1,67
Corn	930	82,52	13,84	1,38
Wheat	153	13,58	2,28	0,22
Rye	17	1,51	0,25	0,02
Barley	27	2,39	0,40	0,04
B. Vegetables	2.680	100	39,87	3,97
Potato	2.010	75,0	29,90	2,98
Beans	138	5,15	2,05	0,20
Other vegetables	532	19,85	7,91	0,78
C. Fodder crops	2.036	100	30,29	3,02
Alfalfa	319	15,67	4,75	0,47
Other cattle fodder	1.712	84,33	25,54	2,54
D. Fallow land	879	100	13,08	1,30
II. Orchards	2.334	100		3,46
III. Meadows	20.502	100		30,42
IV. Grasslands	37.821	100		56,13
TOTAL	67.379			100

Source: Statistical Office of Montenegro, Census of Agriculture (the relevant year), the calculation of data by the author

Arable land is mostly using for sowing harvest. In area, where corn was grown in are region's dominant cereal (930 hectares or 82, 52%). However, the total area under maize fields and gardens accounts for 13, 84% and 1, 38% of the total agricultural area. Thus, areas sown with maize are small and primarily determined by the amount of rainfall during the growing season, especially the government deficit in July and August, when the corn is in the process of maturing grain. Stable yields of maize in the region can be ensuring irrigation of arable land. However, if one takes into account the temporal and spatial distribution of water suitable for irrigation is insufficient, and that they mainly used for irrigation of vegetable crops, then small amounts of water remain so available for irrigation area under maize. Wheat is the most abundant plant other crops in the region (153 ha or 13, 58%). The total area under wheat fields and gardens accounts for 2, 28% and 0, 22% in total agricultural area. Despite cultivars of use and considerably modern agricultural practices in the basin Berane, wheat yields significantly determined by the agro-climatic conditions. However, the cultivation of wheat in the region decreases significantly due to the high cost of its production and labor shortages, and because of the simple reason that it is cheaper to buy bread in the shops than "look at the wheat field and worry about what will be her next race and the effort and costs"(Jaćimović, 1971). The grain structure similar changes have occurred in the rye and barley. Thus rye harvested areas in the region amounted to 17 ha, or 1, 51%, barley 27 ha or 2, 39% of the total area under cereals. The total area under rye fields and gardens accounts for 0,25% and 0,40% barley or rye with barley and 0,02% at 0,04% of the total agricultural area. Both cultures tolerate cold, drought, and even moisture, and succeed where other cultures would be difficult to adapt. In addition, are the area under these crops has reduced, as are grown for their own use and in small areas and less fertile soils. Greatest importance in are diet of the region, in spite of changes in the way of growing a fodder with natural grasslands and arable land under livestock fodder. Moreover as are, fodder for feeding cattle, sheep and much less use and cornstalks (Rajović and Bulatović, 2013 d).

In 2005 the area of fodder production was 2.036 ha or 30, 29% of the total arable or 3, 02% of the total agricultural area. Sown area of l alfalfa amounted 319 ha or 15, 67% of total area under livestock fodder or 4, 75% compared to arable or 0, 47% of total agricultural land. In the same period, the area under other livestock forage crops (clover, vetch, a mixture of grasses) encompassed an area of 1.717 ha or 84, 33% of total area under forage crops or livestock 25, 55% compared to arable or however, 2, 54% of total agricultural land. The main reason for under-sown areas under cattle fodder is due to poor implementation, technology inappropriate select varieties and safeguards. Areas under natural grasslands amount 58.323 ha. The surface of 20.502 ha of meadows and pastures 37.821 ha share in total agricultural area with 30, 42% and 56, 13%. Dying of sheep and goat farming in this region, more meadows and pastures win a variety of shrubs and weed communities. Weed vegetation occurs in a large number of species in agricultural areas, along roads and boundaries. The representatives are Nettle (Urtica dioica), dandelion (Taharacum officinale), spurge (Euphorbia cuparissias), wild oat (Avena tatua), thistle (Cirsium arvense), Buttercup

(Ranunculus pepens), Burdock (Lappa majok), black mallow (Malva silvestris) and others. A combination of mechanical, chemical and biological methods can be suppressed weeds only during a rotation crop rotation, however, but in the next, he reappeared. Vegetable farming is one of the most intensive field crop production, due to the effort and the realized production. Total area under vegetable crops in 2005 amounted to 2.680 ha or 39, 87% of the total arable land, and 3, 97% of agricultural land. Potato is the official statistics dominant vegetable crop. Under these vegetable crops there were 2.010 ha or 75,0% compared to the total sown area under vegetable crops, or 29,90% of total arable or 2,98% of agricultural land. The basic problem is even larger sown area under potato is the fact that despite the use of quality planting materials (the Dutch seed potato and homemade potato), the unfavorable rainfall patterns in the second half of the growing season. Beans grow best in fertile soil and loose, particularly at the upper flood plains of the region. Traditionally sown as intercrop maize, but the penetration of sunlight hinders its development. It has caused the so-called bean planting "Pure culture". The total area under bean in 2005 amounted to 138 ha, or 5, 5% of the total area under vegetable crops, or 2, 05% of the total arable land, or 0, 20% of the total agricultural area. Other vegetables (onion, cabbage, cucumbers, pumpkins, peas ....) are very widespread in the region. Area planted to these kinds of vegetables amounts to 532 ha or 19, 85% of the total area under vegetable crops, or 7, 91% of the total arable land, or 0, 8% of the total

agricultural area. The introduction of new varieties, improved agricultural technology and organization of production, planted area under "other vegetables" may increase because there are natural - environmental conditions (Rajović, 2013).

Forests in northeastern Montenegro, according to the census of 1979 the forest reserves 63.718 ha are occupied, which means that 42, 87% of the territory of the region is covered with forest vegetation (Table 4). According to statistics from 2005 the forest area as compared to 1979, decreased to 128.629 ha (area under forest cover 62.432 ha, and the forest coverage of 42, 02%). Reduced forest area, are not the result of a planning ways of management, lack of timely measures to protect against erosion, fire. Natural conditions in the region caused the structure of forest communities. The alluvial plains and terraces fluvial glacial Lima are characterizing by extremely low vegetation. Forests are mostly mixing, while the most common types are hydrophilic woods willow, poplar, alder, elm, oak, oak, beech, birch, maple. The belt of beech is most common in the form of your images: four: beech (at lower elevations), mountain beech forest at altitude 1.000 – 1.300 m above sea level, sub-alpine beech forest at altitudes greater than 1.800 m with spruce-dominated forests. With some of our mountainous stretch of the mountain forests moonlike and pine. Above this band represented the expanse of white and red pine. Some forest stands and makes the dwarf pine; whose propagation exceeds 2.000 m above sea level, or junipers, stops above 2.200 m (Rajović, 2007 b).

Table 4. Surface and structure of forest reserves in 1979

Forest stand	Area in ha	%	Timber in m^3	%	Annual increment in m^3	Annual growth in m^3/ha
Decorated	54.643	85,76	11.515.192	82,95	208.721	1,81
Disorganized	9.075	14,24	2.367.324	17,05	31.193	1,31
TOTAL	63.718	100	13.882.516	100	239.914	3,12

Source: Statistical Office of Montenegro List of growing stock in 1979, Edition "Studies and Analyses," Titograd, 1983.

Table 5. Forest structure in 2005

Surfaces	ha	%
High economic forests	27.196.24	43,56
Low economic forests	2.657.17	4,26
Protective forests	17.972.64	28,80
Fallow land	7.742.19	12,40
Bushes and shrubs	6.858.05	10,98
TOTAL	62.432.29	100

Source: Statistical Office of Montenegro, Census 2005, forest reserves, calculation of data by the author

Regarding the breeding categories, arranged dominated (high, low and safety) of forest. Specifically, of the total forest area in the northeastern part of Montenegro, the beautiful trees waste 85, 76% (54.643 ha), and the disordered 14, 4% (9.075 ha). The total density, estimated at 13.882.516 m3, i.e. 82, 92% (Table 5). In well is 11.515.192 m^3 or 82, 95% of the total wood mass and disordered 2.367.324 m^3 (17, 05% of total wood pulp).

Annual volume increment amounts to 239.914 m3. At the same time, the increment tended forests 208.721 m^3, while the disordered is 31.193 m^3. The average value of annual increments were relatively small, amounting to 3, 12 m^3/ha, beautiful trees – 1, 81 m^3/ha, unregulated -1.3 m^3/ha.

The structure of forests are most widespread economic forests that cover 27.196.24 hectares or 43,56%, low economic forests 2.657.17 ha or 4,26%, protective 17.972.64 ha or 28,80%, 7.742.19 ha of uncultivated land or 12,40% and scrub and bush 6.858.05 ha or 10,98%. According to statistics from the forest reserves in 2.000 shows a gross weight of timber felled was 76.873 m^3, of which the trees is accounted for 25.122 m^3 and 51.751 m^3 of coniferous trees. Of the total cuts to technical wood accounted for 53.474 m^3, 13.525 m^3 of firewood and scrap 9.824 m^3. The annual increment of forests estimated 190.934 m^3. So the volume of harvest is less than the increment of mass. To forest vegetation played a proper function for economic development in the region, it is necessary in the future to pay special attention to the preservation and reproduction of forest reserves, especially of high economic and protective forests. This implies the use of rational, reconstruction of devastated forests and extensive forestation, especially places that have been exposing to erosive processes. Careful cutting plan, taking into account are environmental and economic criteria. Wood processing industry in this part of northeastern Montenegro, we must pay special attention. Specifically, based on the use of forest resources, this industry is, so far, its development directed to finalize the primary production (timber, wood panels...). Direction main wood processing, must be defined production function, i.e. ensuring all products from the forest, which can be valorizing through wood volume production and other forest products. In addition, as the main forest product occurs virgin wood, either in the unprocessed (sawmill logs, firewood, lumber) or processed form (furniture, cellulose). The other products are some of the woods, which are gaining increasing importance: venison, fish, snails, berries and seeds, mushrooms, resins, essential oils, juices, roots, leaves, lichens, moss peat, stone, gravel, sand and ..., for which there is no prohibition on the collection. Before a hundred years or more livestock has largely rested on the use of forest products as an energy food, especially oak, beech acorns and wild fruit and chestnut. Today this method of feeding livestock remained largely a memory. From the economic point of flora forest (62.432 ha) and pastures (37.821 ha) is enriched with various kinds of medicinal plants and edible mushrooms. Especially important are certain types of mushrooms, wild strawberries, raspberries, cornelian cherries, rose hip, blueberries, juniper berries. In the region there are about 60 medicinal plant species. Most of them ranked highly in traditional medicine, pharmaceutical production, which is very important for are the tourism development. Medicinal herbs rich in its diversity, physiological and pharmacological effect, and a healthy quantity of raw materials, offers immense possibilities in the development health and educational tourism. From early spring until autumn at are latest, in the forests, meadows (20.502 ha), are growth of many types plants, most of which are edible and medicinal. Many of them with are highest nutritional values: Klamath weed, thyme... Wormwood used as a tea. Thanks to the widespread forests, pastures and meadows landscapes of the region are diverse and picturesque, providing significant environmental and tourism values and northeastern Montenegro seems very attractive. Meadows and pastures are covering with juicy mountain meadow grass and flowers, so that together with forests, a special landscape-decorative value region. Belt noise is particularly interesting as a living space of venison, birds, fish and insects that. Is the pearl of the unique natural beauty and tourist aspects unimpaired nature? With has significant resources and a predisposition to the development of different forms of tourism, such as fishing, hunting, adventure, puts the adrenaline. The first place is taken by hunting and fishing. In this regard, the region has a very

pure nature and abundance of flora and fauna. The forests are rich in tiny and big game from that aspect is very interesting tourist destination for hunting tourism. Travel offers could include individual and group package tour for hunting small and big game: bear, deer, mountain goats, wild pigs, wolves, fox, rabbit, squirrel, grouse, partridge, wild duck, marten, badger, and others. Rivers and lakes are rich in fish (trout, grayling, brown cataracts, minnow, bullhead, and northern pike). For improving fish stocks, in particular salmonid stocks continued to implement stocking. By artificial stocking, one of the measures for improvement of fish stocks and is the determination of brood fish for the salmonid fish species. The most important brood fish were on the River Lim, Plav Lake and River Ljuča. Sport fishing deals with around 800 registered sport fishing and recreation. The large number of people in the region who regularly resides in nature should be with proper training, use it as an important and indispensable factor of its conservation. With sports - fishing, as well as the hunting, we should build primarily rural - tourism offer in the form of accommodation, local specialties kitchens and the like, service with a night in a hotel or hunting and forestry buildings.

4. Conclusion

Diversity and complementarily of natural resources is the main characteristic of the considered region, which is of special importance to its future economic development. Of course, no matter how natural resources in this part of northeastern Montenegro were great, they are not unlimited and inexhaustible. Therefore, their use must be planned and rational. Harmonious economic development in the future will depend on many factors, which are at the end of the twentieth century proved to be limiting. State of the economy at the beginning of the first century twenty is fraught with numerous problems. The concept of transition based on liberalization and privatization, has not brought the expected results. The slow and poorly conducted privatization processes that have caused the economic development Berane found at the very bottom in Montenegro, although once on this ground occupied third place, after Podgorica and Nikšić. The downfall of Berane industry started in the late eighties, when it came to closing the "pulp and paper" later "Beranka" and "New Beranka", which was the biggest economic giant in the region. Stopped is production in 1989. The factory was privatized in 2004, has since been repeatedly moved to production, but it all ended up on trying. Industrial zone "Rudeš" is a collection of dozens of abandoned factory halls, some of which been turned into scrap metal. There is no production in the coalmine "Ivangrad" as in IMG "Bricks". Neither are former leather factory drive "Polimka" not long ago published widely famous leather goods. Among settled collectives and found the factory for retreading "Gumig". Several agricultural cooperatives in Berane, no longer exist. The entire property JSC "Building" The door is in bankruptcy HTC Berane for many years, no one opens. And "Obod" drive on Rudeš has long been out of order. The only bright spot is the factory "Polieks" with the police, who are engaged in the production of explosives and initiators. Besides "Polieks" and several small private companies in Berane are still working. Among them is the "Asphalt base" in Lower Ržanica, "shirt factory" Petnjica and "FIC - Polimlje" a company engaged in wood processing (www.beranetown.net). In Plav municipality companies that are backbone of development, have experienced a failed privatization or are in bankruptcy: "FIC - Bor", "A-SC - Alpet", "A-SC -Plav," agricultural cooperatives "Murina" Confection "Maxim" and the Murina in "Termoplast " from Gusinja "Metals Processing". The sale of part of the assets of the Agricultural Cooperative "Plav" for the recovery of claims of creditors and employees (www.gusinje-foundation.org). A similar phenomenon we see in the municipality Andrijevica. Unsuccessful privatization or bankruptcy characterized once very successful business collectives, "Soko Štark", "Termovent" list, leather factory "Polimka", "Stationery", Agricultural Cooperatives "Vasojevka" agricultural cooperatives "Andrijevica" agricultural cooperatives "Konjuhe". Shut down the company and "Marble". This state of the economy has caused high unemployment that has caused major socio-economic problems in the region. Intensification of socio-economic problems was emphasizing by "ill-treatment" of agriculture. "Given the current state of the economy ... balance and the current model, based on the sale of the company and copious outflow of capital, and its spillover into consumption, escalation global economic crisis has irreversibly exhausted (Bulatović and Rajović, 2007). At this point, we affirm clearly formulated position Grčić (1991), which indicates that the development problems and irrational economic system kept all the technical and scientific narratives, without being able to you any concrete action to implement. Then, and seems now, we were not able to rise above statement. Hence, the conclusion that it is necessary to develop a special economic strategy for innovative regional policy adapted to the hilly - mountainous regions, such as the right and discussed. Forecasting the future development of the region is difficult. Therefore, the conclusion that requires immediate access to defining regional development strategy to build economic competitiveness. By Cvijanović (2004): "There is a widespread perception that the organization is a valuable resource, as important as raw materials, energy, equipment, technology and personnel, and perhaps more important because it brings together all the resources and makes them meaningful .... In practical dealing with organizational issues, still dominate the descriptive and analytical approach, and lacks an active attitude towards the perceived organizational changes. The effectiveness and efficiency of the approach to solving organizational problems ... is diminishing due to the different individual conceptions and approaches". Of course, we are advocating for programs that meet high

scientific criteria, and the willingness to appropriate institutions in the economic development of the considered space invests the necessary funds.

References

[1]   Pavlović M, Šabić D, Vujadinović S. Natural resources as a factor of socio-economic development Polimlja, *Journal Serbian Geographical Society*, 2, 2009, 3-15.

[2]   Gylfason T. Natural resources, education, and economic development, *European Economic Review*, 45(4–6), 2001, 847–859.

[3]   Rajović G. Mini projects in teaching geography, *Pedagogical reality*, 53 (3-4), 2007a, 205-213.

[4]   Barbier EB. Natural resources and economic development, *Cambridge University Press*, 2007.

[5]   Arezki R, Ploeg F. Do Natural Resources Depress Income Per Capita? *Review of Development Economics*, 15(3), 2001, 504–521.

[6]   Ploeg F. Natural Resources: Curse or Blessing? *Journal of Economic Literature*, 49 (2), 2001, 366-420(55).

[7]   Albrecht D. Amenities, Natural Resources, Economic Restructuring, and Socioeconomic Outcomes in Nonmetropolitan America, *Journal Community Development Society*, 35(2), 2004, .36-52.

[8]   Hopwood B, Mellor M, Brien O'G. Sustainable development: mapping different approaches, *Journal Sustainable Development*, 13 (1), 2005, 38–52.

[9]   Rajović G. Natural conditions for the development of agriculture in northeastern Montenegro, *Journal Industry*, 37 (4), 2009, 15-27.

[10] Rajović G. Natural conditions for the development and deployment of tourism in northeastern Montenegro, *Journal Industry*, 38 (4), 2010, 182-203.

[11] Rajović G, Bulatović J. Some economic-geographic factors development of the example rural areas northeastern Montenegro, *Russian Journal of Agricultural and Socio-Economic Sciences*, 9 (9), 2012 a, 3-20.

[12] Rajović G, Bulatović J. Some geographical factors economic development of rural areas in the manicipality of example Andrijevica (Montenegro), *Russian Journal of Agricultural and Socio- Economic Sciences*, 5 (5), 2012 b, 3-16.

[13] Rajović G, Bulatović J. Some Economic -Geographical Factors Development: The Case of Local Communities Kralje, *Journal of Studies in Social Sciences*, 2 (2), 2013 a, 105-133.

[14] Rajović G, Bulatović J. Geographic Favor of Analyzing Rural Space: The Case Rural Local Communities Kralje, *Journal of Sustainable Development Studies*, 3 (2), 2013 b, 136-167.

[15] Rajović G. Climate as the Value of Agricultural of the Example Northeastern Montenegro, *American-Eurasian Journal of Agricultural & Environmental Sciences*, 12 (12), 2012, 1558-1571.

[16] Rajović G. Evaluation og the morphometric attributes for tourism in Gornje Polimlje, *Journal Natura Montenegrina*, 5, 2006, 161-168.

[17] Regulation of Lima (On-line). Available from: http://www.beranetown.net (27.09.2011).

[18] *Regional Business Centre Berane*. Profile of Berane 1, 21-22, Available from: http://www.nasme.me (07.12. 2011), 2004.

[19] *Regional Business Centre Berane*. Profile of Plav 1, 21-22, Available from: http://www.nasme.me (07.12. 2011), 2004.

[20] Rajović G. Structure and trends of soil usage in Montenegro, *Journal Natura Montenegrina*, 9(5), 2010, 205-214.

[21] *Municipality Andrijevica*. Content spatial and urban plan 1, 49-52, Available from: http://www.andrijevica.me (26.12.2011), 2010.

[22] Marković ĐJ. Thermal mineral water Yugoslavia, *Proceedings Geographical Institute of Sciences*, 1979, pp.36-37.

[23] Leko TM, Ščerbakov A, Joksimović MH. Healing waters and climate of the Kingdom of Serbs, *Croats and Slovenes, the Ministry of Public Health*, 1922, 26.

[24] Nikolić P. Dimitrijević D. Coal Yugoslavia, *"Invention"*, Belgrade, 1990.

[25] *Coal Mine Ivangrad*. Available from: http:// www.mans.co.me (24.12 2011).

[26] Boričić R, Lutovac MV, Petrić D. Commune Ivangrad, *Workers University Ivangrad*, Ivangrad, 1967.

[27] Group of author. Review of mineral resources, *Federal Bureau Geo*, Belgrade, 1982.

[28] Lutovac V M. Andrijevica - characteristics and geographic location factors development, Establishment Andrijevica schools and Development 1863-1973, *Primary school "Bajo Jojić"*, Andrijevica,1973.

[29] Rajović G. Geographic View of the Industry Northeastern Montenegro with Special Emphasis on Handicrafts, *Journal of Studies in Social Sciences*, 4 (1) 2013c, 24-51.

[30] Todorović M. Development of modern agricultural complex in Gornji Milanovac, *Proceedings of the Geographic Institute "Jovan Cvijić "*, Serbian Academy of Sciences and Arts, 37, 1985, 41-78.

[31] Rajović G. Agricultural production factors intensification in North-Eastern Montenegro, *Journal of Agricultural and Food Science Research*, 1 (1), 2012, 11-25.

[32] Rajović G, Bulatović J. Structural changes utilization agricultural land and plant production: the case northeastern Montenegro, *International Letters of Social and Humanistic Sciences*, 3, 2013 d, 10-20.

[33] Rajović G, Bulatović J. Structural changes in livestock production: the case northeastern Montenegro, *International Letters of Social and Humanistic Sciences*, 3, 2013c, 37-45.

[34] Rajović G. Infrastructure projects in forestry Upper Polimlja, Proceedings of abstracts, *III May Conference of Strategic Management, University of Belgrade, Technical Faculty in Bor*, Jagodina, on 31 May -02 June, 2007b, 43.

[35] *Statistical Office of Montenegro*. List of Agriculture, Podgorica, 2006.

[36] Jaćimović B. Socio-geographic transformation and land use in Ostružnica in Belgrade Sava basin, *Journal of Geography Institute*, 17, 1971, 217-240.

[37] Statistical Office of Montenegro. List of forest reserves in 1979, *Edition Studies and Analysis*, Podgorica, 1983.

[38] Statistical Office of Montenegro. *List of growing stock*, Podgorica, 2006.

[39] Bulatović J, Rajović G. Environmental Awareness for Sustainability: A Pilot Survey in the Belgrade Settlement Brace Jerkovic, *International Journal of Advances in Management and Economics*, 2 (1), 2013, 20-27.

[40] *State of the economy in Plav municipality*. Available from: http://www.gusinje-foundation.org (25.12.2011).

[41] Bulatović J, Rajović G. Market development in Gornje Polimlje and the necessity of top management introduction in function of new operations, *Serbian Journal of Management*, 2, 2007, 147-155.

[42] Grčić M. Problems of development and deployment of industry in the mountains of Serbia, *Journal Serbian Geographical Society*, 71, 1991, 57-68.

[43] Cvijanović, J. Organizational Changes, *Institute of Economics*, Belgrade, 2004, 7.

23

Time dependent studies on the energy gain of D-T fuel using determination of total energy deposited of deuteron beam in hot spot

S. N. Hosseinimotlagh[1], M. Jahedi[2]

[1]Department of Physics, Colleges of Sciences, Islamic Azad University of Shiraz, Iran
[2]Department of Physics, Science and Research Branch, Islamic Azad University, Fars, Iran

Email address:
hoseinimotlagh@ hotmail.com (S. N. Hosseinimotlagh), marjanjahedi@yahoo.com (M. Jahedi)

Abstract: In fast ignition (FI) mechanism, a pellet containing the thermonuclear fuel is first compressed by a nanosecond laser pulse, and then irradiated by an intense "ignition" beam, initiated by a high power picosecond laser pulse, is one of the promising approaches to the realization of the inertial confinement fusion (ICF). If the ignition beam is composed of deuterons, an additional energy is delivered to the target, in which coming from fusion reactions of the beam-target type, directly initiated by particles from the ignition beam .In this article, the D+T fuel is selected and at first step we compute new average reactivity using three parameter cross section formula in terms of temperature at second step we use the obtained results of step one and calculate the total deposited energy of deuteron beam inside the target fuel at available physical conditions, then in third step we write the nonlinear point kinetic balance equation of D+T mixture and solve numerically these nonlinear differential coupled equations versus time .In forth step ,we estimate the power density and energy gain under physical optimum conditions and finally we conclude that maximum energy deposited in the target from D+T and D+D reaction are equal to 19269.39061 keV and 39198.58043 keV, respectively.

Keywords: Deuteron Beam, Fast Ignition, Gain, Dynamics

1. Introduction

The fast-ignition concept has been described thoroughly in literature as one alternative to direct-drive hot-spot ignition. In this scheme a high-energy, high-intensity laser is used to compress a cold shell containing fusion fuel to high areal densities. A short-pulse, ultrahigh-intensity laser is then used to generate megavolt electrons to heat the core of the dense fuel assembly in a time that is short compared to hydrodynamic time scales. The use of two independent laser drivers to compress the fuel assembly and subsequently heat the core allows for higher target gains, in principle, for the same amount of driver energy. This is because high fuel-areal-density cores can be assembled with slow implosion velocities and ignition is achieved by efficiently coupling the short-pulse beam energy to the dense core. In comparison to conventional hot-spot ignition, the symmetry requirement of the fuel assembly in fast ignition is not as stringent; this relaxes the illumination uniformity and power-balance constraints of the driver.

In 1975 A. W. Maschke suggested the use of relativistic heavy ion beams to ignite an inertially confined mass of thermonuclear fuel [1]. As in conventional inertial confinement fusion, the fuel was assumed to be precompressed by a factor of the order of 100 in order to minimize the energy needed for ignition. Maschke suggested that lasers might compress the fuel, by high velocity impact, or by ion beams other than the ignition beam. Maschke's "fast ignition" scheme was finally abandoned in favor of the more conventional approach to inertial fusion where the implosion supplies the energy for both compression and ignition. In 1994 an important paper by Tabak et al. [2] rekindled interest in fast ignition ,this time using short-pulse lasers to provide the ignition temperature. Later, at the 1997 HIF Symposium, Tabak estimated requirements for heavy-ion-driven fast ignition [3].This approach is now the subject of intense investigation in a number of countries. If ion beams could be made to deliver the energy density needed for ignition, they would have a number of distinct advantages. The reliability, durability, high repetition rate, and high driver efficiency are expected to be advantages of any accelerator driven inertial fusion system. In the case of fast ignition, there are some additional advantages. Moreover, for ions of the appropriate range, the beam energy can be deposited directly in the fuel, eliminating the inefficiency of converting laser light to electrons or ions that then deposit their energy in the fuel. Finally, because of the reduced requirements on illumination symmetry and stability, it may be possible to devise simple illumination schemes using direct drive or tightly coupled indirect drive, or use of single sided ion illumination for indirect drive fuel compression. This could simplify chamber design and, since direct drive and or tightly coupled indirect drive are efficient implosion methods, it could lead to lower driver energy and, as in the laser case, higher energy gain. We consider ion beam requirements for fast ignition in general, where a single short pulse of ions comes in from one direction onto one side of a pre-compressed D+T fuel mass, heating a portion of that fuel mass to conditions of ignition and propagating burn in a pulse shorter than the time for the heated region to expand significantly. The igniter beam (or beams) would be arranged to penetrate the target in such a way that the Bragg peak occurs at the usual target hot spot. Or one might increase the expansion time by tamping the ignition region. In fact Magelssen published a paper in 1984 [4] in which he presented calculations of a target driven by ion beams having two very different energies. The lower energy ions arrived first and imploded the target to a spherical configuration with a rather dense pusher or tamper surrounding the fuel. The higher energy beams were then focused onto the entire assembly, heating both the fuel and the pusher. The combination of the exploding pusher and the direct ion energy deposition heated the fuel to ignition.

In summary, the fast ignition (FI) mechanism, in which a pellet containing the thermonuclear fuel is first compressed by a nanosecond laser pulse, and then irradiated by an intense "ignition" beam, initiated by a high power picosecond laser pulse, is one of the promising approaches to the realization of the inertial confinement fusion (ICF). The ignition beam could consist of laser-accelerated electrons, protons, heavier ions, or could consist of the laser beam itself. It had been predicted that the FI mechanism would require much smaller overall laser energies to achieve ignition than the more conventional central hot spot approach, and that it could deliver a much higher fusion gain, due to peculiarities of the pressure and density distributions during the ignition. It is clear, however, that interactions of electrons and ions with plasma, and most importantly the energy deposition mechanisms are essentially different. Moreover, if the ignition beam is composed of deuterons, an additional energy is delivered to the target, coming from fusion reactions of the beam-target type, directly initiated by particles from the ignition beam [5]. These and other effects had been of course taken into account in later works on this topic [6,7]. In this work, we

choose the D+T fuel and at first step we compute the new average reactivity in terms of temperature for first time at second step we use the obtained results of step one and calculate the total deposited energy of deuteron beam inside the target fuel at available physical condition then in third of step we introduced the dynamical balance equation of D+T mixture and solve these nonlinear differential coupled equations by programming maple-15 versus time .in forth step we compute the power density and energy gain under physical optimum conditions and at final step we analyzed our obtained results .

2. New Calculation of Average Reactivity for D+T and D+D Reactions

In order to fuse two nuclei, they must be brought close to each other such that the strong nuclear force dominates the repulsive Coulomb force. Since this Coulomb force preventing two nuclei from getting closer is an increasing function of the atomic number Z, the most promising fusion reactions are those involving Hydrogen and its isotopes, namely Deuterium and Tritium. The most widely investigated reactions are so-called D+D reaction and D+T reaction which are listed below.

The D+D reaction:

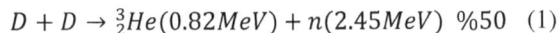

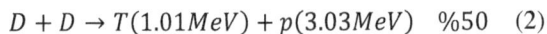

$$D + D \rightarrow {}^{3}_{2}He(0.82MeV) + n(2.45MeV) \quad \%50 \quad (1)$$

$$D + D \rightarrow T(1.01MeV) + p(3.03MeV) \quad \%50 \quad (2)$$

The D+T reaction:

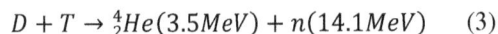

$$D + T \rightarrow {}^{4}_{2}He(3.5MeV) + n(14.1MeV) \quad (3)$$

The rate of thermonuclear reaction is a function of reaction cross section, temperature, and density. For a mixture of two species with densities n_i and n_j, the reaction rate is given by the following expression [6]:

$$R_{ij} = \frac{n_i n_j}{1+\delta_{ij}} <\sigma v> \quad (4)$$

where σ is the reaction cross section and δ_{ij} is the Kronecker symbol .If the fusion is "Thermonuclear", i.e. if the energy provided to the reacting particles is the result of heating the system to a temperature T, then we need to use the averaged $<\sigma v>$. Assuming a Maxwellian distribution, equation (4) becomes,

$$R_{ij} = \frac{n_i n_j}{1+\delta_{ij}} \left[\frac{8}{\pi\mu}\right]^{3/2} \left[\frac{1}{kT}\right]^{\frac{3}{2}} \int_0^\infty \sigma(E)E\exp\left(-\frac{E}{kT}\right)dE \quad (5)$$

By comparing equation (4) and (5) we have

$$<\sigma v> = \left[\frac{8}{\pi\mu}\right]^{3/2} \left[\frac{1}{kT}\right]^{\frac{3}{2}} \int_0^\infty \sigma(E)E\exp\left(-\frac{E}{kT}\right)dE \quad (6)$$

Since the D+T reaction has the highest reaction rate at relatively low fuel temperature, it is the easiest fuel to assemble and burn. However, the presence of neutrons as a product (see D+T reaction) makes it unattractive for fusion

reactors. The formula of fusion cross section for D-D and D-T fusion reactions is given by [8]:

$$\sigma(E_{lab}) = -16389C_3\left(1+\frac{m_a}{m_b}\right)^2 \times \left[m_aE_{lab}\left[Exp\left(31.40\,Z_1Z_2\sqrt{\frac{m_a}{E_{lab}}}\right)-1\right]\left\{(C_1+C_2E_{lab})^2+\left(C_3-\frac{2\pi}{[Exp(31.40Z_1Z_2\sqrt{m_a/E_{lab}})-1]}\right)^2\right\}\right]^{-1} \quad (7)$$

with 3 adjustable parameters (C_1, C_2 and C_3) only. In (7), m_a and m_b are the mass number for the incident and target nucleus, respectively (e.g. m_a = 2 for incident deuteron);E_{lab} (deuteron energy in lab system) is in units of KeV and σ is in units of barn. The numerical values of C_1, C_2 and C_3 for these reactions are listed in Table.1.

Table 1. Numerical values of C_1, C_2 and C_3 for reactions D+T and D+D[8].

	D+T	D+D (p+T and n+^{3}He)
C_1	−0.5405	−60.2641
C_2	0.005546	0.05066
C_3	−0.3909	−54.9932

From this formula, we calculated and plotted the variations of fusion cross sections for these reactions in terms of E_{lab} and our results are shown that in Figure.1.Also by comparing our calculated numerical values with available experimental results (see ref.[8]),we concluded that this formula is very exact.

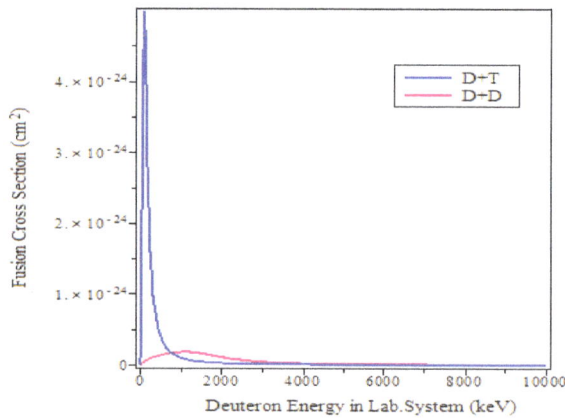

Figure 1. Variations of fusion cross sections versus deuteron energy in lab system for two fusion reactions: D+T and D+D.

Using data in T able.1 and inserting equation (7) into equation(6) and integrate it for these two reactions we can obtain the numerical values of averaged reactivity parameter in terms of temperature and plotted it for D+T and D+D in the Figure.2.Note that $E_{lab} = \frac{m_a+m_b}{m_b}E$,where$E$ is energy in the center of mass frame. By observing Figs.1 and 2 we conclude that the cross section and averaged reactivity of D+T fusion reaction is greater than D+D, also$< \sigma v >_{D+T}$ and $< \sigma v >_{D+D}$ are strongly temperature dependent .From Fig.2, it will therefore be recognize that at resonant temperature in which the probability for occurring fusion is maximized for D+T fusion reaction is approximately 60KeV.

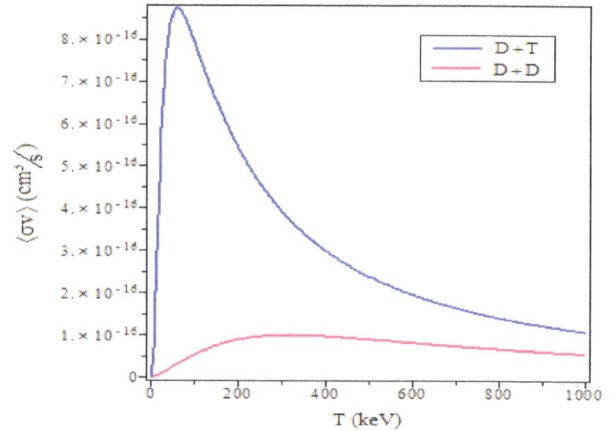

Figure 2. Variations of averaged reactivity in terms of temperature for $< \sigma v >_{D+T}, < \sigma v >_{D+D}$.

3. Gain Requirement

In order to have economically fusion, an attractive energy source is necessary that the energy from thermonuclear reaction exceeds both the energy invested in achieving it and the losses due to radiation, reactor wall, conversion inefficiencies, etc. A simple estimate of the driver efficiency and target gain can be obtained using energy bookkeeping approach (Fig.3).

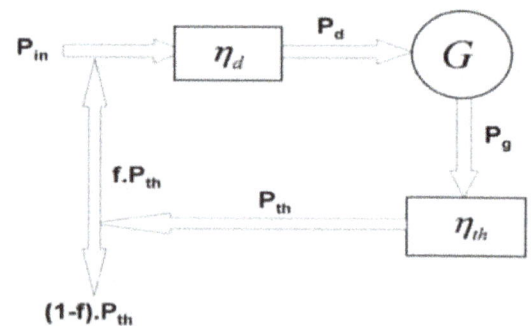

Figure 3. Power cycle.P_{in}, the input power (this could be the electrical power from the outlet).P_d, the power at the output of the driver.P_g,the power fromthermonuclear fusion. P_{th} ,the power after converting the thermal energy to electricalenergy.

Let P_{in} be the electrical power that we intend to invest in achieving fusion and let η_d be the efficiency of conversion of this power to laser light power P_d. If we definethe target gain as the ratio of the power obtained from the fusion reaction to the driver power, then

$$P_g = GP_d = G\eta_dP_{in} \quad (8)$$

taking into account the conversion efficiency from thermal to electrical power, the available electrical power is given by,

$$P_{th} = \eta_{th}P_g = G\eta_{th}\eta_d P_{in} \qquad (9)$$

A fraction $fG\,\eta_{th}\,\eta_d P_{in}$ of this power is used to run the driver and the remaining fraction $(1-f)G\,\eta_{th}\,\eta_d P_{in}$ is send to the customers. The closer factor $(1-f)$ to unity, more electricity is available to customers. This can be achieved using small values of f, however, f has a lower bound determined by the equation,

$$P_{in} = f_{min}G\eta_{th}\eta_d P_{in} \rightarrow f_{min} = \frac{1}{G\eta_{th}\eta_d} \qquad (10)$$

Using a typical η_{th} value of 40% and recycling less than 1/4 of the power to run the driver, we get, $G\eta_d \geq 10$. Solid state lasers for example have an efficiency of 1-10% [9] and to fulfill the available condition, gains of 100-1000 are needed.

4. Deuteron Beam and Deposited Energy in Pre-Compressed D+T Fuel

As we mentioned that in section 1, deuterons have been considered for fast ignition as well [9–17]. Bychenkovs group, considered an accelerated deuteron beam, but decided that deuterons would have too high an energy (7–8 MeV) to form the desired hot spot [11]. Deuterons would not only provide proven ballistic focusing, but also fuse with the target fuel (both D and T) as they slow down [18], providing a "bonus" energy gain. Depending on the target

plasma conditions, this added fusion gain can be a significant contribution [19]. We must notice that the idea of bonus energy for first time is presented by Xiaoling Yang and her group in low temperatures [20]. In this work we use of this idea, to compute the added energy released as the energetic deuterons interact with the target fuel ions in different range of temperatures (0-70keV) that is contained resonant temperature. This added energy increases total energy gain of the system. We use a modified energy multiplication factor φ to estimate the bonus energy in terms of the added "hot spot" heating by beam-target fusion reaction for D+T [18]. The deuteron beam deposited energy and stopping range and time are also calculated for this reaction. Also we determine energy gain time dependent in different temperatures.

F value is the ratio between the fusion energy E_f produced and the ion energy input E_I to the plasma and for D+T fuel is given by [18]:

$$F_{D+T} = n_T \frac{\int_{E_{th}}^{E_I} S(E)dE}{E_I} \qquad (11)$$

where E_I and E_{th} are, the average initial energy and the asymptotic thermalized energy of the injected single ion for each reaction, respectively [18,21,22] and

$$S(E) \equiv \sum_k K_k [< \sigma v(E) >_b]_{Ik}(E_f)_{Ik}/\left(\frac{dE}{dt}\right) \qquad (12)$$

Such that for this reaction we have:

$$\frac{1}{n_T}\left(\frac{dE}{dt}\right) = -\frac{Z_I^2 e^4 m_e^{1/2} E \ln \Lambda_{D+T}}{3\pi(2\pi)^{1/2}\varepsilon_0^2 m_I (kT_e)^{3/2}} \times \left[1 + \frac{3\sqrt{\pi}m_I^{3/2}(kT_e)^{3/2}}{4m_k m_e^{1/2}E^{3/2}}\right] \qquad (13)$$

Where m_e and m_I are mass of electron and mass of injected ion, respectively and both of them are in atomic mass unit (amu). $< \sigma V >_{Ik}$ is the fusion reactivity for the injected ion I of species k having atomic fraction K_k in the target, $(E_f)_{Ik}$ is the corresponding energy released per fusion, and T_e is the target electron temperature [22]. Combining equations (11),(12) and (13) we can estimate F_{D+T}.

$\ln \Lambda_{D+T}$ is Coulomb logarithm for D+T fusion reaction

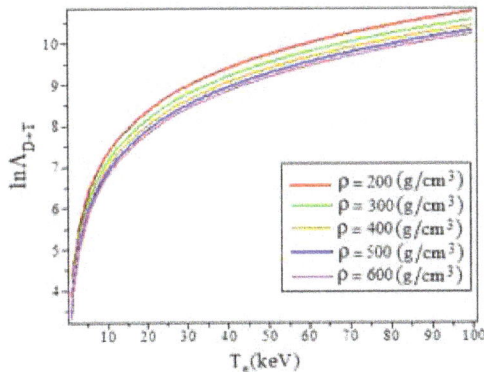

that is given by [20]:

$$\ln \Lambda_{D+T} \approx 6.5 - \ln\left(Z_k\sqrt{\rho}/T_e^{3/2}\right) \qquad (14)$$

Where Z_k is the atomic number of target fuel ions and T_e here is in kilo-electron-volt, ρ is in $\frac{g}{cm^3}$. We plotted the two and three dimensional variations of $\ln \Lambda_{D+T}$ in terms of target density and temperature (see Figure 4).

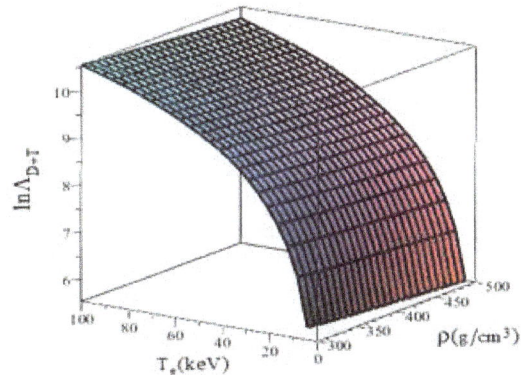

a)

b)

Figure 4. a) The two and b) three dimensional variations of $\ln \Lambda_{D+T}$ in terms of target density and temperature.

From this figure we see that by temperature enhancement and target density reduction, $\ln \Lambda_{D+T}$ is increased. The $(E_f)_{lk}$ in Eq. (12) gives the energy released in the fusion reaction carried by both fast neutrons and charged particles. However, for hot spot heating, the ρR of the hot spot is too low to significantly slow down the neutrons, so only charged particles contribute. Thus, for D+T reaction in the target(see reaction (3)) only the 20% of the fusion energy carried by the alphas is useful for heating while for the D+D reaction about 63% of the total is useful(see reactions (1) and (2))[20].Therefore, to prevent confusion, we introduce a new factor φ to represent the energy multiplication for the hot spot heating by the charged particles, and then $\varphi_{D+T} = 20\%\ F_{D+T}$ for D+T fusion and for occurring D+D fusion in D+T mixture $\varphi_{D+D} = 63\%\ F_{D+D}$.

In summary, the total energy that could be deposited into the target due to combined deuteron ion heating and beam-target fusion for D+T and D+D, respectively becomes:

$$\varepsilon_{D+T} = E_I(1 + \varphi_{D+T}) \qquad (15)$$

$$\varepsilon_{D+D} = E_I(1 + \varphi_{D+D}) \qquad (16)$$

so it is seen that parameters φ_{D+T}, φ_{D+D} play the major role in the "bonus energy" for deuteron driven fast ignition in these two reactions. To avoid confusion, please note that the ε here is the total energy deposited by the ion beam plus any contribution from its beam-target fusion in the hot spot, but not the total input energy to the target that is often cited in energy studies and represents the total laser compression plus fast ignition energy delivered to the total target, also the deuteron stopping range and stopping time can be calculated by following equations [20]:

$$R_S = \int_{E_{th}}^{E_I} v_D\, dE / \left(\frac{dE}{dt}\right) \qquad (17)$$

$$t_S = \int_{E_{th}}^{E_I} dE / \left(\frac{dE}{dt}\right) \qquad (18)$$

Where, $\left(\frac{dE}{dt}\right)$ is calculated from equation (13) for D+T, and the deuteron velocity is $v_D = \sqrt{\frac{2E}{m_D}}$.

For calculating the total energy deposited into the target of D+T at first step we substitute equation (14) into equation (13),then at second step the obtained result is substituted into equation (12) and at third step the results of second step are inserted into equation (11) and we compute F_{D+T} and F_{D+D} for D+T reaction, at forth step we use of these parameters for determination of φ_{D+T} and φ_{D+D}, finally the obtained results from forth step inserted into equations (15) and (16) thus we can obtain the numerical values of ε_{D+T} and ε_{D+D} in D+T mixture for $300 \leq \rho(\frac{g}{cm^3}) \leq 500$, $0 \leq T_e(keV) \leq 70$ and deuteron energy E, with range of $0 \leq E(MeV) \leq 10$. Also under these conditions we can calculate the deuteron stopping range and stopping time by using equations(17) and (18).Figure.5 shows the results of our calculations

of φ_{D+T}, φ_{D+D} , ε_{D+T} and ε_{D+D} in the case of $\rho = 500(g/cm^3)$ and in Table.2numerical values computed for maximum of φ_{D+T}, φ_{D+D} , ε_{D+T} and ε_{D+D} are given for studying D+T reaction at different temperatures and target density $300 \leq \rho(g/cm^3) \leq 500$.

We see clearly that from Fig.5,variations of, ε_{D+T} and ε_{D+D} by increasing temperature is such that we have the maximum value of them in resonance temperature of 60 KeV. We must be notice that the temperature 60 keV is a resonant temperature for D+T reaction such that in this temperature we have maximum probability for occurring fusion reaction. Also from Table.2 we find that the maximum total deposited energy of ε_{D+T} and ε_{D+D} inside hot spot and also the maximum of φ_{D+T} and φ_{D+D} by increasing target density from 300 to 500 (g/cm^3) are raised very slowly.The total energy deposited inside the hot spot (ε_{D+T} and ε_{D+D}) is also function of deuteron energy beam (E_I)such that by increasing energy from 1 to 10000KeV is increased.Fig.5 shows that the values of multiplication factors φ_{D+T}, φ_{D+D} increased by increasing deuteron energy (please note that our calculations show that in temperature lower than 5keV at first by raising the deuteron energy, parameters φ_{D+T} and φ_{D+D} are increased and then slowly decreased).In all temperatures ε_{D+D} is higher than ε_{D+T}.From Fig.6 we see that by increasing temperature stopping time ($t_{S_{D+T}}$) is increased because by raising temperature particle energy is increased then the stopping time become large. Since that by increasing temperature stopping range is increased then particle can move more distance (see Fig.7). The effective parameter that decreased stopping time and stopping range, is n_T. Our calculations show that by changing n_T from 10^{22} to $10^{24} cm^{-3}$, $t_{S_{D+T}}$ and $R_{S_{D+T}}$ are decreased by the order of 10 to 100. By increasing target density from 300 to 500 (g/cm^3), $R_{S_{D+T}}$ and $t_{S_{D+T}}$ are increased, but this changes for $t_{S_{D+T}}$ is not very sensitive.

5. Balance Equations of Deuterium-Tritium Mixture

The following system of equations is used to describe the temporal evolution of plasma parameters averaged over the volume (the density of deuterium ions and tritium are n_D and n_T, respectively .n_α is density of thermal alpha-particles, E is plasma energy, for D+T nuclear fusion reaction:

$$\frac{dn_D}{dt} = -\frac{n_D}{\tau_P} - n_D n_T \langle \sigma v \rangle_{D+T} + S_D \qquad (19)$$

$$\frac{dn_T}{dt} = -\frac{n_T}{\tau_P} - n_D n_T \langle \sigma v \rangle_{D+T} + S_T \qquad (20)$$

$$\frac{dn_\alpha}{dt} = -\frac{n_\alpha}{\tau_\alpha} + n_D n_T \langle \sigma v \rangle_{D+T} \qquad (21)$$

Energy balance equation is written as

$$\frac{dE}{dt} = -\frac{E}{\tau_E} + Q_\alpha n_D n_T \langle \sigma v \rangle_{D+T} - P_{rad} \quad (22)$$

S_D and S_T are the source terms which give us the fuel rates; τ_α, τ_P, and τ_E are lifetimes of thermal alpha-particles, deuterium and tritium, and the energy confinement time, respectively, also the energy of the alpha particles are: $Q_\alpha = 3.52 MeV$. The radiation loss P_{rad} is given by:

$$P_{rad} = P_{brem} = A_b Z_{eff} n_e^2 \sqrt{T} \ for D + T \quad (23)$$

Where $A_b = 4.85 \times 10^{-37} \frac{Wm^3}{\sqrt{KeV}}$ is bremsstrahlung radiation coefficient. No explicit evolution equation is provided for the electron density n_e since we can obtain it from the neutrality condition $n_e = n_D + n_T + 2n_\alpha$, whereas the effective atomic number, the total density and the energy are written as

$$Z_{eff} = \frac{\sum_i n_i z_i^2}{n_e} = \frac{n_D + n_T + 4n_\alpha}{n_e} \quad (24)$$

where, Z_i is the atomic number of different ions. The fusion energy gain is defined as:

$G(t) = \frac{E_f(t)}{E_{driver}}$ where $E_f(t)$ is equal to the energy due to the number of occurred fusion reactions in target in terms of time and E_{driver} is the required energy for triggering fusion reactions in hot spot and is equal to 4MJ[23].Also the fusion power density for D+T reaction is given by $P_{D+T} = n_D(t)n_T(t) < \sigma v >_{D+T} Q_{D+T}$ where $Q_{D+T} = 17.6 MeV$ .We solve time dependent nonlinear coupled differential equations (19) to (22) with the use of computers (programming, Maple-15) under available physical conditions. Our computational obtained results are given , in Figs.8 to 10 and Table.3.

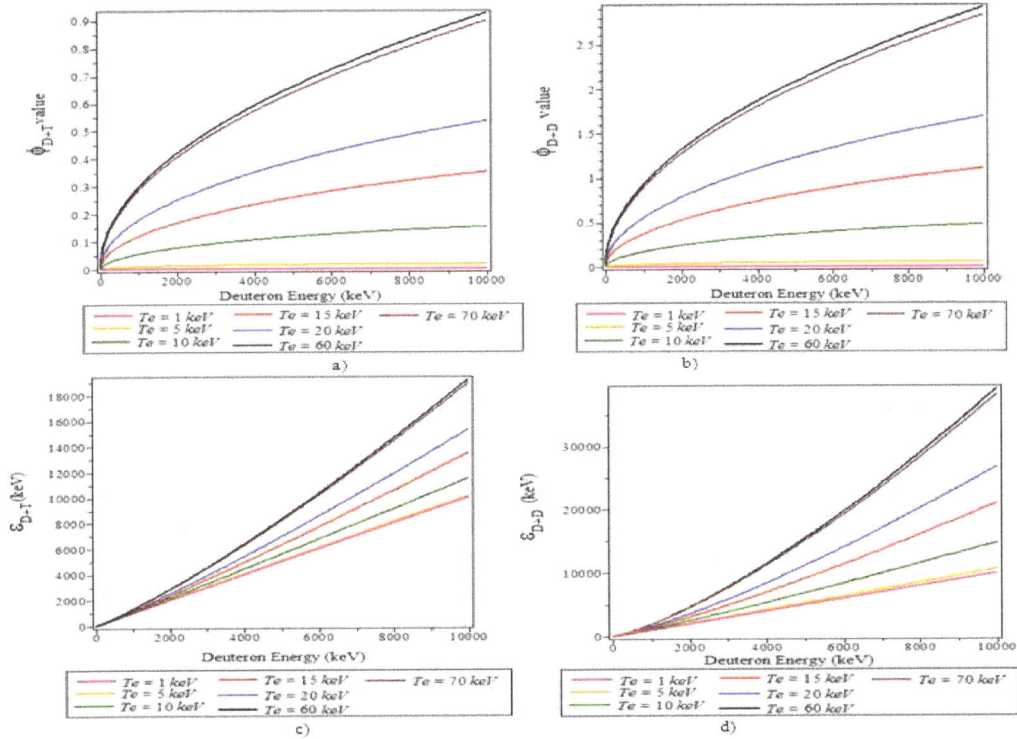

Figure 5. *The two dimensional variations of a) φ_{D+T} b)φ_{D+D} c) ε_{D+T} d) ε_{D+D} in terms of deuteron energy in different temperatures in D+T mixture for $\rho = 500 \left(\frac{g}{cm^3}\right)$*

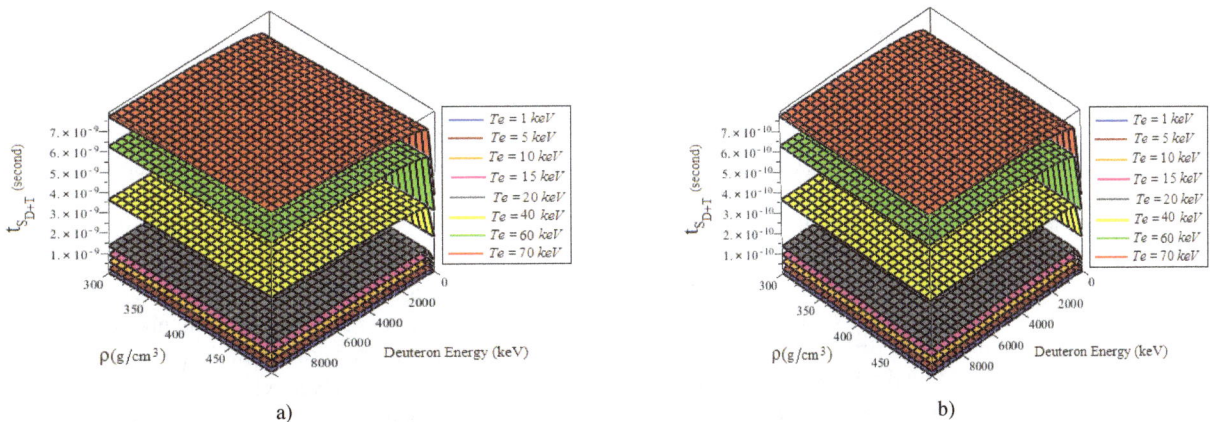

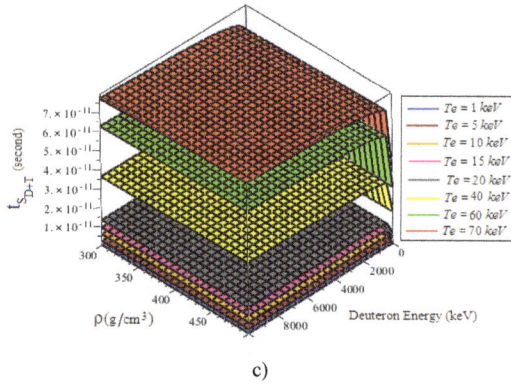

c)

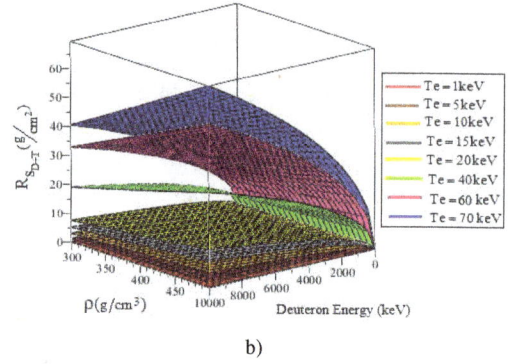

b)

Figure 6. The three dimensional variations of stopping time versus target density and deuteron energy in different temperatures in D+T mixture at a)$n_T = 10^{22}(cm^{-3})$, b)$n_T = 10^{23}(cm^{-3})$, c)$n_T = 10^{24}(cm^{-3})$.

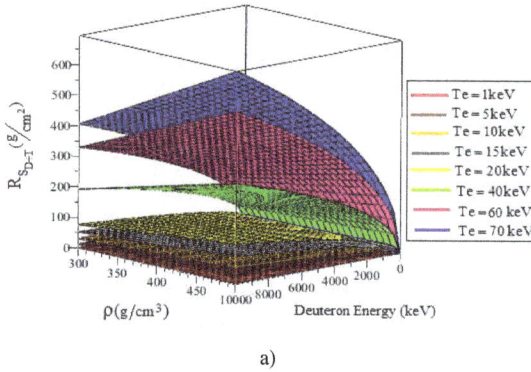

a)

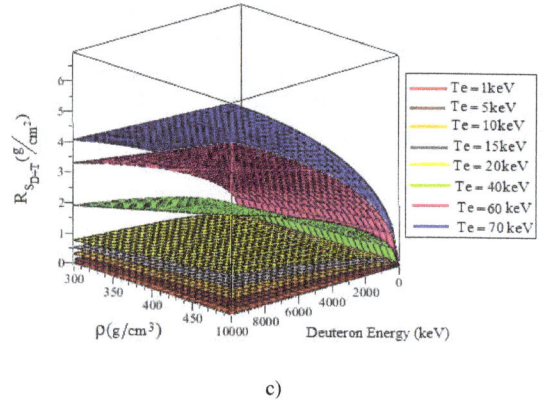

c)

Figure 7. The three dimensional variations of stopping range versus target density and deuteron energy in different temperatures in D+T mixtureat a) $n_T = 10^{22}(cm^{-3})$, b) $n_T = 10^{23}(cm^{-3})$, c) $n_T = 10^{24}(cm^{-3})$.

Table 2. Maximum numerical values of total energy deposited and multiplication factors in D+T mixture at different temperature for $300 \leq \rho(\frac{g}{cm^3}) \leq 500$.

		D+T			
$\rho(g/cm^3)$	$T_e(keV)$	$\varepsilon_{D+T_{max}}(keV)$	$\varepsilon_{D+D_{max}}(keV)$	$\varphi_{D+T_{max}}$	$\varphi_{D+D_{max}}$
300	10	11463.92692	14611.36978	0.146392691	0.461136978
300	20	15185.35591	26333.87112	0.518535591	1.633387112
300	60	19020.60063	38414.89199	0.902060063	2.841489199
300	70	18763.52452	37605.10224	0.876352452	2.760510224
400	10	11494.18969	14706.69751	0.149418969	0.470669751
400	20	15278.61416	26627.63459	0.527861416	1.662763459
400	60	19159.04394	38850.98842	0.915904394	2.885098842
400	70	18891.14962	38007.12129	0.889114962	2.800712129
500	10	11518.539	14783.39786	0.151853900	0.478339786
500	20	15353.29364	26862.87495	0.535329363	1.686287495
500	60	19269.39061	39198.58043	0.926939061	2.919858043
500	70	18992.73249	38327.10736	0.899273249	2.832710736

From Figs.8 to 10 we see clearly, increasing temperature from 1 keV to 70 keV the variations of deuterium and tritium density in terms of time ($n_D(t), n_T(t)$) are decreased since that by increasing time the consumption rate of $n_D(t)$ and $n_T(t)$ are increased but variations of $n_D(t)$and $n_T(t)$, in all temperature are similar and as we see in Figs.8 to 10 they coincide each other. Thus, by increasing temperature 1 keV to 70 keV the variations of alpha density ($n_\alpha(t)$) versus time at first by increasing time is increased and then decreased while the production rate of fusion plasma energy ($E_f(t)$) is increased until in resonant

temperature ,60 KeV, is maximized because in this temperature the highest number of D+T fusion reaction is occurred. Also, our calculations show that enhancement of the injection rate of deuterium and tritium(S_D and S_T) from 10^{22} to $10^{24}cm^{-3}$ the rate of variations of $n_D(t)$ and $n_T(t)$ in terms of time are increased while $n_\alpha(t)$ and $E_f(t)$ are raised. Therefore, for having optimum value of power density and fusion energy gain using the above discussions we select injection rate of deuterium and tritium and resonant temperature respectively equal to $10^{24}cm^{-3}$ and T_e =60keV(see Table.3).

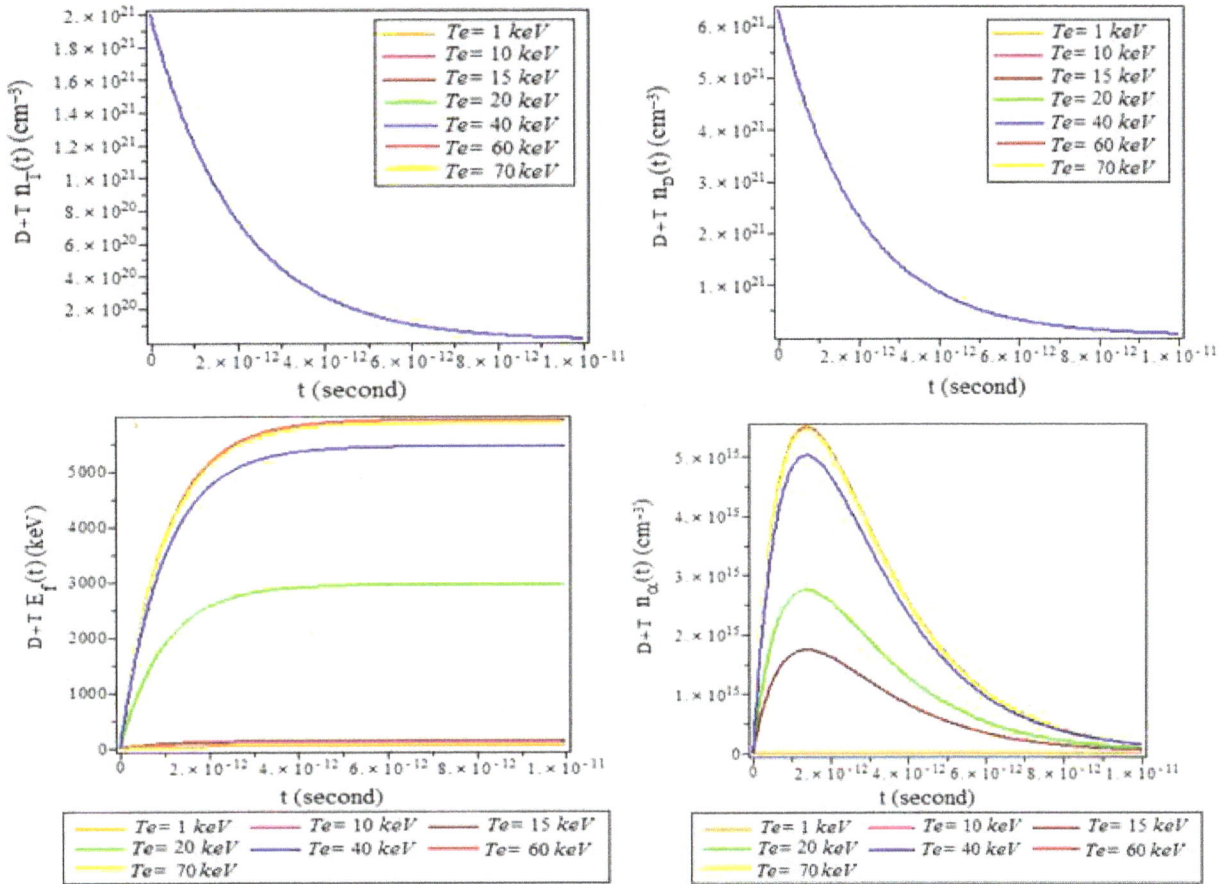

Figure 8. *The two dimensional variations of deuterium , tritium ,alpha particles densities and plasma energy in terms of time at different temperatures for D+T mixture under choosing $S_T = 0.20 \times 10^{22}(cm^{-3})$ and $S_D = 0.63 \times 10^{22}(cm^{-3})$.*

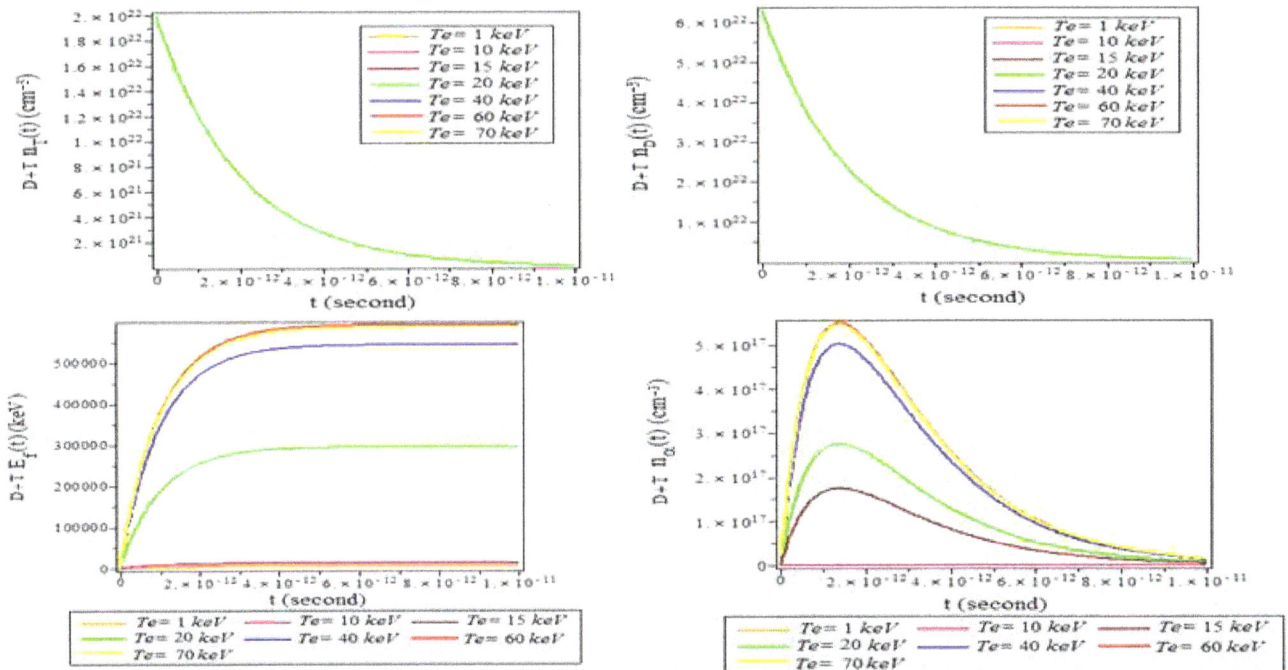

Figure 9. *The two dimensional variations of deuterium , tritium , alpha particles densities and plasma energy in terms of time at different temperatures for D+T mixture under choosing $S_T = 0.20 \times 10^{23}(cm^{-3})$ and $S_D = 0.63 \times 10^{23}(cm^{-3})$.*

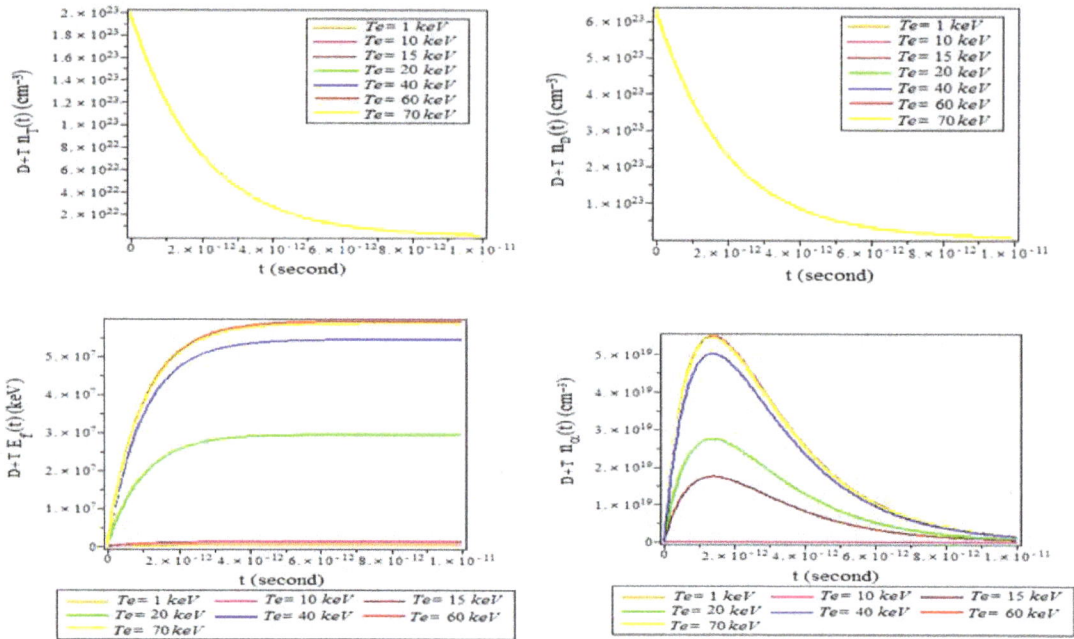

Figure 10. The two dimensional variations of deuterium, tritium , alpha particles densities and plasma energy in terms of time at different temperatures for D+T mixture under choosing $S_T = 0.20 \times 10^{24}(cm^{-3})$ and $S_D = 0.63 \times 10^{24}(cm^{-3})$.

Table 3. Time dependent of numerical values for fusion power density and target energy gain.

D+T $S_D = 0.63 \times 10^{24}(cm^{-3})$, $S_T = 0.20 \times 10^{24}(cm^{-3})$			
T_e (keV)	t (s)	$P_{D+T}(t)(\frac{W}{cm^3})$	$G_{D+T}(t)$
20	10^{-20}	1549.2E17	0.11802E-22
20	10^{-11}	702.75E13	1.1798E-15
20	60	619.61E-6	0.572432E-7
20	110	619.61E-6	0.99075E1
40	10^{-20}	2827.6E17	0.21788E-22
40	10^{-11}	1282.1E13	2.1777E-15
40	60	1131.4E-6	1.0566E-6
40	110	1131.4E-6	1.8288E1
60	10^{-20}	3092.4E17	0.23718E-22
60	10^{-11}	1401.9E13	2.3705E-15
60	60	1236.9E-6	1.1501E-6
60	110	1236.9E-6	1.9906E1
70	10^{-20}	3073.8E17	0.23485E-22
70	10^{-11}	1393.5E13	2.3473E-15
70	60	1229.5E-6	1.1389E-6
70	110	1229.5E-6	1.9711E1

6. Conclusion

From this work, we conclude that resonant temperature for D+T mixture is 60 keV and in this temperature we have maximum energy gain. Our computational results show that numerical values of stopping time ($t_{s_{D+T}}$) are increased very strongly by raising temperature , while slowly increase by raising target density ρ. Another effective parameter on the value of stopping time is the deuteron energy such that by increasing this energy stopping time is increased. But by raising n_T in D+T reaction numerical values of stopping time strongly decreased. The important parameter that effect on the value of stopping range , is target density, by increasing target density $R_{s_{D+T}}$ is strongly increased. The multiplication factors and total deposited energies are time independent because in calculating these parameters n_T is omitted. By observing the related figures and tables we will conclude that the maximum variations of particle densities , plasma energy ,power density and target energy gain are occurred in resonant temperature. From table.3 we have maximum gain equal to 19.906 at resonant temperature 60 keV under optimum conditions $S_D = 0.63 \times 10^{24}(cm^{-3})$, $S_T = 0.20 \times 10^{24}(cm^{-3})$,t=110s.Finally, we concluded that the maximum energy deposited in the target from D+T and D+D reaction are equal to 19269.39061keVand 39198.58043keV,respectively. So, deposited energy can reduce laser driver energy.

Nomenclature

$\sigma(E_{lab})$ Fusion cross section
C_1, C_2, C_3 3 adjustable parameters
m_a Mass number for the incident nucleus
m_b Mass number for the target nucleus
E_{lab} Deuteron energy in laboratory system
Z_1 , Z_2 Charge numbers of the colliding nuclei
$< \sigma v >$ Reactivity
m_r Reduced mass
k_B Boltzmann constant
T Temperature
ε Energy in the center of mass frame
P_{in} Input power
P_d Driver power output
P_g Thermonuclear fusion power
P_{th} Thermal power
E_I Average initial energy of the injected single ion
E_{th} Asymptotic thermalized energy of the injected single ion
F The ratio between the fusion energy produced and the ion energy input to the plasma

K_k Atomic fraction

$(E_f)_{Ik}$ Energy released per fusion for the injected ion I of species k

m_e The mass of electron

$\ln \Lambda_{D+T}$ Coulomb logarithm for D+T fusion reaction

m_I The mass of the injected ion

T_e The target electron temperature

Z_k The atomic number of target fuel ions

ρ Target density

φ_{D+T}, φ_{D+D} Energy multiplication factor for the hot spot heating by the charged particles in D+T and D+D

ε_{D+T}, ε_{D+D} The total energy that could be deposited into the target due to combined deuteron ion heating and beam-target fusion for D+T and D+D

R_S The deuteron stopping range

t_S The deuteron stopping time

v_D The deuteron velocity

E The deuteron energy

n_D The density of deuterium ions

n_T The density of tritium ions

n_α The density of thermal alpha-particles

S_D, S_T The source terms which give us the fuel rates

τ_p The lifetimes of deuterium and tritium

τ_α The lifetime of thermal alpha-particles

τ_E The energy confinement time

Q_α The energy of the alpha particles

P_{rad} The radiation loss

A_b Bemsstrahlung radiation coefficient

n_e The electron density

Z_{eff} The effective atomic number

Z_i The atomic number of the different ions

$G(t)$ The fusion energy gain in terms of time

$E_f(t)$ The energy due to the number of occurred fusion reactions in target in terms of time

E_{driver} The required energy for triggering fusion reactions in hot spot

$P_{D+T}(t)$ The fusion power density in terms of time for D+T reaction

References

[1]   A. W. Maschke, Proceedings of the (1975) Particle Accelerator Conference, page 1875, IEEE report NS-22, June (1975)

[2]   M. Tabak, et al., Phys. Plasmas 1, 1626 (1994)

[3]   M. Tabak, et. al. Int. HIF Symposium (Heidelberg) paper (1997)

[4]   T. C. Magelssen, "Targets Driven by Dual-Energy Heavy Ions", Nuclear Fusion24, 1527(1984).

[5]   M. Sherlock et al., Phys. Rev. Lett. 99, 255003 (2007).

[6]   S. Atzeni, M. Temporal, and J. Honrubia, Nucl. Fusion 42, L1 (2002).

[7]   S. Atzeni et al., Nucl. Fusion 49, 055008 (2009).

[8]   Xing Z. Li, Qing M. Wei and Bin Liu, Nucl. Fusion 48, 125003 ,5pp (2008).

[9]   G. Velarde, Y. Ronen, and J. Martinez-Val. Nuclear fusion by Inertial Confinement: A comprehensive treatise.

[10] Logan, B. Grant Bangerter, Roger O.Callahan, Debra A.Tabak, Max Roth, Markus Perkins, L. John Caporaso, George ,Lawrence Berkeley National Laboratory (2005).

[11] V. Bychenkov, W. Rozmus, A. Maksimchuk, D. Umstadter, and C.Capjack, Plasma Phys. Rep. 27, 1017 (2001).

[12] N. Naumova, T. Schlegel, V. T. Tikhonchuk, C. Labaune, I. V. Sokolov,and G. Mourou, Phys. Rev. Lett. 102, 025002 (2009).

[13] V. T. Tikhonchuk, T. Schlegel, C. Regan, M. Temporal, J.-L. Feugeas, P.Nicolaï, and X. Ribeyre, Nucl. Fusion 50, 045003 (2010).

[14] M. L. Shmatov, J. Br. Interplanet. Soc. 60,180 (2007).

[15] M TEMPORAL J Honrubia, S Atzeni. Phys. Plas.9,3102(2002)

[16] A. Maksimchuk, S. Gu, K. Flippo, D. Umstadter, and V. Y. Bychenkov, Phys. Rev. Lett. 84, 4108 (2000).

[17] H. Schwoerer, S. Pfotenhauer, O. Jackel, K. U. Amthor, Ziegler, R. Sauerbrey, K. W. D. Ledingham, and T. Esirkepov ,Nature London439,445 (2006)

[18] C. Bathke, H. Towner, and G. H. Miley, Trans. Am. Nucl. Soc. 17, 41 (1973).

[19] M. L. Shmatov, J. Br. Interplanet. Soc. 57,362(2004).

[20] Xiaoling Yang, George H. Miley, Kirk A. Flippo, and Heinrich Hora, PHYSICS OF PLASMAS 18, 032703 (2011)

[21] D. J. Rose and M. Clark, Jr., Plasmas and Controlled Fusion MIT Press, Cambridge, MA(1965).

[22] G. H. Miley, Fusion Energy Conversion American Nuclear Society, Hinsdale, IL(1976).

[23] S .Pfalzner, "An Introduction to Inertial Confnement Fusion", Published by CRC Press Taylor & Francis Group(2006).

24

The feasibility study of a grid connected PV system to meet the power demand in Bangladesh

Mohammad Shuhrawardy, Kazi Tanvir Ahmmed

Department of Applied Physics, Electronics and Communication Engineering, University of Chittagong, Chittagong, Bangladesh

Email address:
mswardy@cu.ac.bd (M. Shuhrawardy), tanvir@cu.ac.bd (K. T. Ahmmed)

Abstract: This paper represents the feasibility study of a grid connected PV system with battery backup in the south-east part of Bangladesh. In Bangladesh, only 53% of the total population gets access to grid power and yet, the current consumers cannot be provided with uninterrupted and quality supply of electricity due to the inadequate generation compared to the national demand. The world is gradually moving towards sustainable renewable energy sources due to diminishing fossil fuel energy resources and increasing demand for power. Most of the power stations in Bangladesh are based on fossil fuels. Fossil fuels are not environment friendly and are responsible for global warming. So a renewable grid connected power system with battery backup can be a better option to provide continuous power in a load shedding prone country like Bangladesh and also to reduce the emission of greenhouse gases. Our proposed system is simulated using HOMER optimization tool and the simulation results and analysis of the system are presented in the paper.

Keywords: Renewable Energy, Non-renewable Energy, Fossil Fuels, Grid Connected System, Global Warming, Independent Power Producers' (IPPs), Rental Power Producers (RPPs)

1. Introduction

Electric power supply is the key for economic development of any country. To sustain the economic development huge amount of electricity is required in all sectors. However in Bangladesh there is a huge gap between the supply and demand of electricity. In United States the energy consumption rate is 11.4 kW per person where in developing countries, like India, the per person energy use rate is closer to 0.7 kW. Bangladesh has the lowest consumption rate with 0.2 kW per person [1]. There are many different sources of energy that are naturally available throughout the world in different forms. Depending on regeneration energy can be categorized into two main different sources: renewable and non-renewable energy sources [2]. The sources that can be replenished or which are available in plenty in nature are renewable energy sources [3]. By far they are the cleanest and more environment-friendly sources of energy available on this planet, as they do not cause any natural imbalances. Solar energy, wind energy, biomass energy, tidal force, hydropower and geothermal energy are the major instances of renewable energy sources. The sources that cannot be replaced or can be replaced only very slowly by natural processes are non-renewable energy sources. The fossil fuels- natural gas, oil and coal, are the main examples of non-renewable energy sources.

Today's global energy production is mainly, in fact 80.6% dependent on fossil fuels [4]. But the fossil fuels are limited and their use is responsible for the global climate change due to the emission of greenhouse gases like carbon dioxide. WHO (World Health Organization) estimates that climate change is already causing more than 150,000 deaths a year [5]. Being concerned about the negative impact of conventional energy sources on environment and to provide clean and sustainable energy, there is an increasing demand of energy from renewable energy sources.

In this paper we focused our research on the feasibility study and cost analysis of a grid connected PV system in Bangladesh. We performed our analysis in University of Chittagong which is situated in the south-east part of Bangladesh. The HOMER (Hybrid Optimization Model for Electric Renewable) optimization tool is used for simulating the proposed system.

2. Electricity Scenario of Bangladesh

Electricity, the most convenient form of energy, is a vital ingredient to upgrade the socio-economic condition and to alleviate poverty [6]. Bangladesh is a densely populated country with 142.3 million people [7]. Only 53% of the country's population has access to grid electricity with 265 kWh per capita generations, which is very low compared to other developing countries in the world [8]. But it is not possible to meet the electricity demand for the current consumers due to the insufficient production of electricity around the country. This is because the government is unable to ensure the supply of natural gas, the primary fuel used to produce electricity in Bangladesh. Many power plants are idle due to the shortage of gas supply and this creates a struggling situation of electricity generation. On the other hand, most of the public power stations of the country have become very old and they are operating lower than their rated capacity. So load shedding is a regular matter in the country [6]. At present the maximum generation is 6066 MW against the average demand of 7518 MW with a load shedding of 1452 MW [9].

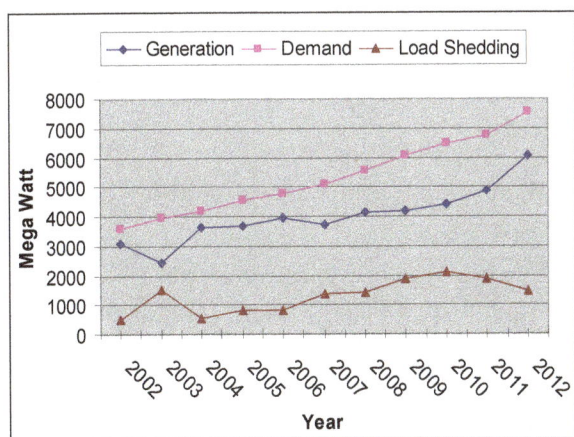

Figure 1. Electricity Demand, Generation and Load Shedding Scenario in Bangladesh from 2002 to 2012 [9], [10], and [11].

Due to the growth in almost all sectors, electricity demand in Bangladesh is increasing day by day. To face the increasing demand and because of time limitations for coal mining, gas supply constrains and lack of other resources, Government has initiated some high-cost temporary solutions, on an emergency basis, such as rental power and small independent power producers' (IPPs), which are mainly diesel or high sulphur fuel oil (HSFO) based. As a result, in just two years (from 2010 to 2012), there was a substantial change in the generation capacity profile. Natural gas share fell from 88% to 67% whereas high sulphur content fuel oil rose from 3% to 22%. The share of diesel, which is more expensive, trebled from 2% to 6%. The generation profile is now dominated by less efficient reciprocating engines which now represent 39% of the generation capacity [12]. Under contracts between the Power Development Board (PDB) and the rental power producers (RPPs), the PDB guarantees access to fuel by the

RPPs at lower rates charged by the Bangladesh petroleum corporation (BPC) to other consumers. On the contrary, the PDB sells electricity at prices lower than the prices the PDB spends to purchase from generators. A BIDS (Bangladesh Institute of Development Studies) study estimated that the average cost of un-served energy is $0.344 per kilowatt-hour (kWh), while the average bulk tariff on electricity is $0.052 per kWh (1 USD = 77.76 BDT) [13, 14].

Bangladesh has a vast potential for renewable energy which can be used to meet the power shortage. Out of various renewable energy sources solar energy can be effectively used in the country. The geographical position of Bangladesh is very favorable for the utilization of solar energy. The amount of annual solar radiation in Bangladesh varies from 1840 to 1575 kWh/m^2, which is 50-100% higher than in Europe [15, 16]. Taking an annual average solar radiation of 1800 kWh/m^2, only 0.00083% of the incident radiation will be enough to meet the present electricity demand of the country.

3. A Grid Connected PV System with Battery Backup

A grid connected power system is a power system that involves getting power from solar energy and/or other renewable energy sources, such as wind and hydro, and using the grid as backup. But in countries, like Bangladesh, where load shedding is a common this type of power system fails to provide continuous electricity to the users when there is no energy coming from the sun or other renewable energy sources and a blackout from grid occurs. Hence we proposed a grid connected PV system with battery backup, which is able to provide uninterrupted power supply at any condition. A grid connected system with battery backup is mainly composed of solar panels, a battery bank, a charge controller, an inverter/charger (converter), a utility grid and a two way electric meter (Net meter). The overall components of the system are shown in figure 2.

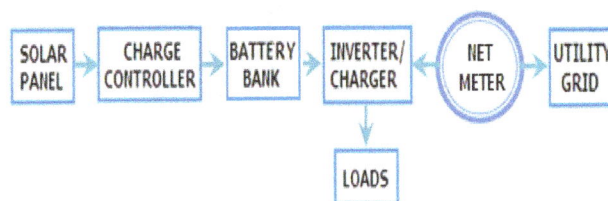

Figure 2. Grid connected PV system with battery backup.

4. Load Profile of the Proposed Area

Our work provides a design of a grid connected PV system with battery backup for the IT building of University of Chittagong. The building consists of two departments and an ICT centre. An estimation of the electrical appliances used by the building is given in the table below. The over all loads of the building is about 80kW. But all loads do not work at a time, while each load working at a particular time.

Table 1. Typical electrical appliances used by the IT building of University of Chittagong

Appliance item	Typical Power Rating (W)	Quantity
Ceiling fan	80	160
Tube light	40	200
Bulb	60	40
Computer	400	120
Refrigerator	110	3
Air conditioner	1800	4
Water pump	746	1
Laser printer	436	6
Other loads	3000	-

The office starts at 8.00 am and the class at 9.00 am and both closes at 5.00 pm. Normally, load demand is high at afternoon in the area under investigation. The load demands after the sunset is very low, only the refrigerators and few lights serve the load. In winter load requirement is small and in summer it is high. The load demand is high in the month of May and low in June because of rain vacation thorough out the whole month. Figure 3 shows the hourly load profile for a day in the month of May for the proposed area. The average daily load profile for the whole year is shown in figure 4. On an average the daily load demand is 304 kWh/day and the peak demand is 69 kW.

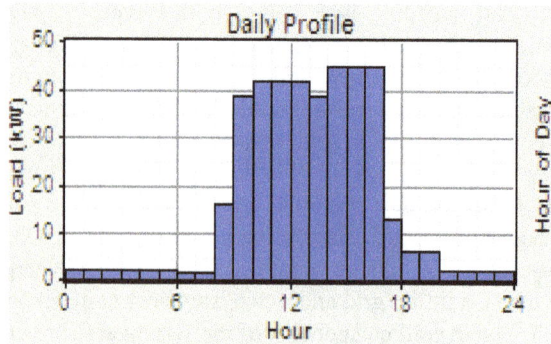

Figure 3. Hourly load profile for a day in the month of May for the proposed area.

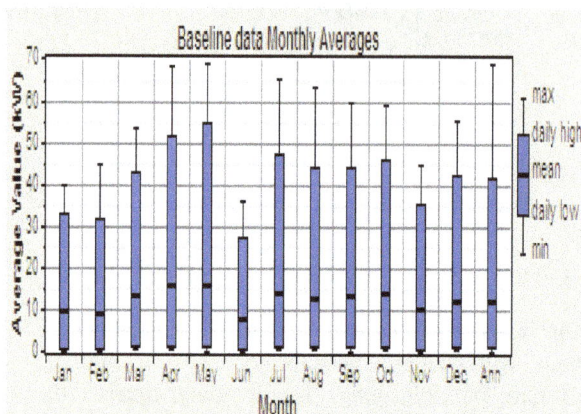

Figure 4. Monthly average load data for the proposed area.

5. Solar Radiation

The latitude and longitude of the site under investigation are 22° 21′ N and 95° 50′ E, respectively. HOMER uses these data to introduce the average daily radiation and clearness index for twelve months (Figure 5). The annual average solar radiation is 5.054 kWh/m^2/d. The maximum daily radiation is in the month of May, which is best suitable for our site as the load demand is also high is this month. The clearness index is the highest in the month of January and the lowest in the month of July.

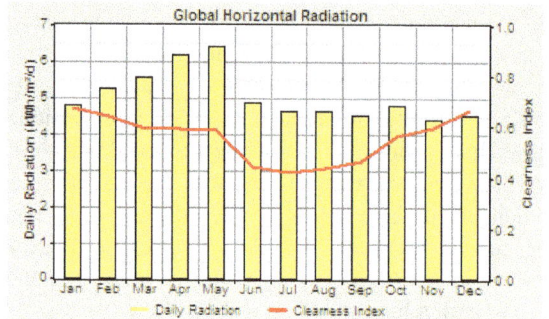

Figure 5. Average daily solar radiation and clearness index.

6. Implementation of the HOMER Model

6.1. System Configuration and Components

The HOMER schematic of the proposed system is given in the figure 6. The components considered are PV system, converter, battery, grid and loading system. Table 2 shows the component wise cost of the system and all cost are estimated considering the market price of Bangladesh.

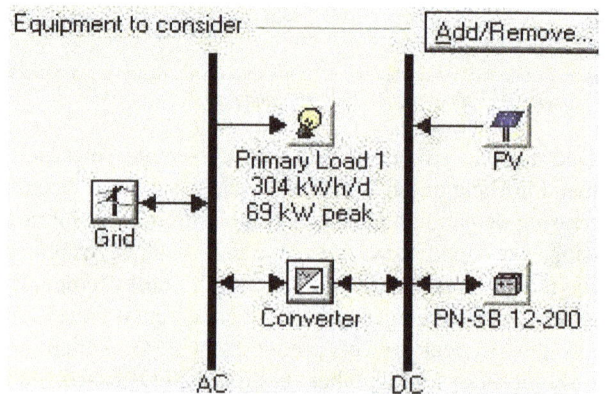

Figure 6. The HOMER schematic of the proposed system.

Table 2. Cost of different components of the proposed system.

Component	Capital cost ($)	Replacement cost ($)	O & M cost ($/yr)
PV system (1kW)	3500	2800	0
Converter (1kW)	850	730	0
Battery(12V, 200 Ah)	800	680	0

6.1.1. PV System

For the system the PV panels considered are the crystalline solar panels as they have high efficiency, about 21.5% [17], and long lifetime. At present the production cost of the best crystalline-silicon (c-Si) PV module in China is $0.5/W and will fall to $0.36/W by the end of 2017[18]. The capital and replacement costs for 1kW PV system are considered to be $3500 and $2800 respectively. The capital cost includes PV modules, structural components, the cost of installation and the charge controller cost, as HOMER does not model the battery charge controller as a separate component. An efficiency of 90% for solar panel and an efficiency of 95% for charge controller are considered. So the derating factor is 85.5%. No maintenance cost was considered for the PV system because crystalline solar panels require little or no maintenance. The lifetime of the PV panels are taken as 25 years and no tracking is considered in the PV system. Such a size of PV panel for the system is considered so that it can provide at least 50% of the total load demand.

6.1.2. Converter (Inverter/Charger)

Converter is one of the key components of a PV system. It is also called an inverter/charger. It is an electrical component that can works both as an inverter (DC to AC) and rectifier (AC to DC) depending on the direction of flow of power. It can also operate simultaneously with an ac generator. For the system a grid connected true sine wave power inverter/charger is considered as it produces the closest to a pure sine wave of all power inverters and in many cases produces cleaner power than the utility company itself [19]. The capital cost of inverter is $850/kW and replacement cost is $730/kW. It has no operational and maintenance cost. The inverter's efficiency is 95 % and lifetime is 15 years, which means that it has to be replaced once in 25 years period of system lifetime.

6.1.3. Battery

The battery provides backup power during power outage on the grid in a cloudy day or at night. Today different types and classifications of batteries, each with specific design and performance characteristics suited for particular applications, are manufactured. In this paper deep cycle AGM battery is considered, as they are maintenance free and have superior performance and longer lifetime, where most people use flooded solar battery. Because they are completely sealed, unlike flooded battery, they can't be spilled, do not need periodic watering, and emit no corrosive fumes, the electrolyte will not stratify and no equalization charging is required [20], [21]. The battery considered is Phaesun PN-SB 12-200 with nominal voltage of 12V and capacity of 200Ah (2.24 kWh). The capital and replacement cost of each battery are $800 and $680, respectively. The number of batteries considered is to give at least 3 hour backup continuously.

6.1.4. Grid Input

In Bangladesh the average selling price of grid power is lower than the average generation cost, selling price is about 60% of the average generation cost [22]. At present the proposed area uses the category "F" tariff offered by the grid, which has two rates - $0.077/kWh for off peak period and $0.120/kWh for peak period. We considered the same prices as sellback prices for off peak period and peak period respectively. The demand rate is $0.579/kWh/month. We considered monthly net metering scheme in the simulation. Net metering is a billing system by which the utility allows one to sell power to the grid at the retail rate [23]. At the end of the billing period one will be charged for the net amount purchased (total purchases minus sales).

6.2. Economics and Constraints Analysis

The project lifetime is considered to be 25 years and the annual interest rate is taken as 6%. No fixed operation and maintenance costs are considered as they are given in the individual components. Operating reserve as percentage of the current load is 10% and this reserve as a percentage of solar power output is 20%. Annual capacity shortage is set to zero.

7. Optimized Simulation Result

After simulation HOMER omits all infeasible system configurations and shows a list for the feasible combinations sorted by net present cost (NPC). The simulation results data for our proposed system obtained from HOMER is shown in figure 7.

		PV (kW)	PN-SB 1...	Conv. (kW)	Grid (kW)	Initial Capital	Operating Cost ($/yr)	Total NPC	COE ($/kWh)	Ren. Frac.
					400	$0	9,363	$119,687	0.084	0.00
			40	40	400	$66,000	11,742	$216,104	0.152	0.00
		45		40	400	$191,500	5,095	$256,636	0.166	0.53
		45	40	40	400	$223,500	6,699	$309,136	0.200	0.53

Figure 7. Optimized simulation results from HOMER.

Our proposed optimized result contains PV panel of 45 kW, converter of 40 kW and 40 batteries for storage. The power demand supplied by the PV system is 56% and the other 44% is purchased from the grid (Figure 8). There is no unmet electric load and capacity shortage is zero. Excess electricity fraction is almost zero percent. The amount of PV energy that is used to serve the load is 92% and the other 8% PV energy is sold to the grid around the year. The renewable fraction is 0.535 and the maximum renewable penetration is 113%.

Production	kWh/yr	%		Quantity	kWh/yr	%
PV array	71,894	56		Excess electricity	0.920	0.00
Grid purchases	56,256	44		Unmet electric load	0.00	0.00
Total	128,150	100		Capacity shortage	0.00	0.00

Consumption	kWh/yr	%		Quantity	Value
AC primary load	110,960	92		Renewable fraction	0.535
Grid sales	9,999	8		Max. renew. penetration	113 %
Total	120,959	100			

Figure 8. Electrical Outputs of simulation result.

The monthly average electric power produced by PV system and grid is shown in the figure 9. The PV production is high in the months of January and February because of higher clearness index and low in the months of July and August for lower clearness index.

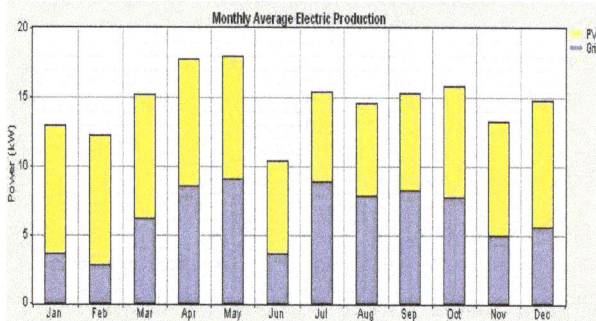

Figure 9. *Monthly average electric production from PV and grid.*

The cash flow summary is given in figure 10. The highest part of the cost is due to the PV panels, although there is no replacement and operating cost in this purpose. All cost associated with grid is considered as operating cost. The total net present cost (NPC) is \$309,136, the capital cost is \$223,500, the operating cost is \$6699/yr and the cost of energy is \$0.200/kWh.

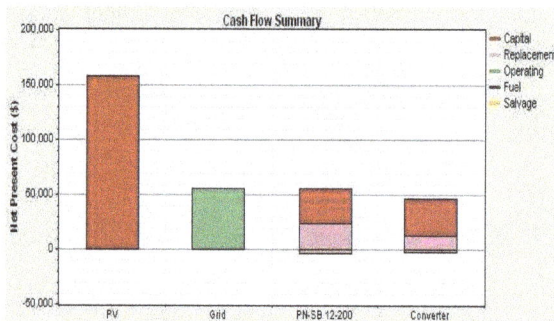

Figure 10. *Cost summary of the proposed system.*

8. Analysis of the Simulation Result

From simulation result (Figure 7), first optimum option is using the grid with government subsidy. The second option is to use grid with battery backup and the third option is using solar plus grid with no battery backup. The fourth option is our proposed option that provides almost continuous power using solar renewable energy plus grid with battery backup. Although the second option can provide continuous power, it will contribute to the overall increase in the load shedding scenario of the country and hence can not be the permanent solution. The first and third options do not provide continuous power. But uninterrupted power supply is must required for all the educational institutions. Completing all the regular activities, performing laboratory experiments, doing research work will be greatly hampered without continuous power. So our proposed system is the ultimate solution to provide uninterrupted power in a load shedding prone country like Bangladesh.

9. Conclusion

The power sector of Bangladesh is mainly dependent on natural gas. But the gas supply is decreasing day by day as no new reserve could be discovered and hence the country is suffering a severe power crisis. To minimize the power shortage Government has taken short term plan, on an emergency basis. Under this plan, rental and quick rental power plants are installed using diesel and high sulphur fuel oil (HSFO). The average cost of un-served energy from quick rental power plants is \$0.344/kWh. The cost of energy for our proposed system is \$0.200/kWh and it will be feasible to install our proposed system in Bangladesh. So our government should increase power generation from renewable sources instead of other conventional sources. If we can generate significant amount of power from renewable energy sources then it will not be necessary to hike the price of electricity so frequently and in this process emission of significant amount of CO_2 and other green house gases can be reduced.

References

[1] A. Zaman, M. A. Rahman and S. Islam, "Design of a hybrid power system for rural area", Journal Basic Science and Technology, 1(2), 1-4, 2012.

[2] http://renewablegreen.net/?p=124 [Accessed on 5 th March, 2013]

[3] http://readanddigest.com/what-are-the-different-types-of-energy-sources/ [Accessed on 19 th May, 2013]

[4] Renewable energy policy network for the 21st century (REN21). Renewables 2012, Global Status Report.

[5] World Wide Fund for Nature (WWF). The Energy Report 2011, 100% Renewable Energy by 2050.

[6] Unnayan Onneshan Research Report. Electricity Scenario in Bangladesh, November 2011.

[7] Population & Housing Census 2011, Preliminary Results, July 2011, Bangladesh Bureau of Statistics, Statistics Division, Ministry Of Planning, Government of the People's Republic of Bangladesh.

[8] M. R. Hamid, "Photovoltaic based solar home systems-current state of dissemination in rural areas of Bangladesh and future prospect", International Journal of Advanced Research in Electrical, Electronics and Instrumentation Engineering, Vol. 2, Issue 2, February 2013.

[9] Annual Report 2011-2012, Bangladesh Power Development Board.

[10] "Power and Energy Sector Road Map: An Update", June 2011, Finance Division, Ministry of Finance, Government of the People's Republic of Bangladesh.

[11] Annual Report 2007-2008, 2008-2009, 2010-2011, Bangladesh Power Development Board.

[12] Oxford Institute for Energy Studies. "Natural gas in Pakistan and Bangladesh: Current issues and trends", June 2013.

[13] http://www.dhakatribune.com/economy/2013/jun/18/rental-power-slows-base-load-power-stations-bids [Accessed on 10 th July, 2013]

[14] http://www.thedailystar.net/beta2/news/quick-rental-power-helps-fuel-gdp-growth-study/ [Accessed on 7 th July, 2013]

[15] M. A. Khan, M. Shamsuddoha, A. A. Helal and A. Hassan, "Climate change mitigation approaches in Bangladesh", Journal of Sustainable Development, vol. 6, no. 7, 2013.

[16] M. R. Islam, and T. H. M. S. Rashid, "Prospects and potential analysis of solar and biomass energy at Pabna district, Bangladesh: A realistic way to mitigate district energy demand", International Journal of Engineering and Advanced Technology (IJEAT), Volume-2, Issue-2, December 2012.

[17] http://energyinformative.org/best-solar-panel-monocrystalline-polycrystalline-thin-film/ [Accessed on 10 th July, 2013]

[18] http://www.greentechmedia.com/articles/read/solar-pv-module-costs-to-fall-to-36-cents-per-watt [Accessed on 10 th July, 2013]

[19] http://www.freesunpower.com/inverters.php [Accessed on 12 th February, 2013]

[20] http://www.bdchargers.com/batterytypes.php [Accessed on 15 th February, 2013]

[21] http://www.altestore.com/store/Deep-Cycle-Batteries/Batteries-Sealed-Agm/c436/ [Accessed on 2 nd April, 2013]

[22] Power Division Annual Report, 2010-2011(In Bengali), Ministry Of Power, Energy and Mineral Resources, Bangladesh.

[23] S. Talebhagh and H. K. Kareghar, "Applying grid-connected PV systems as supplement to public electricity network in rural area of Iran", International Journal Of Advanced Renewable Energy Research, Vol. 1, Issue. 7, pp. 419- 428, 2012.

25

Economic load dispatch with the proposed GA algorithm for large scale system

Hamed Aliyari[1, *], **Reza Effatnejad**[2], **Ardavan Areyaei**[1]

[1]Electrical Engineering Department, Science and Research Alborz branch, Islamic Azad University, Alborz, Iran
[2]Electrical Engineering Department, Karaj branch-Islamic Azad University, Alborz, Iran

Email address:

hamedaliyary@gmail.com (H. Aliyari), Rezaeffatnejad@yahoo.com (R. Effatnejad), ardavan.aryaei@gmail.com (A. Areyaei)

Abstract: Economic load dispatch (ELD) have been applied to obtain optimal fuel cost of generating units. Genetic Algorithm (GA) is a global search technique based on principles inspired from the genetic and evolution mechanism observed in natural biological systems. This paper presents a novel stochastic Genetic Algorithm approach to solve the Economic Load Dispatch problem considering various generator constraints and also conserves an acceptable system performance in terms of limits on generator real and reactive power outputs bus voltages, shunt capacitors/reactors, transformers tap-setting and power flow of transmission lines. The ELD problem in a power system is to determine the optimal combination of power outputs for all generating units which will minimize the total fuel cost while satisfying all practical constraints. To show its efficiency and effectiveness, the proposed GA algorithm is applied to some types of ED problems containing non-smooth cost functions of 13 and 40 generating units systems (large scale systems). The experimental results show that the proposed GA approach is comparatively capable of obtaining higher quality solution.

Keywords: Non-Smooth Cost Functions, Genetic Algorithm, Economic Dispatch, GA, IEEE Tests Systems

1. Introduction

Conventionally, economic load dispatch problem allocates loads to plants at minimum cost while meeting the constraints. It is an optimization problem which minimizes the total fuel cost of all committed plants while meeting the demand and losses. The optimal power system operation is achieved when both the objectives of power systems i.e. cost of generation and system transmission losses simultaneously attain their minimum values. Economic load dispatch reflects the optimal electrical output of generation facilities, to fulfill the system load demand, at the lowest possible cost, while providing power in a robust and reliable way. Economic load dispatch problem is one of the fundamental matters in power system operation. In essence, it is an optimization problem and its main objective is to cut-down the total generation cost, without breaching any constraints [1]. Preceding efforts on solving ELD problems have made use of various mathematical programming and optimization techniques [2]. Numerous techniques have been established to help solve the ELD problems, such as Particle Swarm Optimization [3], Artificial Bee Colony Algorithm [4], Genetic Algorithm [5], Pattern Search Algorithm [6], Neural

Networks [7], Evolutionary Programming [8], and Harmony Search Algorithm [9]. Each of the employed techniques may have some advantage and disadvantages. For instance, Particle Swarm Optimization (PSO) is well recognized for its capability to permit each particle to maintain a memory of the best solution it has discovered in the particle's neighborhood swarm. Furthermore, PSO is easy to device, and effective [10]. However, the algorithm might experience inequality constraints difficulties. Recently, Ant Colony Optimization (ACO)[11] has turn into a candidate for many optimization applications [12] such as the combinatorial optimization travelling salesman problem (TSP), quadratic assignment problem (QAP), and optimal design and scheduling problem of thermal units [13].

This paper presents an innovative approach based on Ant Colony Algorithm was chosen for solving the load-flow problem.

2. Economic Load Dispatch

The objective of Economic Load Dispatch is to minimize the operating cost of each generating unit in the system. Thus, an optimal generated output can be acquired from the solution. Economic Load Dispatch can be calculated by using the following equations[15-19]

$$optimum \cos t = \sum_{i}^{Ng} Fi(Pi) \qquad (1)$$

Where *cost* is the operating cost of power system and the objective function is to minimize the cost. *Ng* is the number of units. *Fi(Pi)* is the cost function and *Pi* is the power output of the unit *i*. *Fi(Pi)* is usually approximated by a quadratic function of its power output *Pi* as:

$$Fi(Pi) = a_i + b_i P_i^2 + b_i P_i + c_i \qquad (2)$$

Where *ai*, *bi*, and *ci* are the cost coefficients of unit *i*. The above equation is subjected to both the equality and inequality constraint as follow:

Real power balance constraint is given by:

$$\sum_{i}^{Ng} Fi(Pi) = P_D + P_L \qquad (3)$$

Real power generation limit is given by:

$$P_{i\min} \leq P_i \leq P_{i\max} \qquad (4)$$

Where *PD* is the total load demand in MW, *PL* is the total transmission loss of the system in MW; *Pimin* and *Pimax* are the minimum and maximum generation limit of *Pi*. Next, The search of the optimal control vector is performed using into account the real power flow equation which present the system transmission losses (*PL*). These losses can be approximated in terms of *B*-coefficients as [20]:

$$Pf = \left(1 - \frac{\partial P_L}{\partial P_g}\right)^{-1} \qquad (5)$$

These losses are represented as a penalty vector [21-24]given by:

$$P_L = \sum_{i}^{N} \sum_{j}^{N} P_i B_{ij} P_j \qquad (6)$$

The transmission loss of a power System *PL* can be calculated by the *B*-Coefficients method [25] and given by:

$$P_L = \sum_{i}^{N} \sum_{j}^{N} P_i B P_j + \sum_{i}^{N} B_{oi} + B_{oo} \qquad (7)$$

Where *B* is an *ng×ng* coefficients matrix, *B0* is an *ng*-dimensional coefficient column vector and *B00* is a coefficient.

3. Genetic Algorithm

3.1. Basic Principle of Genetic Algorithm

Genetic Algorithms (GA)[26,19] are direct, parallel, stochastic methods for global search and optimization, which imitate the evolution of the living beings, described by Charles Darwin. GA is part of the group of Evolutionary Algorithms (EA). The evolutionary algorithms use the three main principles of the natural evolution: reproduction, natural selection and diversity of the species, maintained by the differences of each generation with the previous. Genetic Algorithms works with a set of individuals, representing possible solutions of the task. The selection principle is applied by means of a criterion, giving an evaluation for the individual with respect to the desired solution. The best-suited individuals create the next generation.

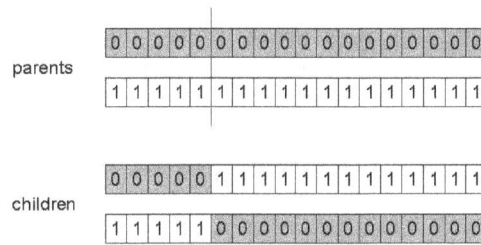

Fig 1. Behavior of Genetic Algorithm.

3.2. Genetic Algorithm with Arithmetic Crossover

In order to formulate the algorithm for economic/environmental ED problem, let the chromosome of the k-th individual Ck be defined as follows:

$$c_k = [P_{k1}, P_{k2},...,P_{kn}] \qquad (13)$$

Where
k : 1, 2, . . . , popsize
n : 1,2….number_of_gene

popsize means population size, number_of_gene is the number of unit in our experiment

Pki is the generation power of the n-th unit at k-th chromosome

Reproduction involves creation of new offspring from the mating of two selected parents or mating pairs. It is thought that the crossover operator is mainly responsible for the global search property of the GA. We used an arithmetic crossover operator that defines a linear combination of two chromosomes [23]. Two chromosomes, selected randomly for crossover, C_i^{gen} and C_j^{gen} may produce two offspring, C_i^{gen+1} and C_j^{gen+1} , which is a linear combination of their parents i.e.,

$$C_i^{gen+1} = a.C_i^{gen} + (1-a)C_j^{gen} \qquad (14)$$

$$C_j^{gen+1} = a.C_j^{gen} + (1-a)C_i^{gen} \qquad (15)$$

where

C_i^{gen} : an individual from the old generation

C_i^{gen+1} : an individual from new generation

a : is the weight which governs dominant individual in reproduction and it is between 0 and 1

The mutation operator is used to inject new genetic material into the population and it is applied to each new structure individually. A given mutation involves randomly altering each gene with a small probability. We generate a random real value which makes a random change in the m-th element selected randomly of the chromosome.

3.3. Proposed GA Algorithm for Solving ED Problem

In this study, an innovative approach based on Genetic Algorithm was chosen for solving the load-flow problem, i.e. the two groups of genes are pretended in a parallel mood, in a way that one group moves in the values decreasing direction and the other one moves in the values increasing direction. After giving the primary values, the two groups converge to the minimum answer found.

The initialization method that is utilized in the following algorithm is an innovative method. The increasing primary amount is chosen in a way that at first, all the units should take their minimum amount and after that, we act according to the conditions 3 and 4 and the following algorithm:
A. A gene is chosen.
B. According to the probability function of one of the units, one of them is selected. Then the constant value (defined according to the accuracy and cpu time of the program, which can change based on the type of the operation), is added in accordance to the primary value that was defined for it previously. The amount of constant value should be selected in a way that it increases the power value of each unit, because each unit has already got its minimum power value. If this action does not violate condition "4" and that the amount of $\sum_i^{Ng} F(P_i)$ is less than P_D, then adding the constant value is accepted   and proceeding to the level "C" is appproved, otherwise the constant value is not acceptable and this level should be repeated.
C. If the condition "3" is not violated, then proceeding to the next level is approved, otherwise going back to the level "B" is mandatory.
D. The primary amounting is done.
E. Chose another gene, i.e. go to level "A". (The number of genes is determined at first.)

We should do the same guess algorithm in a decreasing way, in which we give the maximum amount of units to them and we do the same steps in the decreasing route.

In order to get to the optimum result in accordance to the guess algorithm, in the part that we are getting the best results, the more the number of these results (genes), the more the accuracy and the less the speed of running the

code(CPU time), so the existing results give the prices of each unit.

Actually a large amount of data is obtained at the end of the process. We have to get to the most optimum level within the data, by means of GA algorithm based on arithmetic crossover and standard mutation separately.

Based on this method, the optimization was applied in 13 units [27], 40 units [28] systems (assuming power losses P_L to be zero). It should be mentioned that in all cases, constraints such as speed and the forbidden work zones of each generator is considered.

4. Simulation Results

To assess the feasibility of the PSO approach, the studies of ED were compared with many optimization methods such as GA, TS, PSO, and ACO, implemented in MATLAB (7.6.0.6324(2008a) version). These programs were run on a Pentium Dual core, 2.5 GHz personal microcomputer with 3 GB RAM under Windows XP. In each case study, 100 independent runs are carried out for each optimization method. In addition, 100 different initial trial solutions are used for each method.

The GPSO is applied to two ED problems with 13 and 40 generating units. The input data for 13 generating units system are given in [27] with 2520MW load demand. Also, the input data for 40 generating units system are given in [28] with 10,500MW load demand. The global solutions for these systems are not discovered yet. The best local solutions reported until now for 13 and 40 generating units are 24,169.92 \$/h [29] and 121,741.98 \$/h [30], respectively. After performing 100 trials, the best results for P_is, in the 13 units system and 40 units system, in order for the best answer to be found, are shown in the Tables 1 and 2 respectively.

Table 1. Best result obtained by proposed GA for 13-unit system.

unit	Pi(min) MW	Pi(max) MW	Pi MW
1	0	680	582
2	0	360	307
3	0	360	304
4	60	180	150
5	60	180	152
6	60	180	160
7	60	180	170
8	60	180	151
9	60	180	145
10	40	120	91
11	40	120	88
12	55	120	112
13	55	120	108
Total power output	(MW)		2521.05807120690
Total generation cost	(\$/h)		24167

Table 2. *Best result obtained by proposed GA for 40-unit system.*

unit	Pi(min) MW	Pi(max) MW	Pi MW
1	36	114	102.998723543689
2	36	114	48.8339865534452
3	60	120	81.9858291219816
4	80	190	184.910084237253
5	47	97	94.2352602174419
6	68	140	105.422756354223
7	110	300	276.389659153576
8	135	300	270.466620780899
9	135	300	298.060182246735
10	130	300	147
11	94	375	237.691701220395
12	94	375	206.462568526111
13	125	500	284.649567876062
14	125	500	299.585612037311
15	125	500	337.915510467145
16	125	500	276.073863005347
17	220	500	477.22822388695
18	220	500	498.847725762535
19	242	550	541.67218476475
20	242	550	420.642175667156
21	254	550	512.202032512945
22	254	550	530.777820055416
23	254	550	541.468907284464
24	254	550	521.954612690208
25	254	550	550
26	254	550	549.808717554535
27	10	150	18.4444706057755
28	10	150	17.5935208491752
29	10	150	12.786627396382
30	47	97	92
31	60	190	190
32	60	190	176.788840003834
33	60	190	186.052294384747
34	90	200	200
35	90	200	190.129452886447
36	90	200	200
37	25	110	102.62394666683
38	25	110	63.3481518226166
39	25	110	108.692272455022
40	242	550	543.705916841164
	Total power output	(MW)	10499.45
	Total generation cost	($/h)	121168.9255

Table 3. *Convergence results for 13-unit system.*

Method	Best Cost
Load demand	2520 MW
Propose GA	24167
[31]	24169.89
GA	24186.02
PSO	24171.70
ACO	24174.39
TS	24180.31

Table 4. *Convergence results for 40-unit system.*

Method	Best Cost
Load demand	10500 MW
Propose GA	121168.9255
[31]	121532.41
GA	121996.40
PSO	121800.13
ACO	121811.37
TS	122288.38

5. Conclusions

The problem of economic load dispatch with non-smooth cost functions has been investigated in this paper. We used GA algorithm with arithmetic crossover and standard mutation for arrived at this goal. A comparison analysis has been done for different intelligent techniques with respect to the total minimum generation cost that shown at the table 3 and the table 4. The present algorithm is able to solve the economic dispatch problem under partially deregulated environment. The performance of the developed algorithm has been demonstrated on an 13 units systems and 40 units systems test system.

According to the results of this research, and the comparison between other researches results, and the significant difference between them, it is obvious that this method can be a proper method for operating gigantic power systems.

References

[1] S. M. V. Pandian and K. Thanushkodi, "Solving Economic Load Dispatch Problem Considering Transmission Losses by Hybrid EPEPSO Algorithm for Solving Both Smooth and Non-Smooth Cost Function," International Journal of Computer and Electrical Engineering, vol. 2, 2010.

[2] A. J. Wood and B.F. Wollenberg, "Power Generation, Operation and Control", New York: John Willey & Sons, Inc., 1984

[3] S. M. V. Pandian and K. Thanushkodi, "Solving Economic Load Dispatch Problem Considering Transmission Losses by Hybrid EPEPSO Algorithm for Solving Both Smooth and Non-Smooth Cost Function," International Journal of Computer and Electrical Engineering, vol. 2, 2010.

[4] S. Hemamalini and S. P. Simon, "Economic Load Dispatch With Valve-Point Effect Using Artificial Bee Colony Algorithm," presented at XXXII National Systems Conference, 2008.

In the Tables 3 and 4, there is a brief comparison between our suggested method and some other innovating methods proposed by other researchers and some standard methods based on the smart algorithms(that we discussed above) which are the results of other researchers.

[5]   Y. Labbi and D. B. Attous, "A Hybrid GA-PS Method to Solve The Economic Load Dispatch Problem," Journal of Theoritical and Applied Information Technology, 2005.

[6]   Al-Sumait, J. S., Sykulski, J. K. and Al-Othman, A. K. (2008) 'Solution of Different Types of Economic Load Dispatch Problems Using a Pattern Search Method', Electric Power Components and Systems, 36:3, 250 – 265

[7]   A. Y. Abdelaziz, S. F. Mehkhamer, M. Z. Kamh, and M. A. L. Badr, "A Hybrid Hopfield Neural Network - Quadratic Programming Approach for Dynamic Economic Dispatch Problem," 2008.

[8]   H. T. Yang, P.C. Yang, and C. L. Huang, "Evolutionary Programming Based Economic Dispatch for Unit with Non-Smooth Fuel Cost Functions," IEEE Trans. Power System, vol. 11, no 1, pp. 112-118, Feb 1996

[9]   R. Arul, D. G. Ravi, and D. S. Velusami, "Non-Convex Economic Dispatch with Heuristic Load Patterns Using Harmony Search Algorithm," international Journal of Computer Applications, vol. 16.

[10] B. H. Chowdhury and S. Rahman, "A Review of Recent Advances in Economic Dispatch," IEEE Trans. Power System, vol. 5, no. 4, pp. 1248-1259, Nov 1990.

[11] R. Effatnejad , H.aliyari, H.Tadayyoni, A.Abdollahshirazi, "Novel Optimization Based On The Ant Colony For Economic Dispatch" ,International Journal on Technical and Physical Problems of Engineering (IJTPE); Iss. 15, Vol. 5, No. 2, Jun. 2013

[12] N. H. F. I. Ismail Musirin, Mohd Rozely Kalil, MUhammad Khayat Idris, Titik Khawa Abdul Rahman, Mohd Rafi Adzman, "Ant Colony Optimization (ACO) Technique In Economic Load Dispatch," in Inrternational MultiConference of Engineers and Computer Scientist 2008, Hong Kong, 2008, p. 6.

[13] D. Nualhong, et al., "Diversity Control Approach to Ant Colony Optimization for Unit Commitment Problem," in TENCON 2004. 2004 IEEE Region 10 Conference, 2004, pp. 488-491 Vol. 3.

[14] P . SUREKHA, S.SUMATHI, "Solving Economic Load Dispatch problems using Differential Evolution with Opposition Based Learning",WSEAS TRANSACTIONS on INFORMATION SCIENCE and APPLICATIONS, Issue 1, Volume 9, January 2012

[15] F. N. Lee and A. M. Breipohl, "Reserve constrained economic dispatch with prohibited operating zones," IEEE Trans. on Power Syst., vol. 8, no. 1, pp. 246-254, 1993.

[16] J. Y. Fan and J. D. McDonald, "A practical approach to real time economic dispatch considering unit's prohibited operating zones," IEEE Trans. on Power Syst., vol. 9, no. 4, pp. 1737-1743, 1994.

[17] B. H. Chowdhury and S. Rahman , "A review of Recent Advances in Economic Dispatch", IEEE Transactions on Power Systems, November 1990, Vol.5, No. 4, pp. 1248-1259.

[18] Y. S. Haruna, Comparison of Economic Load Dispatch Using Genetic Algorithm and Other Optimization Methods, M. Eng. Degree Thesis, A. T. B. University, Bauchi, Nigeria, 2003.

[19] A. G. Bakistzis, P. N. Biskas, C. E. Zoumas, and V. Petridis, "Optimal power flow by enhanced genetic algorithm, "IEEE Trans. Power Systems, vol. 17, no. 2, pp. 229-236, May 2002.

[20] H.T. Yang, P.C. Yang, C.L. Huang," Evolution programming based economic dispatch for units with non-smooth fuel cost functions", IEEE Trans Power Syst 1996;11(1):112–8.

[21] C. Chen, "Non-convex economic dispatch: a direct search approach" ,Energ Conver Manage 2007;48:219–25.

[22] L.S. Coelho, V.C. Mariani, "Combining of chaotic differential evolution and quadratic programming for economic dispatch optimisation with valve-point effect", IEEE Trans Power Syst 2006;21(2):989–96.

[23] O. Abedinia, N. Amjady, K. Kiani, H.A. Shayanfar, A. Ghasemi, "MULTIOBJECTIVE ENVIRONMENTAL AND ECONOMIC DISPATCH USING IMPERIALIST COMPETITIVE ALGORITHM", International Journal on Technical and Physical Problems of Engineering (IJTPE), Iss. 11, Vol. 4, No. 2, Jun. 2012.

[24] H. Shayeghi, A. Ghasemi, "MOABC ALGORITHM FOR ECONOMIC/ENVIRONMENTAL LOAD DISPATCH SOLUTION", International Journal on Technical and Physical Problems of Engineering (IJTPE), Iss. 13, Vol. 4, No. 4, Dec. 2012

[25] S. K. Wang, J. P. Chiou, C. W. Liu, " non -smooth/non-convex economic dispatch by a novel hybrid differential evolution algorithm" ,The Institution of Engineering and Technology, 1, (5), pp. 793–803, 2007

[26] D. C. Waters and G. B. Sheble, "Genetic algorithm solution of economic dispatch with valve point loading," IEEE Trans. PWRS, vol. 8, pp. 1325- 1332, 1993.

[27] N. Sinha, R. Chakrabarti, P.K Chattopadhyay," Evolutionary programming techniques for economic load dispatch" ,IEEE Trans Evolut Comput 2003;7(1):83–94.

[28] H.T. Yang, P.C. Yang , C.L. Huang," Evolution programming based economic dispatch for units with non-smooth fuel cost functions", IEEE Trans Power Syst 1996;11(1):112–8.

[29] C. Chen, "Non-convex economic dispatch: a direct search approach" ,Energ Conver Manage 2007;48:219–25.

[30] L.S. Coelho, V.C. Mariani, "Combining of chaotic differential evolution and quadratic programming for economic dispatch optimisation with valve-point effect", IEEE Trans Power Syst 2006; 21(2):989–96.

[31] S. K. Wang, J. P. Chiou, C. W. Liu," non -smooth/non-convex economic dispatch by a novel hybrid differential evolution algorithm" ,The Institution of Engineering and Technology, 1, (5), pp. 793–803, 2007

26

Impact of different factors on biogas production in poultry dropping based biogas plants of Bangladesh

Mohammad Shariful Islam[1], Asif Islam[2], Dipendra Shah[3, *], Enamul Basher[1]

[1]Electrical & Electronic Engineering, Bangladesh University of Engineering & Technology (BUET), Dhaka, Bangladesh
[2]Planning & Development Division, Power Grid Company of Bangladesh (PGCB) Ltd., Dhaka, Bangladesh
[3]NIS Department, Huawei Technologies Co. Ltd., Kathmandu, Nepal

Email address:
sabuz03@gmail.com (M. S. Islam), asif038@gmail.com (A. Islam), dipendrashah@gmail.com (D. Shah),
enamul_basher@eee.buet.ac.bd (E. Basher)

Abstract: Bangladesh, with poultry farms number of 215000 and poultry population of 200 million, has a potential of 1.33TWhr electricity per year. Despite of this huge potential, only few number of biogas plants have been deployed till date. Besides, most of these plants are unable to produce the expected amount of biogas as most of the digesters of these plants are made locally and they don't have any scheme for monitoring and controlling temperature, pH, bacterial population in digester, mixture of different substrate, hydraulic retention time, total solid (TS), periodic agitation, periodic loading and unloading of substrate etc. But, there is few bigger size plants which contain imported digesters equipped with monitoring and controlling schemes, hence they produce bigger amount of biogas. In this study, an analysis has been done focusing the impacts of above mentioned factors on biogas production by comparing biogas production between locally made digesters and imported digesters. The result reveals that up to 75% more biogas is generated in imported digesters than the locally made digesters.

Keywords: Digester, Anaerobic Digestion, Psychrophilic, Mesophilic, Thermophilic, Hydraulic Retention Time

1. Introduction

In Bangladesh only 3% of the people enjoy the facility of natural gas coming to their homes through pipe lines. The lucky few mostly live in the cities. Most of the Bangladesh's rural people depend on biomass, crop residues, plant debris, animal dung and wood for fuel creating deforestation, flood, soil erosion etc. Women and children, on whom the burden of collecting fuel falls, suffer the most. They are the worst victims [1] of indoor air pollution such as smokes in the kitchens. Biogas technology is one of the best means [2] to provide natural gas to the largest number of rural people. It can provide them with pollution free, efficient energy for cooking and at the same time protect them from diseases by giving them a cleaner environment. Biogas technology can be used to implement a sustainable waste management program suitable for rural areas, as wastes of all sorts are transformed into biogas or slurry. Besides, biogas can be used to produce electricity for poultry farm itself to reduce farming cost as well as household use. According to the Bangladesh Bureau of Statistics (BBS) the number of poultry farms [3, 4] in Bangladesh was 49825 in 2007-08. Out of which 33225 were broiler farms, 10099 were layer, 227 were hatchery, 5524 were duck and 750 were duck farms. The annual growth rates of those farms were 28%, 18%, 20%, 22% and 9% respectively. According to International Finance Corporation (IFC), the number of poultry farms in Bangladesh is 215000. From the poultry droppings and litter, biogas can be produced [5]. This biogas can be used for daily cooking, lighting or electricity generation. Different poultry birds give different amount of droppings. Hence different amount of biogas is produced from different type of birds [6]. Breeder birds give approximately 245gm of droppings, whereas other birds give around 100gm of droppings per day. Almost thirty thousand tons of droppings are projected to be generated. If all these droppings can be used for generation of biogas then a great amount of energy [7-10] could be produced. The usual solid contents of the droppings are about 20%. Biogas produced from per kg of solid content is about 0.4 m^2. 1kWhr of electricity can be produced from 22.4 CFT of biogas according to BCSIR. About 82 million CFT of biogas can be produced per day which is equivalent to 30 billion CFT of biogas per year. These amounts of biogas can produce 3.65 GWhr of electricity per day which is equivalent to 1.33 TWhr per year [9]. For 14 hours of operation this biogas can run a 260MW generation unit. Biogas plant can supply required electricity to the poultry farm itself and to the neighboring localities where national grid power is not available. This will also reduce CO_2 emission, alleviate environmental pollution, bad odor produced in poultry farms, create job scopes for people of the adjacent area, mitigate the stress on the national grid and can generate revenue selling organic fertilizer. Besides, efficient management of produced biogas, electricity and heat wastage will provide the maximum benefit out of the available resources. Despite all these prospects only a limited number of biogas plants are deployed in the poultry farms of Bangladesh. Besides, plants already deployed are not managed properly [11] which reduces the production of biogas hence the efficiency. There are different factors which influence the production of biogas e.g.-temperature, pH, bacterial population in digester, mixture of different substrate, hydraulic retention time, total solid (TS), periodic agitation, periodic loading and unloading of substrate etc. This study is aimed at to reveal the influence of these factors on biogas production in biogas plants deployed in poultry farms.

2. Methodology

For technical and economical success insuring stability process and maintaining required quality and quantity of gas is a prerequisite [12]. To maintain these prerequisite following conditions must always be fulfilled:

- Anaerobic condition in the digester must be maintained.
- Temperature must me controlled to the permissible values.
- Availability of nutrients for bacteria;
- Substantial amount of nutrients bacteria must always be present in the digester.
- Substrate must be loaded and unloaded periodically.
- C/N ration must always be observed and work accordingly.
- Correctly chosen proportion of solids content and proper agitation;
- Absence of inhibitors of the digestion process

2.1. Temperature

Temperature has the most pronounced impacts on anaerobic digestion. It has three main temperature ranges: from 10-25°C (psychrophilic conditions), from 30-37°C (mesophilic conditions) and from 48-55°C (thermophilic conditions) [13]. Psychrophilic digestion process is very slow hence only mesophilic and thermophilic digestion processes are used in practice. At very low or high temperature the activities of bacteria population is almost

stopped consequently the digestion process is becomes very long. Hence the production of biogas is reduced. The methane content becomes very low. As a result the produced biogas cannot be used as fuel for electricity generation and cooking [12-15].

2.2. pH Value

The pH of the digester is a function of the concentration of volatile fatty acids produced, bicarbonate alkalinity of the system, and the amount of carbon dioxide produced. The optimum range of pH for biogas production is between 7.0 and 7.2 [16]. But the substantial biogas can be produced for the pH range of 6.6 to 7.6. Biogas production reduces many fold for the pH value of less than 5 as the bacteria population decrease significantly under the circumstances.

2.3. Bacterial Population in Digester

Digester must always contain a certain amount of bacteria for acceptable amount of biogas production. Any dip in bacteria population impedes the production of biogas. Addition of enzymes enhances the bacteria population which in turns increases biogas production [13].

2.4. Mixture of Different Substrates

As different substrate produce different amount of biogas, the mixture of them increases the biogas production and methane content [10].

2.5. Hydraulic Retention Time (HRT)

The Hydraulic retention time (HRT) is also known as hydraulic residence is the time required for complete digestion of the substrates in Digester. Hydraulic retention time is the volume of the digester divided by the influent flow rate.
HRT = Volume of Digester / Influent Flow Rate
where using SI UnitsVolume is in [m^3] and Influent flow rate is in [m^3/h]. HRT is usually expressed in hours (or sometimes days).

2.6. Total Solid (TS)

Total solid contained in a certain amount of materials is usually used as the material unit to indicate the biogas-producing rate of the materials. Most favorable TS value desired is 8%.

2.7. Daily Substrate Input

The daily substrate input depends on the HRT of the digestion process. The HRT can be 10 days to 80 days.
If digester HRT is 10 days, than daily substrate load will be 1/10 from the overall volume of substrate in digester. For biogas plants working in thermophilic range, daily substrate load can constitute up to 1/5 of overall substrate volume in digester.

2.8. Substrate Digestion Time

Choice of substrate digestion time depends also from the type of substrate used. Cattle substrate takes 10-15 days, pig manure 9-12 days, poultry manure 10-15 days and 40-80 days manure mixed with vegetation waste for complete digestion [14].

2.9. C/N Ratio

For optimal C/N ratio different substrates are mixed for which maximizes the production of biogas.

2.10. Regular Agitation

For maximizing the production of biogas regular agitation of substrate in digester is required. The regular agitation frees the produced biogas entrapped in the substrate, mixes the fresh substrate with bacteria population, prevents scum and sediment, prevents temperature gradient inside the digester, provides the homogeneous mixture of bacteria population and prevents the formulation of voids.

3. Study Area

In Bangladesh two types of digesters are used for biogas production in poultry farms [9]. The large portions of these digesters are locally produced and are used for smaller size of biogas plants. But in larger size of plants the digesters are imported from abroad.

Locally produced digesters are made of concrete and are of low costs. But no control system is available for these digesters. Hence they produce less biogas. On the other hand, imported digesters are made of steel, bigger in size and cost higher. But as they have control system for controlling above factors they generate larger amount of biogas.

There are numerous biogas deployed in different parts of Bangladesh. Paragon Poultry Limited (PPL), one of the biggest poultry farms in Bangladesh, has deployed biogas plants at Gazipur, Mymensigh, Rangpur and Sylhet. One plant is ready to operate at Panchagarh also. We have surveyed biogas plants in Gaziupur, Mymensingh and Rangpur, and also collected information about Sylhet and Panchagarh plants. Felix Energy Service (FES), a Danish company for setting biogas power plant, in collaboration with Computerized Service Centre (CSC) of Narayanganj has set up more than 20 biogas plants in poultry farms of different parts of Bangladesh. A few of them has been surveyed and also information has been collected for remainder of them. Local companies and FES have set up the biogas plants; FES and CSC have provided all the equipments. There is another biogas plant set up by individuals in Rangunia, Chittagong for household electricity generation. The plants generate 6.4 m³ to 4000 m³ of biogas. Out of 18 (Figure 1) plants first 16 are using digesters with local materials and last two are using imported digesters. One plant is using local equipments which includes generator. 13 are using local equipments and assembled generators. 4 of them are using imported

machineries. Three of them are using 2 generator units. Rests 15 are using 1 generator unit. Table 1 shows the detail of these plants.

Only 3 plants are using mixing tanks, all of them are of PPL. Most of them are using 1 digester, but Rangpur is using 4 digesters and Gaziupur using 2 digesters. Only three are using separate holding tanks and only two of them are using how water tank for temperature control.

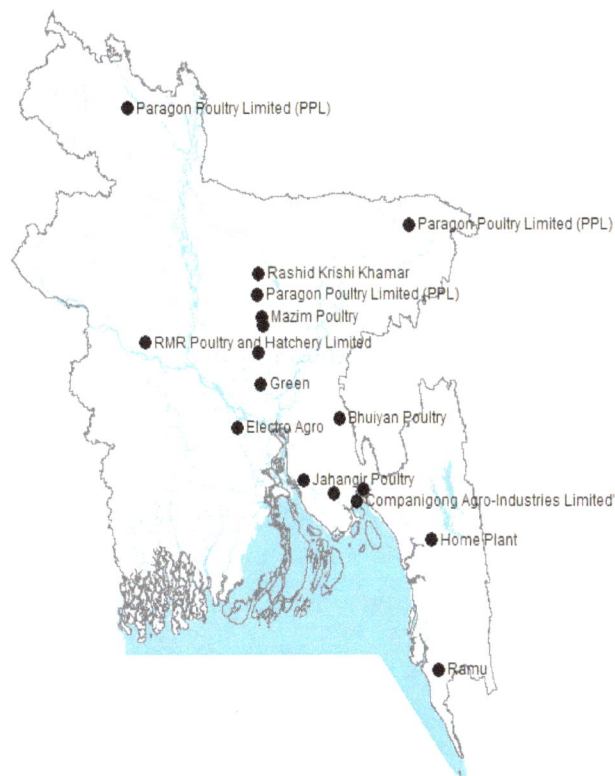

Figure 1. Locations of different biogas plants in different parts of Bangladesh which are studied

4. Results and Analysis

Digesters can be locally built or can be imported. For both the cases the cost for per cubic meter of biogas generation is almost equal. The plant of 1200 cubic meter capacity has a bigger cost because they have designed a bigger size digester for future expansion. For locally converted generators the overall costs remain constant and much smaller than the plants with imported equipments. The generation cost per kWhr of electricity is given by the Figure 2.

The digester cost remains fairly constant for all scale of generation. Generator cost for locally converted biogas generators remain small, but for imported generator the cost is fairly high. Other costs remain fairly constant all over.

The value of pH between 6.8 and 7.2 gives the best production of biogas. Anything outside of this range reduces biogas production. Hence it is very important to maintain the pH. Usually pH of a digester is self-regulating, hence any aberration of it tend to be converged to the reference value on its own. But there could be cases where pH needed to be controlled if too much deviation is observed. Usually lime water is introduced to the digester to maintain pH.

For locally made digesters there are no arrangements for periodic measurement of pH which could reduce the biogas production but for the cases of imported digesters there are arrangements for periodic pH measurement, but there are no automatic pH control arrangements, hence the onus is fully on the operator to maintain the pH. For this research, biogas plants of Gazipur and Mymensingh have been visited which use imported digesters. The pH is measured and maintained regularly. Note that not every day's pH is tabulated. Only November and December's pH are found and it seen that the pH always remains 7 over the mentioned time (Figure 3).

Temperature of digester is the most important parameter for optimal biogas production. The average temperature over the year varies between 17°C to 30°C. The maximum and minimum temperature can vary over a larger range. But the temperature should remain as close to 38 as possible for maximum production of biogas (Figure 4).

biogas will vary with the variation of temperature. The variation of temperature will reduce biogas yield and methane content of the gas. Hence the air and H_2S removal unit need to work more time than usual. The running time of generator will be reduced, in turn it will reduce electricity yield. So for rated load it will increase current and reduce voltage, rotational speed and frequency. The energy of cooling water and exhaust gas can be used to control the temperature of the digester. The cooling water of the generator can be directly used to control the temperature of the digester. But this water along is not enough for the digester in an extremely cool weather. Hence the energy of the exhaust gas is used with a heat exchanger for producing more hot water. The cooler the weather the more temperature difference between the input and output of the heat exchanger will be observed. As the imported digesters have the arrangements for temperature control the subsequent figure shows the mentioned parameters for them over the year.

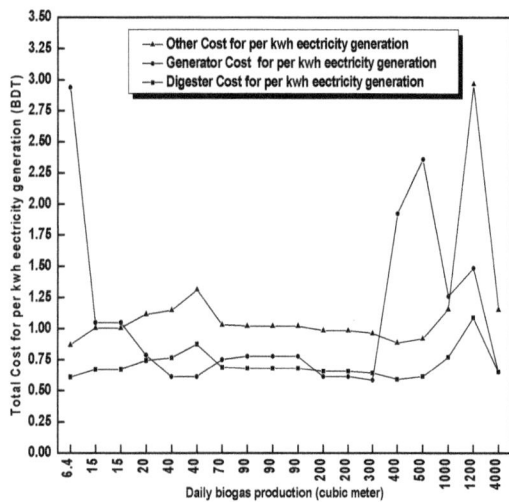

Figure 2. Total Cost for per kWhr electricity generation

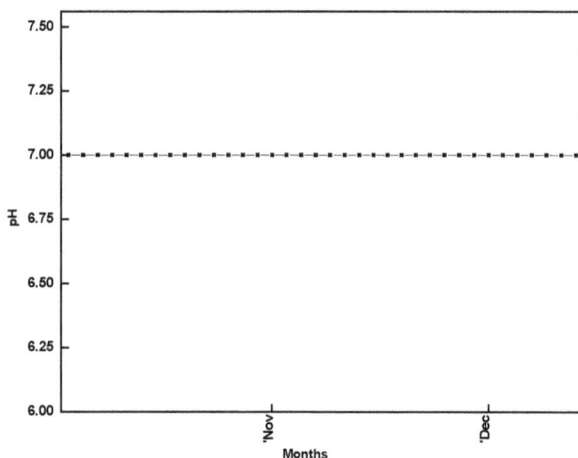

Figure 4. Average, maximum and minimum temperature throughout the year

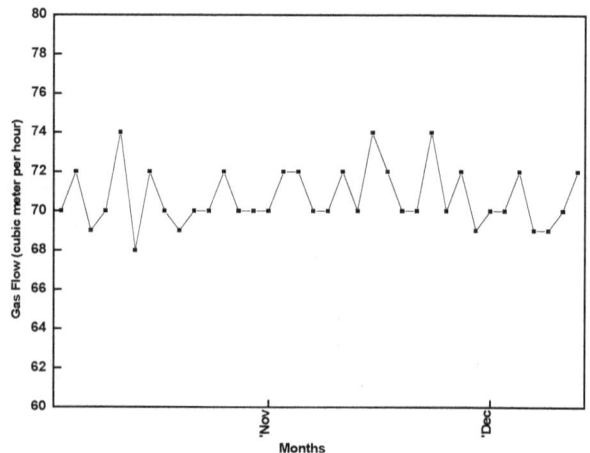

Figure 3. pH of an imported digester over the different months

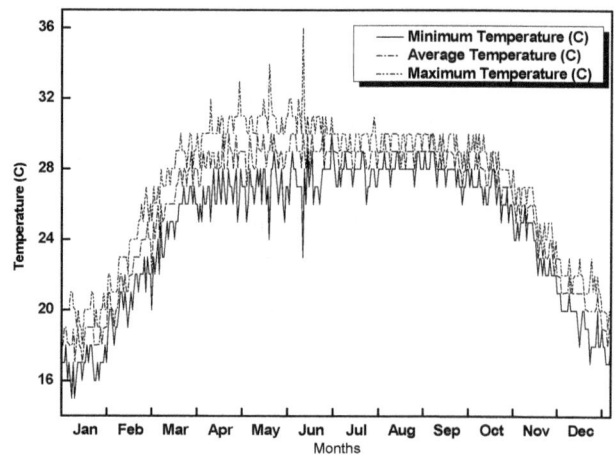

Figure 5. Gas flow over the year

So if temperature is not controlled the production of

Figure 5 shows that despite the temperature variations, as the temperature is controlled the biogas flow remains almost constant. The voltage remains constant at 400V over the year, but current varies with variation of the load as per Figure 6.

The frequency remains constant 50Hz over the year and rotational speed remain within 1% of the rated value over the year as shown in the Figure 7.

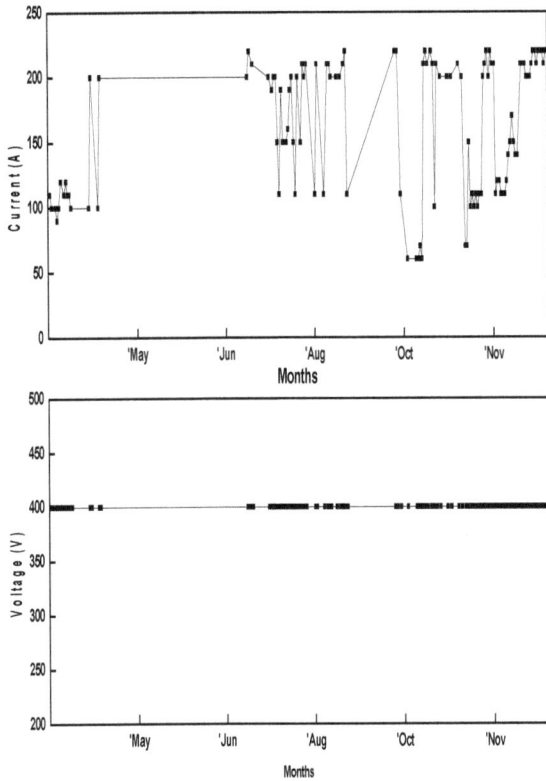

Figure 6. Voltage and current of the biogas plant with imported digester

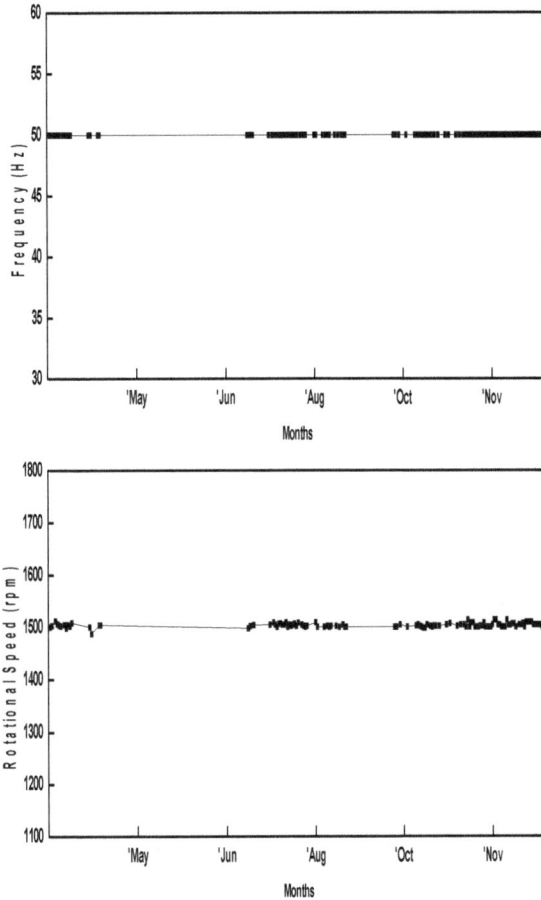

Figure 7. Frequency and rotational speed over the year for imported digester

It shows (Figure 8) that during the cooler part of the year more heat required for maintain the temperature of digester hence the temperature difference is higher, but during the hotter part of the year the temperature difference is lesser as less heat is required for maintaining the temperature of the digester.

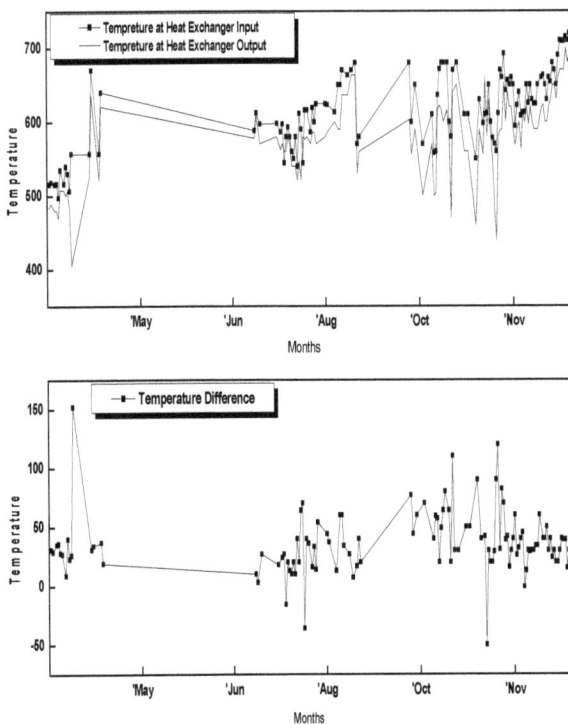

Figure 8. Temperature of exhaust at heat exchanger input and output and temperature difference

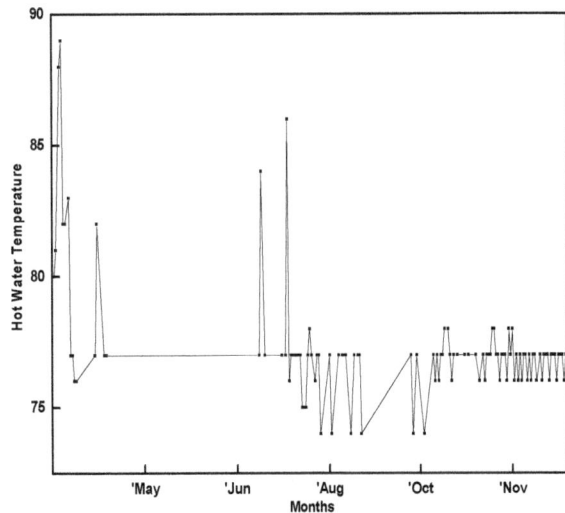

Figure 9. Hot water temperature over the year

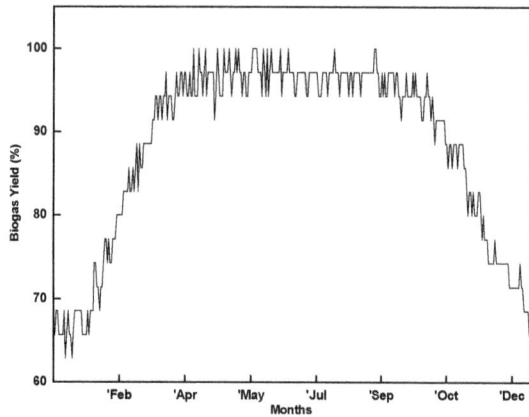

Figure 10. Biogas yield for local digesters

The hot water tank temperature (Figure 9) is maintained around 77 over the year. There cases of overshoot as well undershoot. Now for locally built digesters the temperature and pH are not controlled. Hence less biogas is generated. The biogas yield and methane content vary over the year, consequently electricity generation also varies. Neither biogas yield or methane content nor electricity generation is monitored for local digesters. But methane content of biogas is monitored for Rangpur Poultry Farm at Rangpur, but not tabulated. The methane content during the hotter part of the year varies among 60% to 70%. But during the cooler part it can fall as low as 40%. The variation of methane content as the percentage of maximum methane content can be shown by the Figure 10.

The calculation shows that the overall biogas generation reduced by 12%. For extreme cold year the generation can be reduced by up to 20%.As the biogas yield decreases by 12% to 20% the generation cost increases. We shall use 12% reduction of biogas yield to calculate the generation cost of per kWhr electricity. Figure 11 below shows the normalized cost for per unit electricity generation.

The advantage of local digester is no longer valid as the generation cost increased for per kWhr electricity generation.

Periodic loading and unloading and unloading are very important for maximizing the biogas production. The periodic agitation of digester and mixing tank increases biogas production. Motors and pumps are used for imported digester for loading and unloading of manure. For agitation, pumps and motors are used. At PPL manure is loaded everyday at 8 am. Before loading, unloading is also done. Everyday 20 ton manure is pumped into the mixer, after mixing for 30 minutes the manure is pumped into digester for 2 hours (Figure 12 and Figure 13).

Taking all the factors described above into account we are in the position calculate the production of biogas from two different cases. The cases are:

1) Case 1: Imported digester with all types of controlling schemes
2) Case 2: Local digester with no controlling schemes.

Table 2 shows the comparison of biogas yield between mentioned two cases. It clearly appears that the digesters with controlling schemes yields75% more biogas. More benefits means,

1) The plants which run for 14 hours can run for 24 hours now.
2) More biogas means less generation cost for per kwh.
3) Uninterrupted power supply for 24 hours which cannot be provided by even the national grid.
4) More power from biogas means less from diesel or other sources which reduces green house gas emission.

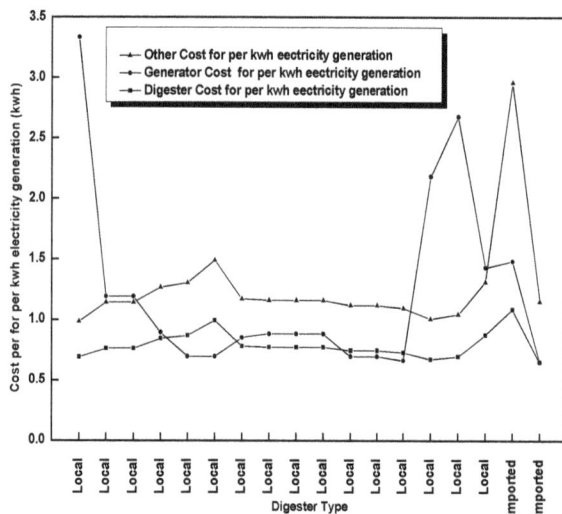

Figure 11. Cost for per kWhr electricity generation taking the pH and temperature into account

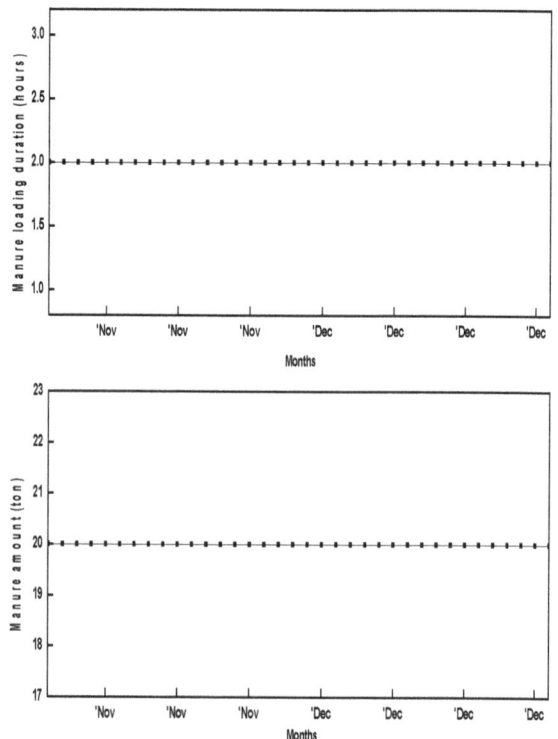

Figure 12. Manure loading duration and manure amount

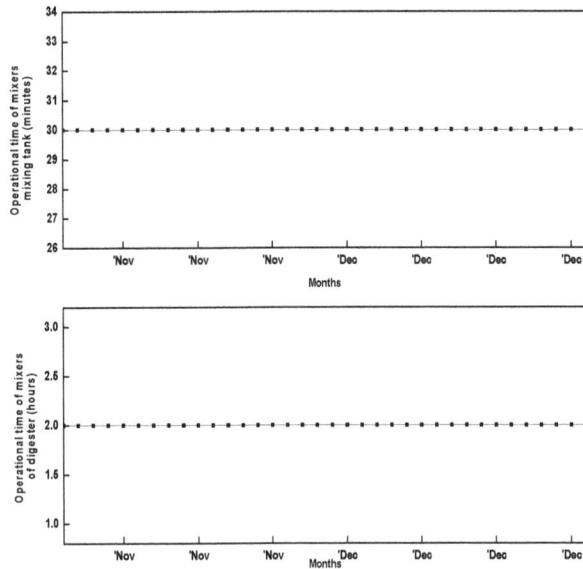

Figure 13. Operational time of mixers mixing tank (minutes) and Operational time of mixers of digester (hours)

But for generating more gas we need to deploy the controlling schemes which will,

1) Incur more cost for equipments and control module.
2) Will reduce operational cost as less operator will be required
3) Will reduce the size of the digesters as digestion with temperature control reduces the Hydraulic Retention Time (HRT) from 40 days to 20 days.

The cost of per kWhr electricity generation will be reduced for additional gas generation. Taking additional gas generation and additional investment required into account

the figure (Figure 14) below shows the comparison between two cases mentioned above.

So if additional control systems are deployed in locally made digesters then 38% money is saved. The digester with less $20m^3$ sizes doesn't have any advantage of using control system as most of the energy is lost in the control system.

4. Conclusion

This research work represents the impact of different factors on production of biogas in different biogas plants of Bangladesh. The operation of different size of biogas plants throughout the country starting from 6.4 m^3 to 4000 m^3 have been analyzed in this paper. The impacts of temperature, pH, agitation periodic loading and unloading [17] on biogas yield are analyzed based on the collected data from survey, internet and other sources.

It has been found that most of the biogas plants don't have the schemes for controlling the temperature, pH, agitation, loading and unloading apart from few of the biogas plants which are imported. These imported biogas plants are well equipped to control temperature, pH, agitation, loading and unloading etc. Hence they produce up to 75% more biogas than the locally produced biogas plant. But for locally produced biogas plants additional controlling schemes can be provided making an additional investment. Although providing controlling schemes incurs additional costs but the increment in gas production suppresses the costs and the capital cost for per unit of electricity generation decreases by up to 38%.

Table 1. Detail of biogas plants studied

S/N	Daily Gas Production (m^3)	Total Size (kW)	Name	Digester Type	Address	Mixing Tank	Digesters	Holding Tanks	Hot Water Tanks
1	6.4	2.4	Home Plant	Local	Rangunia, Chittagong	0	1	0	0
2	15	2	Jahangir Poultry	Local	Laxmipur	0	1	0	0
3	15	2	Green	Local	Solamasi, Keraniganj	0	1	0	0
4	20	2.5	Mohammadia Poultry	Local	Chowpally, Laxmipur	0	1	0	0
5	40	5	Ramu	Local	Cox's Bazar	0	1	0	0
6	40	6	Vision Poultry	Local	Sonagazi, Feni	0	1	0	0
7	70	8	Companigong Agro-Industries Ltd.	Local	Companiganj, Comilla	0	1	0	0
8	90	10	Bhuiyan Poultry	Local	Comilla	0	1	0	0
9	90	10	RMR Poultry and Hatchery Limited	Local	Laxminathpur, Rajapur, Pabna	0	1	0	0
10	90	10	Nazim Poultry	Local	Mauna, Gazipur	0	1	0	0
11	200	24	Electro Agro	Local	Gojaria, Shibchar, Madaripur	0	1	0	0
12	200	24	Fortuna	Local	Targas, Gazipur	0	1	0	0
13	300	34	Vulua Royal Chicks	Local	Sonapur, Noakhali	0	1	0	0
14	400	50	Rashid Krishi Khamar	Local	Trishal, Mymensingh	0	1	0	0
15	500	60	Paragon Poultry Ltd. (PPL)	Local	Sylhet	0	1	0	0
16	1000	100	Paragon Poultry Ltd. (PPL)	Local	Chandaner Hat, Gangachara, Rangpur	1	4	1	0
17	1200	122	Paragon Poultry Ltd. (PPL)	Imported	Chamiadi, Valuka, Mymensingh	2	1	1	1
18	4000	264	Paragon Poultry Ltd. (PPL)	Imported	Baniarchala, Bhavanipur, Gazipur	1	2	1	1

Table 2. Amount of biogas yield from local and imported digesters

Case No.	Manure Amount (t/day)	Total Solid (%)	Total Solid	Electricity Yield (kWhr/day)	Equivalent biogas (m³/day)	Biogas yield (m³/kg)
1	20	0.3	6000	6300	4018	0.7
2	0.08	0.2	16	10	6	0.40

Acknowledgements

The authors would like to acknowledge the contribution of Paragon Poultry Limited (PPL), Felix Energy Services, and Computerized Service Centre for their data of biogas plants. They are also grateful to CES (Centre for Energy Studies), IAT (Institute of Appropriate Technology) and CERM (Centre for Environment & Resource Management) of BUET (Bangladesh University of Engineering & Technology) for their support.

References

[1] A. Srivasata and R. Prasad, "Triglyceride Based Diesel Fuels," Renewable and sustainable Energy Reviews, Vol. 4, No. 2, pp. 111-133, 2000.

[2] C. C. Akoh and B.G. Swanson, "Base Catalyzed Transesterification of Vegetable Oils," Journal of Food Processing & Preservation, Vol. 12, pp. 139-149, 1988.

[3] M. Mirhosseini, "Assessing the Wind Energy Potential Locations in Province of Semnan in Iran," Renewable and Sustainable Energy Reviews, Vol. 15, No. 1, pp. 449-459, 2011.

[4] A.K Hossain and O. Badr, "Prospects of Renewable Energy Utilization for Electricity Generation in Bangladesh," Renewable and Sustainable Energy Reviews, Vol. 11, Issue 8, pp. 1617-1649, 2007.

[5] A.K.M.S Islam, M. Islam, and T. Rahman, "Effective Renewable Energy Activities in Bangladesh," Renewable Energy, Vol. 31, Issue 5, pp. 677-688, 2006.

[6] M.R. Islam and M. R. A. Beg, "Renewable Energy Sources and Technology Practice in Bangladesh'" Renewable and Sustainable Energy Reviews, Vol. 12, Issue 2, pp 299-343, 2008.

[7] S. I. Khan and A. Islam, "Performance Analysis of Solar Water Heater," Smart Grid and Renewable Energy, Vol. 2,

No.4, pp. 396-398, 2011.

[8] T. R Oke, "Initial Guidance to Obtain Representative Meteorological Observations at Urban Sites: Instruments and Observing Methods," World Meteorological Organization, Report No. 81, WMO/TD-No. 1250, 2006.

[9] BPDB, "Annual Report 2006–07," Bangladesh Power Development Board, Dhaka, 2007.

[10] M. Asaduzzaman, A.H.M.M. Billah, "Energy for the Future, in Centre for Policy Dialogue, Emerging Issuesin Bangladesh Economy: A Review of Bangladesh's Development 2005–2006," University Press Limited, Dhaka, 2008.

[11] S. C. Bhattacharyya, G. R. Timilsina, "Energy Demand Models for Policy Formulation: A Comparative Study of Energy Demand Models," The World Bank, Policy Research Working Paper – 4866, 2009.

[12] S. I. Khan, A. Islam, A. H. Khan, "Energy forecasting of Bangladesh in gas sector using LEAP software," Global Journal of Researches in Engineering, Vol. 11, No. 1, pp. 15–20, 2011.

[13] Z. M. Hasib, J. Hossain, S. Biswas, A. Islam, "Bio-Diesel from Mustard Oil: A Renewable Alternative Fuel for Small Diesel Engines," Modern Mechanical Engineering, Vol. 1, No. 2, pp. 77–83, 2011.

[14] "Annual Report 2008", Bangladesh Oil, Gas and Mineral Corporation (Petrobangla), Dhaka, 2009.

[15] A. Roy, "Reliable Estimation of Density Distribution in Potential Wind Power Sites of Bangladesh," International Journal of Renewable Energy Research, Vol.2, No.2, pp. 219-226, 2012.

[16] Z. Wadud, D. J. Graham, R. B. Noland, "Gasoline Demand with Heterogeneity in Household Responses," The Energy Journal, Vol. 31, No.1, pp 47–73, 2010.

[17] M. Mansha, S. Javed, M. Kazmi and N. Feroze, "Study of Rice Husk Ash as Potential Source of Acid Resistance Calcium Silicate," Advances in Chemical Engineering and Science (ACES), Vol. 1 No. 3, pp. 147-153, 2011.

27

Transitioning the Dominican Republic: regimes, niches and scenarios

Daniël Amrish Lachman

Institute for Graduate Studies and Research and Mechanical Engineering Discipline, Anton de Kom University of Suriname; FHR Lim A Po Institute, Paramaribo, Suriname

Email address:
danny_lachman@yahoo.com

Abstract: Energy security in the Dominican Republic is far from acceptable; black-outs, high tariffs, politicized decisions etc. are common. Furthermore, the future outlook seems worse due to effects of the global economy, climate change, oil prices, further degradation of the existing system, etc. A transition towards sustainable alternatives is therefore mandatory. In this paper a combination of existing concepts and approaches is used to indentify possible roadblocks and windfalls for such a transition in the Dominican Republic. This combination starts with defining the unit of analysis, after which actors in the socio-technical energy system are charted through literature research and interviews. Next, using social network analysis, regimes and niches are identified to depict the unit of analysis in a more useful manner for managing transitions. The step hereafter consists of creating internal and external scenarios based on critical uncertainties to insure transition management efforts against uncertainty. Moving to Transition Management, robustness analysis is then used to evaluate strategies and policies in all combinations of these internal and external scenarios to get to an optimum set of strategies and policies which are used to form a normative scenario. This will be used to get stakeholders behind the transition effort. The results are a clear overview of the energy system, impediments and opportunities regarding transitions, possible futures, and the validity of strategies and policies in different scenarios for the Dominican Republic.

Keywords: Energy System Transition Management, Socio-Technical Systems, the Dominican Republic

1. Introduction

The world is on an unsustainable path when it concerns the energy usage which forms the basis for almost all activity undertaken by mankind. There are three main reasons for this: the majority of the energy supply is provided by finite fossil fuels; transformation of these fossil fuels into readily usable energy for the consumer generates greenhouse gases that contribute in a significant way to global warming and subsequently to climate change; demand for energy services is growing primarily due to population and economic growth in developing countries.

Hence, the last decades a lot of effort has been put into place in academia, politics, business and other groups to create so-called "sustainable energy systems", meaning energy systems which have a negligible negative impact on the environment on short, middle and long term which would otherwise result in climate change. However, transitioning towards such energy systems has proven to be difficult due to the fact that existing unsustainable energy systems in use are firmly embedded in society in terms of sunk costs en vested interests; their socio-technical nature – a term that encompasses the technological, social, political, regulatory, and cultural aspects of electricity supply and use (Sovacool 2009) – has influenced and formed engineering practices, academia, legislation, institutes, behavior, spatial planning, among other things, all of which need to change (some more than others) if a new energy system is taking center stage. Another reason is also found in the fact that such a transition on a global scale will cost enormous sums of money and consume years (if not decades) and thus prove to be difficult, in particular for developing countries. Therefore, a lot of research regarding "energy system transition" has emerged and intensified in the last decades, spawning several academic courses, journals and articles, and various approaches to understand and manage transitions.

Developing countries in particular are in need of energy system transitions (Lachman 2013). This paper scrutinizes roadblocks and windfalls regarding energy system transitions in the Dominican Republic and examines the borders that make up the realm of energy system transition management. To do this, a methodology consisting of existing approaches has been used which addresses important transition aspects such as regimes, niches, landscape factors, uncertainty, etc., all of which will be explained in this paper.

This paper is structured as follows: in the next section, the methodology is explained in detail. The section thereafter reflects on and discusses this methodology. The fourth section applied this methodology in the case of the Dominican Republic. The last section concludes this paper and provides a number of recommendations, and is followed by an alphabetical list of references.

2. Methodology

The methodology is aimed at understanding the construction and dynamics of socio-technical systems, and how to manage these in such a way that leverage points are identified and utilized to ignite and accelerate the transition of these systems towards desirable alternatives. Its steps are: conducting interviews and performing a literature study in order to be able to perform a social network analysis (to obtain a clear picture of the unit of analysis, viz. the energy sector, and its construction), and to identify landscape factors (influences originating from outside the energy system) which are categorized into predetermined elements and critical uncertainties (the first are those factors that have expected developments, while the latter comprise factors whose developments are unpredictable). The next step is re-arranging the charted social network in such a manner that sets of regimes and niches become apparent. Next, scenarios are constructed (including a normative scenario) using the earlier mentioned landscape factors and other indicators. The final step consists of using robustness analysis and the normative scenario to guide transition management. The steps are detailed in the following subsections.

These steps are intertwined: information obtained through interviews and desk research is used in social network analysis and to identify landscape factors. These results are used to depict the unit of analysis in a more useful manner (more on this in section 2.4) for managing transitions and the latter is used to create scenarios for this unit of analysis (detailed in section 2.5). These scenarios are used to insure transition management efforts against uncertainty both within the unit of analysis as well as in external (landscape) influencing factors. Validating strategies, policies and actions against each scenario and iteratively tracking towards which scenario the present is heading is embedded in transition management (see section 2.6). Thus in this paper, the social network analysis and scenario creation form the foundation which transition management builds upon.

2.1. *Define the Unit of Analysis*

It is important to start with carefully defining the unit of analysis. Since this paper focuses on the transition from one energy system to another, the unit of analysis will be a socio-technical energy system by definition, but the questions remains as to how large the scope will be, in other words, whether the energy system of a region, country, group of countries, etc., will be placed under scrutiny. Properly defining the unit of analysis will prevent the omission of important actors in the socio-technical energy system, and make a proper distinction between internal and external (regarding the energy system) influencing aspects (which will be called "landscape factors" from subsection 2.2 on). Partly due to these reasons, a proper definition will also influence which questions need to be asked in the interview process (which is discussed in the next subsection). It should be noted as a cautionary warning that a larger unit of analysis will automatically result in a longer and more difficult process (because of more actors, variables, data, etc.).

2.2. *Conduct Interviews*

After the unit of analysis is defined, interviews need to be conducted. To do this in a proper and efficient manner, the following actions need to be undertaken:
- actors that are part of the socio-technical energy system need to be identified, for instance by researching websites, newspapers, magazines, etc.;
- a questionnaire needs to be conceived which is flexible in structure (i.e. the interviewer is allowed to ask additional questions, alter the sequence of questions, apply small changes to the questions, etc.), but also grounded in the local context (Bulmer and Warwick 1993). The questionnaire needs to address the following topics:
  o the strength of ties between the different actors within the energy system
  o landscape factors (named "driving forces" in scenario planning literature), which refer to events and trends occurring outside the confines of the energy system but which do have an impact on the system (Geels 2002, Shove 2012). Furthermore, questions need to unravel whether these landscape factors belong to the category of predetermined elements (forces of change whose development and impact over time can more or less be estimated), or critical uncertainties (unpredictable driving forces that will have an important and sudden impact on a particular area of interest). These driving forces can be divided into social, technological, economic, environmental and political factors that form the structure in the landscape (the contextual environment) from which trends and events emerge (van der Heijden 2005)
  o driving forces within the energy system (also categorized in predetermined elements and critical uncertainties)

○ leading indicators which are used to anticipate towards which scenario the present socio-technical energy system is heading (Conway 2004)

The interview process will be characterized by non-probabilistic sampling since the interviewer is directly targeting actors in the socio-technical energy network, rather than using a sampling frame from which actors will be randomly picked. Furthermore, to enhance the search for relevant actors, sampling will also be characterized by the snowball effect, which refers to gaining access through the initial respondents to other relevant observational units (Bryman 2004).

It is possible that the interviewer is unable to identify all actors within the energy system. However, if the research to determine the actors has been done properly, it can be assumed that those unidentified actors play a marginal role within the energy system – and even in the niches of that system (which will be discussed in more detail in the next subsection) – and thus their impact can be neglected.

2.3. Social Network Analysis

Utilizing the information obtained through research and interviews, the strength of the ties between different actors in the socio-technical energy system is made visual. The definition of tie strength is of the utmost importance. It can be defined in a quantitative sense (e.g. by using the frequency of recent contact as the definition of tie strength) or in a qualitative manner, i.e. through evaluation of qualitative information obtained through interviews and other research (Granovetter 1983). Tie strength is then rated according to a numerical scale with values that associate with a range from "very weak" to "very strong". This information can be presented in a matrix with a size Y x Y, where Y stands for the amount of actors, and with the tie-values in respective cells (Hanneman and Riddle 2005).

The energy network, consisting of actors and their ties can also be made visual using Graph Theory where actors are represented by nodes that are connected by means of weighted edges, which in turn represent ties (Hanneman and Riddle 2005, Izquierdo and Hanneman 2006), see figure 1.

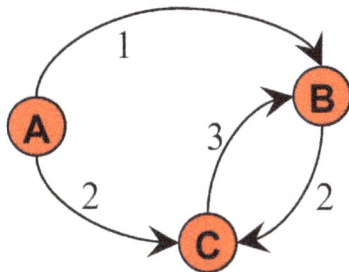

Figure 1. *Actors connected by means of weighted edges (Izquierdo and Hanneman 2006)*

Making an energy network visual as in figure 1 can be done manually or using a numerical program, such as UCINET with NetDraw.

2.4. Identify the Multi-Level Structure

The next step consists of rearranging the visualized energy network in such a manner that different groups of actors are made apparent and place the network in a wider context. This step dives deeper into social network analysis, but particularly borrows from the Multi-Level-Perspective (MLP) which has its origins in Twente school's quasi-evolutionary theory (Geels 2010a).

The Multi-Level Perspective builds on evolutionary thinking (in the sense that it assumes that variation, selection and retention play an important role in the development of systems) and interpretivism that conceptualizes a pattern of long-term change. It consists of a macro, meso and micro level, respectively landscape factors, regimes and niches (Rip and Kemp 1998, Geels 2002):

- Landscape level: the whole set of exogenous impacts on the energy system (like autonomous trends and global events). The energy system itself has little to no influence on the landscape level, but landscape factors can have a significant impact, that can even result in systemic changes (i.e. the rearrangement of the locations of actors within the system);

- Regime level: this level consists of a patchwork of regimes, each consisting of 1 or more dominant actors. The regime concept refers to "the rule-set or grammar embedded in a complex of engineering practices, production process technologies, product characteristics, skills and procedures, ways of handling relevant artifacts and persons, ways of defining problems, etc., which is engrained in institutions and infrastructures". The regime consists of three interlinked elements: (1) a network of actors and social groups, which develops over time; (2) the set of formal and informal rules that guide the activities of actors who reproduce and maintain the elements of the socio-technical system and (3) the material and technical elements (Geels 2004). Regimes co-evolve with each other but also with the environment (landscape).

A change in regimes implies a change in the system in which it functions. Regimes resist systemic change and thus also niches (which are described below). As long as regimes themselves are stable, and the landscape is not unfavorable, regimes create a strong alignment between different elements of the system in which it operates (thus increasing its momentum), thereby making the entire system path dependent / locked in (Raven and Verbong 2007). Even change within regimes follows a dependent path and tends to be incremental;

- Niche level: niches are the spaces where innovative activity takes place and where time-limited protection is offered against dominant selection rules. Niches also tend to be very flexible and adaptive. Thus, they differ from regimes and aim to replace the incumbent regime (which can happen under the right set of conditions), thereby creating new development trajectories. Initially, only technological and market niches were identified; however, Geels (2007) discovered that niches have general relevance.

Figure 2 provides a representation of the Multi-Level Perspective.

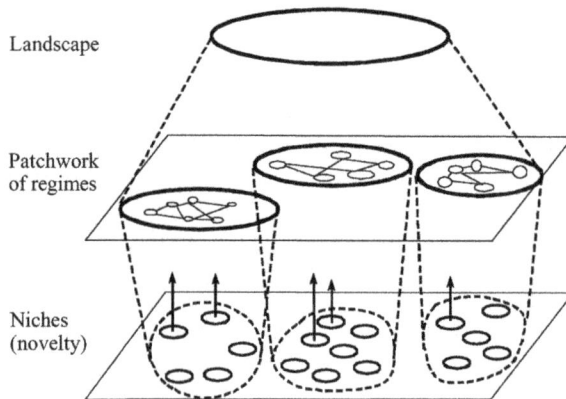

Figure 2. Representation of the multi-level perspective (Geels 2002)

According to the MLP, transitions occur as a result of dynamics at the different levels which reinforce each other creating a "window of opportunity": landscape factors destabilize regimes (actors diverge and start to disagree) while niches, developed in protective spaces, gather momentum to take center stage within the system (Geels 2006, Grin et. al. 2010).

At this point in the methodology, actors need to be grouped in order to visualize regimes and niches. Regimes and niches make up the energy system, and the level remaining is the context (landscape) within which this system exists. This macro level can be described by the external driving forces (consisting both of predetermined elements and critical uncertainties) that have been identified during the research in the first two steps.

With the information now on hand, the multi-level structure of the energy network under consideration can be conceived. It provides a snapshot of the socio-technical system which – due to its rearrangement highlighting regimes and niches placed within the landscape – is an effective manner to present a complex network.

2.5. Create Scenarios

Socio-technical energy systems are complex systems situated in and influenced by a complex landscape. Both the energy system and the landscape have a degree of uncertainty which makes it difficult to manage transitions to sustainable energy systems. This aspect of uncertainty comprises (Schwartz and Ogilvy 1998):

- risks: based on historical data, a probability can be assigned that a certain event will happen;
- structural uncertainty: an event is unique in such a way that (judgmental) probabilities of the event happening are unknown;
- unimaginable events.

It is therefore imperative that a tool is available which enables to deal with the aspect of uncertainty in order to be able to strategize and manage the transition. This is where Scenario Planning comes in the picture; it is a management thinking tool that acknowledges the existence of uncertainty in the organization and its environment by developing a set of scenarios (internally consistent and plausible, but structurally different narratives), which outline the range of possible futures (van der Heijden 2005). It enables to see opportunities and threats in advance and helps in generating more robust strategies, policies and plans (Star and Randall 2007). Scenario Planning has its origins in the Second World War and has proven its worth ever since, as evidenced by the growing number organizations (private companies, governments, and Non-Governmental Organizations etc.) using the methodology (Ogilvy and Smith 2004).

The scenarios are not predictions, but rather hypotheses in the form of rather provocative, structurally different, but internally consistent and plausible narratives, built upon combinations of uncertain, high-impact Driving Forces, about how the future of issues relevant to an organization or individual might unfold (Scearce et. al. 2004). A set of scenarios (each treated with equal weight) outlines the range of possible futures. The scenarios are used to:

1. Set the strategic direction and prepare a rough timetable of events;
2. Be more perceptive of the environment when trying to identify towards which scenario the present is evolving, and anticipate new insights and innovations;
3. Accelerate collaborative learning by providing insight in the environment during the scenario building process;
4. Test existing strategies by challenging assumptions upon which they are built;
5. Rehearse the actions that need to be taken in different environments;
6. Describe goals that need to be achieved (so-called normative scenarios).

There are various approaches to Scenario Planning which can be found in van der Heijden (2005) and Nekkers (2007). In this paper use will be made of the deductive method, which combines extremes of critical uncertainties in a matrix to form the structure upon which scenarios are built (see figure 3).

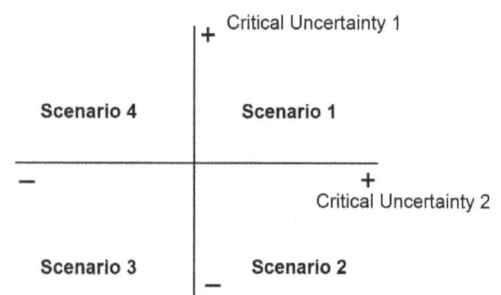

Figure 3. A scenario-matrix (also known as the scenario logic)

The deductive scenario-matrix method has the following advantages over other methods (Scearce et. al. 2004, van der Heijden 2005, Nekkers 2007):

1. due to its nature the method is more likely to produce surprises that challenge existing assumptions;
2. the method is more likely to develop scenarios that cover a wide range of possible futures;
3. the deductive approach is the most analytical Scenario Planning method.

In this phase of the methodology, internal and external scenarios are created to cope with uncertainties which exist in the socio-technical energy system itself and its landscape in order to manage transitions. Regarding internal scenarios, information obtained in the first two steps (desk research and interviews) is scrutinized to identify driving forces of change within the energy system. Similar forces are clustered into larger (underlying) driving forces, which are split into predetermined elements and critical uncertainties. The extremes of the two most uncertain and highest impact driving forces are then plotted on an X-Y axis to create the so-called scenario logic, after which narratives are conceived for each scenario that make up each respective quadrant in the matrix (Scearce et. al. 2004, van der Heijden 2005, Nekkers 2007).

The narratives should link the present state (as described in section 2.4) with the future description, and include the predetermined Driving Forces in each scenario. The scenarios are depicted as short narratives, because these can quickly capture complex matters, embed qualitative information that can not be depicted by means of graphs and tables, make unexpected scenarios believable, and leave a lasting message (van der Heijden 2005). The scenarios need to be internally consistent, structurally different from each other, challenging (they need to display a new and unique perspective), relevant and plausible.

A similar exercise is done for the creation of external scenarios, but instead of using driving forces that exist in the system, driving forces from the landscape level are used (discussed in section 2.4). It is important to note that the researcher pays particular attention during the desk research and interviews to specific local characteristics (on the level of the energy system, region, country, etc.) which can be driving forces that are most relevant for the creation of internal and external scenarios (Lachman 2011).

The researcher has thus created 4 internal and 4 external scenarios which enable to anticipate sudden changes within the socio-technical energy system and within the landscape in which the system functions. Examples of scenarios (which are actually a combination of both internal and external developments) can be found in Lachman (2011).

2.6. Formulate Strategy and Create Management System

In this final step, the focus is on the "Transition Management" (TM) concept, which, like the Multi-Level-Perspective, has its origins in Twente School's quasi-evolutionary theory and is part of what is known as "transition studies". Transition management is a reflexive and participative governance concept that attempts to manage transformative change (i.e. influence the speed and direction of change) towards sustainable development by combining long-term thinking with short term action (thus complementing conventional policy) through a process of searching, experimenting and learning (Loorbach and Rotmans 2006). It is a concept that has gained significant traction in the last decade. Key aspects on TM are:

- Experimenting and learning to guide variation and selection (learning-by-doing and doing-by-learning) while not chasing "silver bullets" (thus keeping all options in consideration and the playing field open). This characteristic of continuous learning places TM in the so-called "Processual Paradigm" school of thought (van der Heijden 2005);
- Obtaining stakeholder (from multiple domains and levels) input through inclusion and involvement;
- Complementing conventional policy (which has a short-term focus) with long-term thinking with the aim of sustainable development;
- Continuous reflection (monitoring, evaluating, improving) on all levels;
- Bringing system innovation alongside system improvement.

Transition management is executed on a strategic, tactical and operational level; these three levels follow a cyclical path consisting of problem structuring and envisioning (strategic level), agenda building and networking (tactical level), experimenting and diffusing (operational level), and executes continuous monitoring, evaluating and adjusting on all levels (Loorbach et. al. 2008).

As part of Transition Management, the Robustness Analysis method is used to determine the best strategy (or set of strategies) and policies to drive the energy system transition. Each internal scenario is placed against each external scenario (both of which were created earlier in the process), resulting in 16 internal-external scenario combinations. Using a matrix (table 1), various strategies and policies can be evaluated in case of each scenario combination. This is done, for instance, by assigning a plus-, zero- or minus-sign, respectively indicating a positive, neutral or negative effect on the energy system transition. This comparison of strategies and policies under each scenario is known as Robustness Analysis (Taylor 1999). An example of a Robustness Analysis exercise can be found in Lachman (2011).

Table 1. *An example of a Robustness Analysis exercise*

		Scenario			
		A	**B**	**C**	**D**
Policy/Strategy	1	-	0	-	0
	2	-	+	0	+
	3	0	+	+	+
	4	-	-	+	0
	5	-	-	-	-

On a strategic level, a normative scenario can be conceived once the most suitable set of strategies and policies are identified through the Robustness Analysis exercise. This scenario describes how these strategies and policies are used to arrive at a plausible and realistic (taking into account present conditions and internal and external driving forces) future state (Nekkers 2007). The normative scenario is used to align and engage multiple stakeholders in a so-called transition arena where vision, goals, roadmaps and milestones are defined based on this normative scenario and the confrontation of different perceptions of and possible directions for the energy system transition (Loorbach and Rotmans 2006, Loorbach 2010).

The results from this strategic level are the framework for the tactical level, where the focus is on conceiving a transition agenda towards the desired goal with the consent of the actors in the transition arena (aligning the actors can be achieved by using the normative scenario and the robustness analysis). Next, at the operational level the execution of the agenda takes place, with a special focus on experiments, rather than "silver bullets", to stimulate learning and thereby guide variation and selection and thus ensure continuous improvement (Loorbach 2010).

To ensure that activities undertaken on all three level remain relevant under changing conditions, adjustments (e.g. of roadmaps, agenda, actions) are done by endlessly tracking leading indicators (this also fosters continuous learning). These indicators point towards which scenario the present is heading, and can be obtained from the earlier conducted interviews and desk research (Lachman 2011). The work done by Hekkert et. al. (2007) also provides some examples of leading indicators used in the Innovation Systems literature. On a larger timescale (e.g. every 5 years), continuous learning, adjustment and improvement is also fed by updating the internal and external scenarios (van der Heijden 2005).

2.7. Visual Representation of the Methodology

The steps in the described methodology are depicted below in a simplified manner.

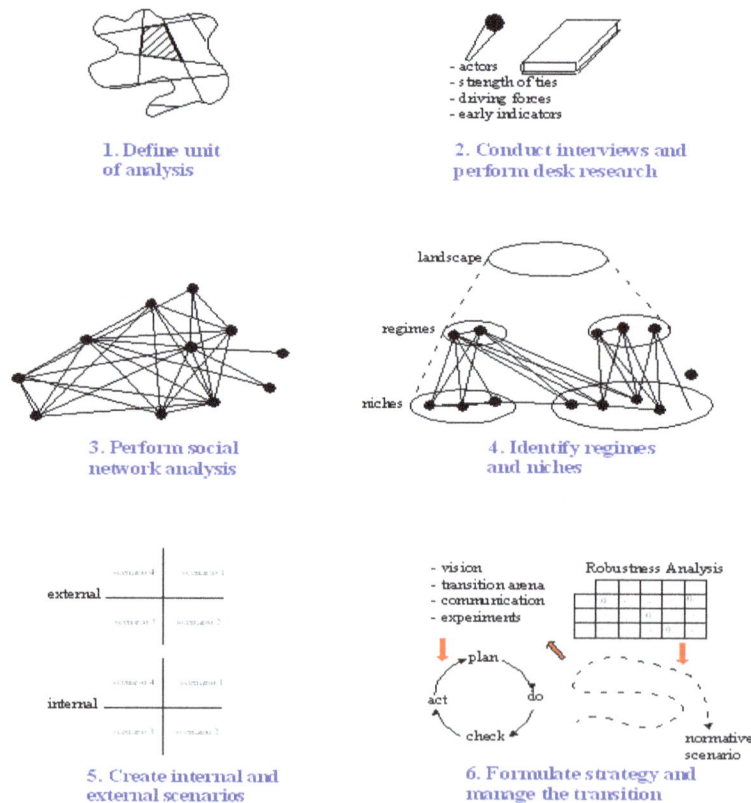

Figure 4. Visual representation of the methodology

3. Reflections on the Methodology

The described methodology is both qualitative and quantitative in nature, in the sense that it gathers qualitative data (obtained through interviews and desk research) and processes this data by assigning various categories and weights, which subsequently are used to perform several numerical exercises, e.g. in the case of social network analysis. After completing the social network analysis, qualitative (and to a lesser extent quantitative) endeavors are performed when creating various types of scenarios which are used in the quantitative robustness analysis method to conceive the normative scenario (which is again a qualitative exercise). It is therefore clearly evident that it is difficult to place the approach strictly under one of these orientations; rather, the methodology shifts from qualitative to quantitative methods, and vice versa (Bryman 2004).

Similar to its research orientation, the methodology is both inductive and deductive in nature; the path to conceive an overview of the socio-technical energy system and the scenarios is clearly inductive, whereas formulating strategies and managing the transition by means of the transition arena, experiments, etc., all derived from the earlier mentioned energy system and scenarios is deductive (Minto 1996). Since the energy system overview, the scenarios, and early indicators need to be updated over time as changes in the landscape become apparent, the subsequent actions on a strategic, tactical and operational level need to be adjusted likewise (van der Heijden 2005). Therefore, this methodology needs to be repeated with a specific frequency in order to adjust transition management strategies and tactics, which implies that this cycle from induction to deduction is required to be repeated during the management of the transition.

From an epistemological perspective, the methodology has traces of phenomenology due to the fact that it is important during the construction of the socio-technical energy network (and its restructuring along the lines of the Multi-Level Perspective) and the scenarios to understand how various social groups, in particular regimes and niches, view the socio-technical energy system and the landscape within which this system operates. Subsequently, the methodology is also interpretivist in nature since it attempts to grasp the subjective meaning behind social action (Bryman 2004) which is formed because of the earlier mentioned perception of reality. This is evident when scenarios (narratives that discuss these meanings behind social actions to render the scenario plausible) are constructed and when shifts in the direction of the unfolding present is becoming apparent through the use of early indicators (which can convey drivers for social action).

When scrutinizing the approach from an ontological point of view, it is important to note the fact that the approach builds upon the notion that the energy system is being regarded as a socio-technical entity. This means that social actors and (technical) artifacts influence each other to such an extent that they shape each others developmental trajectory. Therefore, this clearly indicates that the methodology has a strong constructionist flavor.

It also noteworthy to mention that the methodology is also strongly characterized by the fact that it is an exploratory process at heart; the reasons for this lie in the fact that the approach centers on the observation of phenomena from which the researcher can identify concepts such as driving forces, early indicators, regimes and niches (Bulmer and Warwick 1993). These are in turn are used to create networks and scenarios (which in effect is the inductive portion as described earlier).

The underlying research design is of a mixed nature: there are elements of both a case-study and a longitudinal analysis. The case-study approach is evidenced by the fact

that in-depth research (ranging from literature study, to interviews, robustness analysis, social network analysis, and so on) is done on a particular unit of analysis, which has been carefully chosen with its boundaries properly defined. The longitudinal element of the design is apparent where scenarios are created, since these scenarios formulate of plausible path between the current condition and a particular future state, thereby connecting two points in time.

The described methodology has some advantages; it bridges the gap between understanding the socio-technical energy network and landscape factors (driving forces) on one hand, and concrete activities on a strategic, tactical and operational level to advance a desired transition on the other hand. Secondly, the approach is quite analytical in nature due to its quantitative efforts to create the socio-technical energy network – which also eases comparisons between different energy networks –, lay out its possible (and plausible) developments over time through the use of scenarios, conceive a normative scenario based on the results from a robustness analysis, and managing the transition strategy by keeping track of so-called early indicators, most of which are quantitative in nature. Furthermore, the methodology exercises a form of transition management which is highly flexible and adaptive to (suddenly) changing circumstances through its use of scenarios and the robustness analysis, which is important since the global landscape is increasingly changing, while becoming more complex and unpredictable. Another advantage lies in the fact that the visualizations of regimes and niches in the socio-technical energy system, the various scenarios, an overview (provided by the robustness analysis) of the merits and perils of various strategies, policies and actions in different combinations of internal and external scenarios, can be used in the transition process as powerful communication tools regarding a rather abstract and complex matter.

These benefits notwithstanding, the methodology has some drawbacks and pitfalls; since it consists of several approaches, the drawbacks to each will be discussed (Lachman 2013):

1. The methodology can prove to be quite resource consuming (in particular the stage concerning desk research and interviews) depending of the unit of analysis chosen;
2. With regard to the purposive, non-probabilistic, sampling using unstructured, flexible and even adaptive techniques, the same results will be difficult to obtain in an exactly repeated manner. Therefore, the methodology has a low replicability. However, the validity is regarded to be high, since the method is likely to deliver results relevant to the research questions (Bryman 2004);
3. The interview process can be hampered due to non-response, interviewer variance, and the difficulty of the interviewee to grasp abstract notions such as

landscape, scenarios, driving factors, regimes, and so on (Billiet et. al. 1990);

4. Another pitfall is that the scenarios are often not viewed as the outline of possible futures but as predictions of what may come, thereby nullifying the adaptive ability of transition management (Nekkers 2007);

5. the MLP uses metaphors and imprecise concepts, with the danger of creating ambiguity and being able to categorize phenomena too easily since the concepts have vague boundaries (Smith et. al. 2010);

6. Another drawback of the MLP is its complexity; it might seem straightforward, but attention is required to dynamics between levels and between actors of the same level, resulting in a myriad of events, actors and relations that need to be taken into account, especially when applying the MLP at relatively large transitions. This complicates the conception of computer models (Geels 2010a);

7. TM conveys the idea that a transition can be accomplished through the execution of proper management; transitioning is a managerial task. However, with this assumption, the scope of the transition task is simplified by neglecting the fact that influences exist – both inside and outside the transition management realm –, such as belief systems, political interests, and culture, which obstruct or even prohibit managing transitions according to best management practices and rules (Shove and Walker 2007, 2008).

4. Transition Pathways for the Dominican Republic

In this section, the aforementioned combination of concepts and approaches will be used in the case of the Dominican Republic.

4.1. Unit of Analysis

The unit of analysis will be the total energy sector, consisting of electricity generation, transmission and distribution and (fuels for) the transportation sector. The sector is characterized by a total installed electricity generation capacity of around 3600 MW and consumes about 10 billion kWh. Of its installed capacity, around 84 % is covered by thermal generation (of which 20 % is delivered by one company) and the remaining part by hydro power (which expanded in 2011 with 550 MW) and there are plans for another 600 MW of thermal generation. All of the fossil fuels are imported and in 2010 totaled 7 % of GDP or $ 3.5 billion. Other energy generation technologies are barely existent (e.g. 33 MW of wind power and some bio-fuel initiatives). The sector has an open market where vertical integration is allowed, though hydro power and transmission are run by the state. There

are three public and one private distribution companies (Burgos, 2008).

The Dominican Republic has negligible fossil fuel reserves, and only uses hydro power as its renewable source, though there are other alternative energies in abundance: solar power potential is around 150,000 GWh/year and approximately 4,400 km2 is useful for wind energy generation, equaling 30,000 MW (Elliot et. al. 2001). There are a few micro-scale utilizations of these technologies, though there are plans to expand on that portfolio, but nothing concrete has been set in stone. There have also been talks about a Haiti-Dominican Republic interconnection, though many have deemed it uneconomical (Nexant, 2010).

Total electrification amounts to 95.7 %, whereas rural electrification is close to 90 % and unserved totals 15 %. Technical and non-technical losses are estimated to be between 34 – 40 %, of which illegal connections take up about 18 %. On top of these problems, there are frequent black outs (lasting up to 20 hours), despite the fact that the electricity tariffs are among the highest in the region and electricity subsidies exceed $ 1 billion, while energy consumption is increasing annually with 10 % (Burgos 2008). It is estimated that there is an energy supply deficit of 2,000 – 2,300 MW.

The country has an electricity sector and renewable energy law. The National Energy Commission is the policy agency, one of its main responsibilities being the elaboration of the National Energy Plan. The Electricity Superintendence is the regulatory agency, while the Coordination Agency was created to coordinate dispatch of electricity. The Dominican Corporation of State Electricity Companies is a holding company that brings together all government-owned generation, transmission and distribution companies and associated government programs in the country (OAS 2007). The debt owned by the state's agencies to the generators is quite substantial because of lucrative contracts. Plans to renegotiate these contracts are retaliated by the generators by taking electric capacity offline, thus creating black outs, which in turn causes public outrage.

4.2. Social Network Analysis and Multi-Level Perspective

Data has been obtained from literature, news articles and interviews. The personal interviews taken from people (knowledgeable about aspects regarding the Dominican energy sector) are able to give deeper insight into informal, largely unknown, subtle or increasingly important becoming forces specific to the Dominican context that shape its socio-technical energy system.

A little over 250 articles, 15 reports, 15 books and 4 film documentaries were reviewed. The topics spanned a wide spectrum and included rural electrification, alternative and conventional energy, climate change, geopolitics, the future of transport, social impact of energy and energy efficiency.

A total of 15 persons, from the public, and the private (energy) sector, were interviewed. The amount of interviews was sufficient because during the last couple of interviews no new information, i.e. the saturation point (Strauss and Corbin 1998), was obtained. Questions were asked about actors and their ties, landscape factors (both predetermined elements and critical uncertainties), impediments to transition etc. Using the obtained information, table 2 is conceived which depicts the actors in the Dominican energy system and the strength of their ties (A being the lowest, D being the highest) which is visualized in figure 5.

It can be seen that the generators operating the thermal and hydro power plants take a prominent place within the regime level. This is due to the fact that the government has stakes in their operation and the fact that they are often vertically integrated across the supply chain.

Table 2. Actors and their ties in the energy sector of the Dominican Republic

	Largest Generators	Distributors	Other Generators	Government	RE companies	Fuel distributors
Largest Generators		C	D	B	0	D
Distributors	C		C	C	A	0
Other Generators	D	C		B	0	D
Government	B	C	B		A	B
RE companies	0	A	0	A		0
Fuel distributors	D	0	D	B	0	

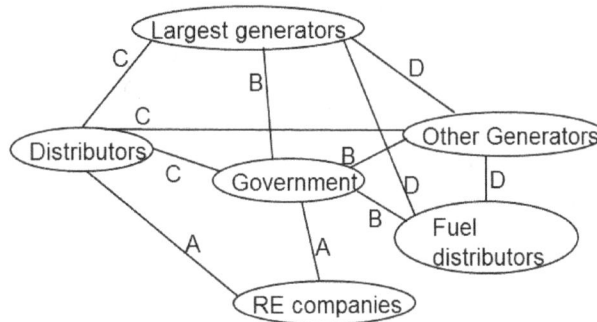

Figure 5. Actors and their strengths in the energy sector of the Dominican Republic

The information in this figure is further rearranged (based on tie strength) into the Multi-Level Perspective format, see figure 6. In this figure concepts, ideas, approaches, etc. that can play an important role in energy systems transitions have also been included in this figure sorted in the niche and regime category.

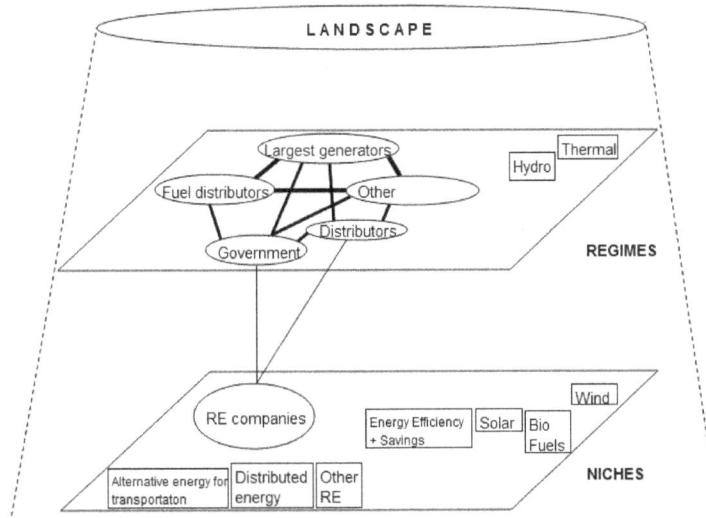

Figure 6. The Multi-Level Perspective applied on the energy sector of the Dominican Republic

As mentioned earlier, the vertical integration of many of the power companies gives these companies (indirect and direct) control of the deployment of renewable and alternative energy technologies, distributed energy systems etc., and since they benefit from the current situation, they have insufficient incentives to speed up deployment. The lack of deployment is also evidenced that mainly thermal (and to a lesser extent hydro) power is considered in energy plans, accompanied by a couple of wind and bio-fuel projects.

4.3. Scenario Planning and Robustness Analysis

From the earlier mentioned desk research and personal interviews, several critical uncertainties have been indentified regarding the future the energy sector in the Dominican Republic is heading to. These critical uncertainties are categorized in internal uncertainties (i.e. internal to the energy sector) and external uncertainties (which belong to the landscape-realm in the Multi-Level Perspective, see figure 6):

Internal Critical Uncertainties
1. The quality of the government regarding (energy) policy making (implementation of existing laws, depoliticizing decisions, ability to deploy unpopular measures);
2. Willingness of the public to accept higher tariffs to improve energy security
3. Willingness of Independent Power Producers (IPPs) to revise existing Power Purchase Agreements (PPAs) which are highly favorable to them;
4. Energy consumption behavior and adoption of energy efficiency and savings guidelines;
5. Perception of new renewable energy projects by affected ethnic / tribal groups;

External Critical Uncertainties
6. Transfer of new energy technologies to the Dominican Republic
7. Climate change the effects thereof on Dominican Republic's energy system and renewable energy potential;

8. Natural disasters and the effects thereof on Dominican Republic's energy system and renewable energy potential;
9. Sudden events, such as the attacks on 11 September 2001, which severely impact the majority of the Dominican economy
10. Oil price behavior;
11. (International) technological breakthroughs and their deployment.

In figure 7 these critical uncertainties are plot according to their level of uncertainty and impact on the energy sector (the numbers refer to the order of these critical uncertainties mentioned above). However, "the quality of the government regarding (energy) policy making" will not be chosen for use in the scenario logic, due to the fact that the ultimate goal of the scenarios will be to formulate resilient and robust strategies, and subsequent policies and actions will mostly fall under the responsibility of the government. Using government quality in the scenario logic will therefore result in normative scenarios, which will describe future governmental action which the government will try to achieve or steer away from (Lachman 2011). The most critical ones (thus those in the upper right corner of the graph) will be used to create internal (public accepting unpopular measures towards energy security and IPPs willing to change PPAs) and external (climate change and impacts on the domestic economy) scenarios, see figure 8.

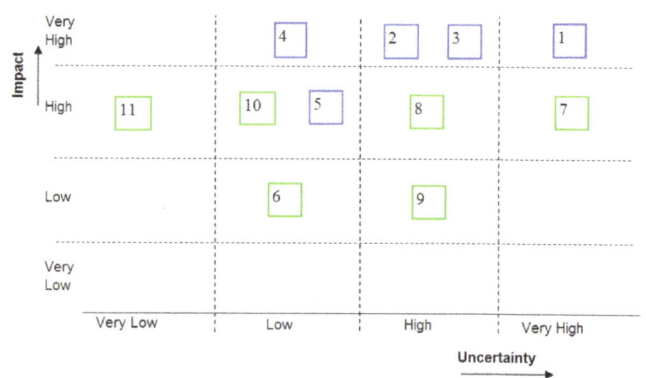

Figure 7. Ranking of the internal (blue) and external (green) critical uncertainties

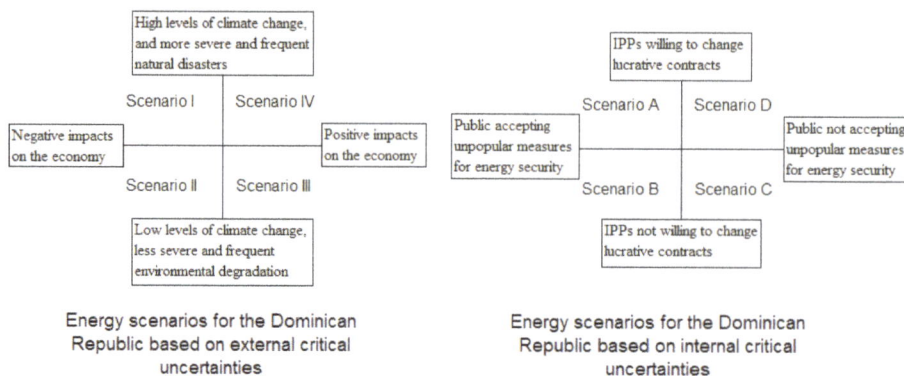

Energy scenarios for the Dominican Republic based on external critical uncertainties

Energy scenarios for the Dominican Republic based on internal critical uncertainties

Figure 8. Internal and external energy scenarios for the Dominican Republic

Various strategies, policies and concrete actions regarding the energy sector can now be validated against each combination of internal and external scenarios using the information from the desk research and interviews. In particular, energy system transition strategies, policies and actions (which take strong account of the actors, tie-

strengths and the multi-level depiction of the Dominican energy sector) can now be tested in different scenarios and thus can transition management insure itself against future uncertainty regarding the energy sector. The robustness analysis is depicted in table 3.

Table 3. Robustness Analysis using created internal and external scenarios

Strategies, policies, actions	Scenario Combination															
	Internal Scenarios:	A				B				C				D		
	External Scenarios:	I	II	III	IV	I	II	III	IV	I	II	III	IV	I	II	III	IV
Expand electricity production through primarily hydro power		-	+	+	-	-	+	+	-	-	+	+	-	-	+	+	-
Expand electricity production through primarily thermal power		-	+	+	-	-	-	-	-	-	-	-	-	-	+	+	-
Expand use of bio fuels for transportation sector		-	0	0	-	-	0	0	-	-	0	0	-	-	0	0	-
Electrify the transportation sector		+	+	+	+	-	-	0	-	-	-	0	-	+	+	+	+
Venture into hydrogen fueled vehicles		-	-	0	0	-	-	0	0	-	-	0	0	-	-	0	0
Provide incentives for less vehicle use (e.g. carpooling etc.)		+	+	+	+	+	+	+	+	+	+	0	0	+	+	0	0
Provide incentives for fuel-efficient vehicles		+	+	+	+	+	+	+	+	0	0	0	0	0	0	0	0
Promote electricity efficiency and energy savings		+	+	+	+	+	+	+	+	+	0	0	0	+	+	0	0
Execute construction of Haiti-Dominican Republic Interconnection		+	0	0	+	+	0	0	+	+	0	0	+	+	0	0	+
Create centralized energy systems		-	+	+	-	-	-	-	-	-	-	-	-	-	+	+	-
Decentralize energy supply		+	+	+	+	+	+	+	+	+	+	+	+	+	+	+	+
Create diversified energy production portfolio		+	+	+	+	+	+	+	+	+	+	+	+	+	+	+	+
Rapid deployment of renewable energy technologies apart from hydro, wind and PV		+	+	+	+	+	+	+	+	+	+	+	+	+	+	+	+
Deploy a firm performance-based rewarding scheme for the governmental bodies in the energy sector		+	+	+	+	+	+	+	+	+	+	+	+	+	+	+	+
Upgrade existing outdated transmission and distribution network		-	+	+	-	-	+	+	-	-	+	+	-	-	+	+	-
Build new transmission and distribution network		-	-	+	+	-	-	+	+	-	-	+	+	-	-	+	+
Mandate inclusion of renewable energy technologies in portfolio of IPPs		+	+	+	+	+	+	+	+	+	+	+	+	+	+	+	+
Gradually stop cross-subsidizing sectors (clearly communicated to stakeholders) while investing the recovered funds in energy saving and efficiency technologies which will lower losses and thus can potentially lower tariffs		+	+	+	+	+	+	+	+	+	+	+	+	+	+	+	+
Immediately stop cross-subsidizing sectors (clearly communicated to stakeholders) while investing the recovered funds in energy saving and efficiency technologies which will lower losses and thus can potentially lower tariffs		-	-	-	-	-	-	-	-	+	+	+	+	+	+	+	+
Unite all governmental bodies with responsibilities in the energy sector into one unit		+	+	+	+	+	+	+	+	+	+	+	+	+	+	+	+
Deploy incentives for efficient energy production, transmission and distribution		+	+	+	+	+	+	+	+	+	+	+	+	+	+	+	+
Place a strong focus on fostering and stimulating niches		+	+	+	+	+	-	-	+	+	-	-	+	+	+	+	+

Legend:
+   positive effect
-       negative effect
0   no / negligible effect

4.4. Transition Management

Once finished with the robustness analysis (the exercise done can be expanded), the execution phase can be entered

using the data from the analysis done thus far, in particular the multi-level perspective, the scenarios and the robustness analysis. The main points are:

- Create a normative scenario for communication purposes and to have a focus for all stakeholders;
- Create a so-called "transition arena" with stakeholders (including actors that belong to the niche level);
- Test important and far-reaching decisions on a strategic, tactical and operational level in each scenario and determine which ones bear the least risk going forward;
- Stimulate the emergence and diffusion of niche-innovations;
- Enhance selection pressure on the regime through economic instruments (e.g., carbon taxes) and regulation (e.g., environmental legislation) (Geels 2012);
- Reiterate the scenario building exercise every 5 years (van der Heijden 2005);
- Use leading indicators to track towards which scenario the present is unfolding, such as, but not limited to:
  o Patterns of climate parameters (e.g. precipitation data, wind velocities);
  o Rate of deforestation;
  o Number of IPPs willing to adjust existing PPAs;
  o The amount of public resistance when implementing unpopular measures at an incremental level;
  o Investment patterns in the Dominican Republic;
  o Shifts in global tourism demand;
  o The frequency and severity of natural disasters;
  o Oil price behavior and forecasting thereof;
  o Time between the deployment of new energy technologies in developed countries and the Dominican Republic;
  o Energy consumption parameters, such as energy intensity;
  o The frequency of sudden (global) events which impact the domestic energy sector.

5. Conclusions and Recommendations

In the analysis of the energy sector in the Dominican Republic it is evident that energy security exists at a low level. Furthermore, energy security appears to be under a greater threat regardless towards which scenario the present is unfolding. The used combination of concepts and approaches, while it has some drawbacks as discussed, has yielded some interesting results, like the identification of not only actors but also ideas, concepts, etc. that belong to either the regime or niche level, the critical internal and external uncertainties and the validity of different decisions in different scenarios (like for instance investing in only a few energy technologies, which is currently the case). thus, it has clearly identified the various roadblocks and windfalls that exist in the energy sector, and which steps need to be taken to deal with these aspects in the wake of both internal and external uncertainty.

Since developing countries are seldom the subject of such analyses, they are paradoxically more in need of them.

It is therefore advocated that similar exercises are done for other developing countries since these cope with specific characteristics not found in the developed world, and thus also not in research focusing on that part of the world. The latter implies that the findings of research on energy system transitions focusing in Western nations can't simply be adopted for developing nations. This paper has highlighted some aspects (important for energy system transitions) when studying regimes, niches and scenarios which are regularly associated with developing countries: poor policy making, vulnerable to climate change, small niche base, weak institutions, etc.

Energy system transition is a difficult task that spans decades. It is therefore advisable to identify the power relations in the energy sector and how to deal with that in the face of looming (internal and external) uncertainty. This combination of concepts and approaches has attempted to provide some insight in this matter for the Dominican Republic in order to help the energy system transition in this country forward in the desirable direction.

References

[1] Billiet, J., Loosveldt, G., Waterplas, L (1990), *Het survey-interview onderzocht, effecten van het ontwerp en gebruik van vragenlijsten op de kwaliteit van de antwoorden.* S.O.I. reeks sociologische studies en documenten, Departement Sociologie, Sociologisch Onderzoeksinstituut, K.U. Leuven, Leuven

[2] Bryman, A. (2004), *Social Research Methods.* Second edition, Oxford University Press, New York

[3] Bulmer, M. Warwick, D. eds. (1993), *Social research in Developing Countries, surveys and censuses in the third world*, reprint 2004, UCL Press, London

[4] Burgos, F.J. (2008). *Regional electricity cooperation and integration in the Americas: potential environmental, social and economic benefits.* Organization of American States, Washington D.C.

[5] Contreras-Lisperguer, R., de Cuba, K. (2008), *The potential impact of climate change on the energy sector in the Caribbean region.* Organization of American States, Washington D.C.

[6] Conway, M. (2004), *Scenario Planning: An Innovative Way to Strategy Development.* Swinburg University of Technology

[7] Elliot, D., Schwartz, M., George, R., Haymes, S., Heimiller, D., Scott, G. (2001), *Wind energy resource atlas for the Dominican Republic*, U.S. Department of Energy, Oak Ridge

[8] Geels, F.W. (2002), "Technological transitions as evolutionary reconfiguration processes: a multi-level perspective and a case study", *Research Policy* 31, pp. 1257-1274

[9] Geels, F.W. (2004), "From sectoral systems of innovation to socio-technical systems: Insights about dynamics and change from sociology and institutional theory", *Research Policy* 33, pp. 897–920

[10] Geels, F.W. (2006), "Co-evolutionary and multi-level dynamics in transitions: The transformation of aviation systems and the shift from propeller to turbojet (1930–1970)", *Technovation* 26, pp. 999–1016

[11] Geels, F.W. (2007), "Analysing the breakthrough of rock 'n' roll (1930–1970). Multi-regime interaction and reconfiguration in the multi-level perspective", *Technological Forecasting & Social Change* 74, pp. 1411-1431

[12] Geels, F.W. (2010a), "Ontologies, socio-technical transitions (to sustainability), and the multi-level perspective", *Research Policy* 39, pp. 495-510

[13] Geel, F.W. (2010b), "A multilevel perspective on system innovation", *Understanding Industrial Transformation. View from Different Disciplines.* Olsthoorn X., Wieczorek A. (eds.), Springer, Dordrecht

[14] Geels, F.W. (2012), "A socio-technical analysis of low-carbon transitions: Introducing the multi-level perspective into transport studies", *Journal of Transport Geography*, 24, pp. 471-482

[15] Granovetter, M. (1983), "The strength of weak ties: a network theory revisited", *Sociological Theory* 1, pp. 201-233

[16] Grin, J., Rotmans, J., Schot, J. (2010), *Transitions to Sustainable Development. New Directions in the Study of Long Term Transformative Change*, Routhledge, New York

[17] Hanneman, R.A., Riddle M. (2005), *Introduction to social network methods*, University of California, Riverside, Riverside

[18] Hekkert, M., Suurs, R., Negro, S., Kuhlmann, S., Smits, R. (2007), "Functions of innovation systems: A new approach for analysing technological change", *Technological Forecasting & Social Change* 74, pp. 413 - 432

[19] Izquierdo, L.R., Hanneman, R.A. (2006), *Introduction to the formal analysis of social networks using mathematica*, available at http://luis.izqui.org/papers/Izquierdo Hanneman_2006-version2.pdf (accessed December 18, 2011)

[20] Lachman, D.A. (2011), "Leapfrog to the future: Energy scenarios and strategies for Suriname to 2050", *Energy Policy*. 39, pp. 5035-5044,

[21] Lachman, D.A. (2013), "A survey and review of approaches to study transitions", *Energy Policy* 58, pp. 269-276

[22] Loorbach, D. (2010), "Transition Management for Sustainable Development: A Prescriptive, Complexity-Based Framework", *Governance: An International Journal of Policy, Administration, and Institutions* 23, pp. 161-183

[23] Loorbach, D., van den Brugge, R., Taanman, M. (2008), "Governance in the energy transition: Practice of transition management in the Netherlands", *International Journal Environmental Technology and Management* 9, pp. 294-315

[24] Loorbach, D., Rotmans, J. (2006), *Managing transitions for sustainable development.* Dutch Research Institute For Transitions, Rotterdam

[25] Magrin, G., Gay García, C., Cruz Choque, D. (2007), "2007: Latin America", *Climate Change 2007: Impacts, Adaptation and Vulnerability. Contribution of Working Group II to the Fourth Assessment Report of the Intergovernmental Panel on Climate Change,* Parry, M. L., Canziani, O. F., Palutikof, J. P., et al. (Eds.), Cambridge University Press, Cambridge, UK

[26] Meisen, P., Krumpel, S. (2009), *Renewable energy potential of Latin America,* Global Energy Network Institute, San Diego

[27] Minto, B. (1996), *The Minto Pyramid Principle: Logic in Writing, Thinking, and Problem Solving.* New and expanded ed., Minto Books International, Inc., London

[28] Nekkers, J. (2006), *Wijzer in de Toekomst. Werken met toekomstscenario's,* Uitgeverij Business Contact, Amsterdam

[29] Nexant (2010), *Caribbean regional electricity generation, interconnection and fuels supply strategy,* Nexant, White Plains

[30] OAS (2007), *Sustainable energy policy initiative report for Latin America and the Caribbean,* Organization of American States, Washington D.C.

[31] Ogilvy, J., Smith, E. (2004), "Mapping Public and Private Scenario Planning: Lessons from regional projects", *Development.* Vol.: 47, No.: 4, Society for International Development, s.l.

[32] Raven, R., Verbong, G. (2007), "Multi-Regime Interactions in the Dutch Energy Sector: the Case of Combined Heat and Power Technologies in the Netherlands 1970-2000", *Technology Analysis & Strategic Management* 19, pp. 491-507

[33] Rip, A., Kemp, R., (1998), "Technological change", *Human Choice and Climate Change.* Eds. Rayner, S., Malone, E., vol. 2. Battelle Press, Columbus, pp. 327–399.

[34] Scearce, D., Fulton, K., the Global Business Network (2004), *What If? The Art of Scenario Thinking for Nonprofits,* Global Business Network, Emeryville

[35] Schwartz, P., and Ogilvy, J.A. (1998), "Plotting your Scenarios", *Learning from the Future. Competitive Foresight Scenarios.* Eds.: Fahey, Liam and Randall, Robert M., John Wiley & Sons, Inc., New York

[36] Shove, E. (2012), "Energy Transitions in Practice. The Case of Global Indoor Climate Change", *Governing the Energy Transition. Reality, Illusion or Necessity?* Eds. Verbong G. and Loorbach D. Routledge, New York

[37] Smith, A., Voβ, J., Grin, J. (2010), "Innovation studies and sustainability transitions: The allure of the multi-level perspective and its challenges", *Research Policy*, 39 (4), pp. 435-448

[38] Sovacool, B.K. (2009), "Rejecting renewables: The socio-technical impediments to renewable electricity in the United States", *Energy Policy* 11 (37), 4500 – 4513

[39] Star, J., Randall, D. (2007), "Growth Scenarios: Tools to Resolve Leaders' Denial and Paralysis", *Strategy & Leadership*, Vol. 35, No. 2, p 56 – 59

[40] Strauss, A., Corbin, J. (1998), *Basics of Qualitative Research,* Sage Publications, Thousand Oaks

[41] Taylor, J.B. (1999), *The Robustness and Efficiency of Monetary Policy Rules as Guidelines for Interest Rate Setting by the European Central Bank,* Stanford University, s.l.

[42] van der Heijden, K. (2005), *Scenarios. The Art of Strategic Conversation,* Second edition John Wiley & Sons, Chichester.

28

Jatropha curcas L: A sustainable feedstock for the production of bioenergy and by products

Kamrun Nahar[1, *], Sanwar Azam Sunny[2]

[1]Department of Environmental Science and Management, North South University, Bangladesh
[2]Department of Mechanical Engineering, University of Kansas, Lawrence, KS, United States

Email address:
nahar@northsouth.edu (K. Nahar), sanwar@ku.edu (S. A. Sunny)

Abstract: Pot experimental studies were carried out in Dhaka, Bangladesh from April 2011 to May 2013, to produce sustainable biomass feedstock of *Jatropha curcas L.* The experiment also focused to evaluate the morphological, physiological and physiochemical parameters of Jatropha including biofuel and seed cake characteristics after fuel extraction. The leaves, petioles and seeds of the plants were collected from the earthen pot to determine the nutrient contents. The current study provides a reliable account of the endogenic concentrations of nutrients present in petiole and their content in leaves and seeds. Experimental results revealed that the morphological parameters responded better in mature plant compared to young plant but the physiological parameter showed variations at 2 growth stages. The different nutrient contents, including the crude protein in the petiole, were higher than the leaves in the young plants when compared to older plant, whereas the reverse was observed at two year old mature plants. The seed kernel contained more nutrients, especially Nitrogen (6.97%) and Crude Protein (43.15%), followed by seed cake and the husk. After maturity, the plant provided about 250 to 300 ml of crude oil per plant and the characteristics of fuel responded better (Flush point-252°C, Ignition point 325.8°C, Specific gravity of 0.9222, Density 0.91992, high Cetane (Ignition Quality) number of 58.7, Sulfur % of 0.128, Iodine (103.67 mg/g) and Saponification (197.88 mg/g) value in comparison to fossil fuel, with higher nutrients content in seed cake residue as byproduct obtained after extraction of oil, which could be used as an excellent organic fertilizer, with nutrients value, N: 3.6%, P_2O_5: 1.9% and K_2O: 1.5%. Also the oil contains high percentage of unsaturated fatty acid (78.74%) resulting in characteristically low levels of free fatty acids, which improves storability. The crude oil without any modification could easily be used in lamp for illumination. The presence of unsaturated fatty acids (high iodine value) allows it to remain fluid at lower temperatures. The low sulfur content indicates less harmful sulfur dioxide (SO_2) exhaust emissions when the oil is used as a fuel.

Keywords: *Jatropha Curcas*, Biofuel, Nutrient Uptake, Seed Cake, Physiochemical Properties

1. Introduction

In the recent past, biofuel derived from plant species has been a major renewable source of bioenergy. The utilization of energy crops as a source of renewable fuels is a concept with great relevance to currant ecological and economic issues at both national and global scales. This non-conventional source of energy will help in removing in regional imbalance in energy use by making energy available in a decentralized manner. The production of biofuel will lower national dependence on foreign oil supplies and will reduce emissions of greenhouse gases. Scientists have identified some plants that bear seeds rich in non-edible vegetative oil. The natural oils when processed chemically show striking similarities to petroleum derived diesel and are called biofuel. Global awareness of the climate changes tagged with gradual increase in Carbon di oxide emission in the atmosphere is creating a relative impact on society of the developed and developing countries.

Bioenergy, which is sometimes made from energy crops, is mainly a liquid or gas derived from biomass of organic plant material, second generation biofuel crops. These crops are not typically used for food purposes but can help supply a portion of the current fuel demand sustainably with minimum environmental impact (Jepsen et al., 2006). The biomass can be converted to fuel and used for various transportation and household purposes. These agro-based fuels are considered an important means of reducing greenhouse gas emissions and increasing energy security by providing an alternative to conventional petroleum based fossil fuel. In recent years, Bangladesh has been suffering from severe energy crisis in urban as well as in rural communities, as a result of its high population growth. Concerns regarding energy security, dependence on foreign resources, and the negative consequences associated with global warming due to continued greenhouse gas emissions have prompted significant interest in the development of low carbon and sustainable energy sources. Biofuels from nonfood bioenergy feedstock, like *Jatropha curcas*, is one such example, which is a second generation energy crop, propagated from seeds, to produce bioenergy and other byproducts for consumption in the context of the developing nation.

Biofuel is a tremendous opportunity to lead the nation to a brighter and better future with respect to energy and can offer environmental benefits such as lower carbon emissions and lower sulfur compared with conventional petroleum-based fuels (Nahar et al., 2011).

Global climate change has a direct effect on agriculture, the primary livelihood of the people of Bangladesh. In Bangladesh especially, water deficiency is a common phenomenon during summer and winter periods. If there were adequate supply of uncontaminated water during these periods, most food crop productions could be ensured. Winter is practically the safest period for crop production as there is little chance of crop failure due to climatic reasons, but the crops suffer from drought during this period. So there is a significant hamper of the growth of agriculture which needs to feed in excess of 162 million people within an area of 147,570 square kilometers, of which a minority is arable land. Further shrinkage is occurring due to erosion, deforestation and other man-made environmental effects. Agricultural production could be enhanced either by supplying adequate water to the crops or by growing drought resistant Plants. This could be done by selecting crops that have less demand for water or have deep root systems sufficient to utilize subsurface water.

They can replace fossil fuels in developed as well as in developing countries like Bangladesh while bringing additional income to poor farmers. As it is an easy to propagate drought resistant plant it can be cultivated on wastelands, without competing with current food production (Heller, 1996; Grimm, 1996). This plant can grow anywhere including soil considered infertile for food production, and can live for about 50 years (Henning, 2010; Openshaw 2000).

Recently, the plant is gaining a lot of importance for the production of biofuel (Biodiesel) in developed as well as in developing countries. The plant can be grown in low to high rainfall areas, light frost areas (short duration) and can be used to reclaim land, as a hedge and/or as a commercial crop. In addition, it can be grown in dry land areas and as well as in flooded lands, in shallow fields, rocky terrains

and marginal and boundary lands. Completely inedible by both humans and animals, it is also an excellent bordering plant, hence prevents animals from wondering into and destroying valuable crops. It can be easily grown from stem cutting, grafting, tissue culture and seeds, and grow extremely fast, producing fruits for about many years. If planted along riverbanks and coastlines, it can prevent erosion and its leaves also enrich the surrounding soil (Nahar et al., 2011). The plant produces seeds containing inedible oil, which is easily convertible into biodiesel. Biodiesel is an alternative and renewable fuel derived from natural oil like jatropha oil. Originally from Central America, *Jatropha* is found throughout the tropics, including much of Africa and Asia (Openshaw 2000; Gubitz et al 1999). Different parts of Jatropha have medicinal values. Each and every part of the plant from roots to the leaves can be used for various purposes.

The plant thrives on different soil types including gravelly, sandy and also under saline conditions (Dagar, 2006). Seed production ranges from about 0.1 tons per hectare per year (t/ha/year) to over 14 t/ha/year, (Heller, 1996; Becker, 2010; Openshaw, 2000; Achten et al., 2008; Jones & Miller, 1991). The seeds contain about 30-35 percent of non-edible oil (Heller, 1996; Deng, 2010; Gubitz et al., 1999; Henning, 2002). One hectare of land, depending on density, can produce 158-396 gallons of oil (Chawla, 2010), as 0.26 gallon of oil can be extracted per 8.8 lbs (Achten, 2008) to 11 -12 lbs of seeds (Jongschaap, 2007). The plant is important for climate change issues as a mature plant or tree absorbs around 18 lbs of carbon dioxide (CO_2) per year. So cultivating Jatropha in one hectare of land can sequester around 20 tons of CO_2 annually (Benard Muok, 2008).

Additionally, the oil derived from the plant seeds directly can be used as a replacement for kerosene cooking fuel, to light lamps and also for energy in engines especially in small farming machineries in rural areas such as pumps for irrigation and machines for generators. The oil can also be converted into biodiesel for use in transportations and also as jet fuel (Cerrate et al 2006, Heller 1996, Achten et al. 2008). 1000 grams of Jatropha oil can produce 980 grams of pure biodiesel. However most optimized practical processes yield around 94% of biodiesel (Alkabbashi et al., 2009), which emits 80% less CO_2 and 100% less SO_2 than fossil diesel (Tiwari et al., 2007). Besides the fuel, it produces many important byproducts after extraction of oil which can be used commercially (Jones & Miller 1991, Nahar and Sunny 2011). Moreover, the important by-product is the seed cake or oil cake, is a good and balanced organic fertilizer containing, Nitrogen, Phosphorus and Potassium including other nutrients (Makkar et al, 2001; Ghosh et al. 2007; Patolia et al. 2007; Wani et al. 2006, Rockefeller Foundation, 1998), which the farmers can use in their fields. Besides that seedcake also contains a lot of energy that can be recovered by digesting it and producing biogas. Biogas digesters are currently a popular energy production unit in rural Bangladesh. In this way both the oil and pressed cake can be used to produce energy/electricity for a rural area which can lead to improved living conditions. Jatropha is considered as a sustainable source of second generation bioenergy feedstock species and the overall supply can be increased with different propagation technologies. In addition, the plant can grow in drought, and the land use patterns in Bangladesh are suitable for its cultivation (Nahar 2011). This can alleviate poverty to a degree and empower women in developing countries as traditionally the seeds have been harvested by women and used for medical treatments and local soap production (Duke 1983; Henning 2002).

So, for potential feedstock, focus has been made on the plant, where tremendous amount of research is being conducted on soil types, farming systems, breeding technologies, cultivation scenarios etc. As a fast growing, non-food energy source it does not require productive arable land that would otherwise be used for food crop production; does not require high amounts of pesticides, fertilizer and irrigation water, result in a lower carbon footprint in the total carbon life cycle basis and provide and equal or higher energy content than the current petroleum-based traditional fuel including jet fuel used by the industry and provides an environmental and socioeconomic benefit to local communities and countries.(OECD/FAO, 2007b).

The planting of *Jatropha* as a living hedge protects the soil from erosion. The roots of the tree bind the topsoil, so it is less vulnerable to the wind erosion-responsible for 28% of soil degradation (Becker and Francis, 2010). In addition to global climate changes and an excess of green-house gases, it addresses deforestation issues as it utilizes infertile, loamy and fallow lands production (Chawla, 2010).

The plant has enormous potential for energy production in Bangladesh. As a multipurpose plant with many attributes, it is considered as potential feedstock for biofuel production which could provide employment, improve the environment and may enhance the quality of rural life.

So, the aim of the present study was to investigate the growth, nutrient uptake and oil characteristics of an important oilseed bioenergy crop, taking in consideration the morphological and physiological quality of the underground and aboveground parts of the plant as well as fuel and seed cake characteristics in pot grown plant with limited soil in roof top condition.

2. Materials and Methods

The Pot experiment was conducted with Jatropha curcas, at author's roof top garden, Dhaka, Bangladesh. Seeds were obtained from local seed mass of Florida, USA. Homogenous seeds were soaked in wet paper towel for 24 hours before sowing and were sown on 1st April, 2011 in poly bags having size of 22 x12 cm and germinated on 7th April. Poly bags were filled with mixed soil and well decomposed farm yard manure in equal proportion in ratio

of 1:1:1. The drainage holes were provided at the bottom of the polybags. 4 weeks seedlings were transplanted in 25 cm height and 30 cm diameter clay pots. The pots were filled with 8 kgs of silty clay loam soil. The soil was collected from Tejgaon series, Bangladesh and the collected soil was air dried and sieved to pass a 2 mm mesh screen. The general physical and chemical characteristics of the soil were: Sand - 5.8%, Silt- 60.2%, clay-34.0%, Field capacity of the soil - 33%, Maximum water holding capacity-46%, Hygroscopic moisture-1.40%, Porosity-49%, Bulk Density-1.27g/cc, Particle Density – 2.57g/cc, PH 7.2, EC-143uS, OM – 1.14, CEC – 17.9 meq/100g soil and N -0.06%. The experiments were arranged to get all the plants in the pots with sufficient sunlight. Nutrient supply, water supply and other intercultural operations were done as and when necessary. Collection of petiole and leaves were done at 12 (1 year) months and 24 (2 years) months after transplanting. Other morphological data and physiological parameters at this two vegetative growth stages including the amount of oil, physiochemical properties of crude oil and nutrients value of kernel, seed husk and seed cake as a byproduct were also recorded. The samples were collected at the end of first year and 2nd year to measure the morphological and physiological parameters during the growing period. Finally at harvesting, sampled for growth (plant height, diameter, leaf and petiole length, fresh and dry wt, root length etc.) and also for the nutrients content. Plant heights were measured as the distance between soil surface and the main-stem apex. Leaf and petiole length were estimated non-destructively by measuring the distance from the point of petiole attachment to the leaf tip.

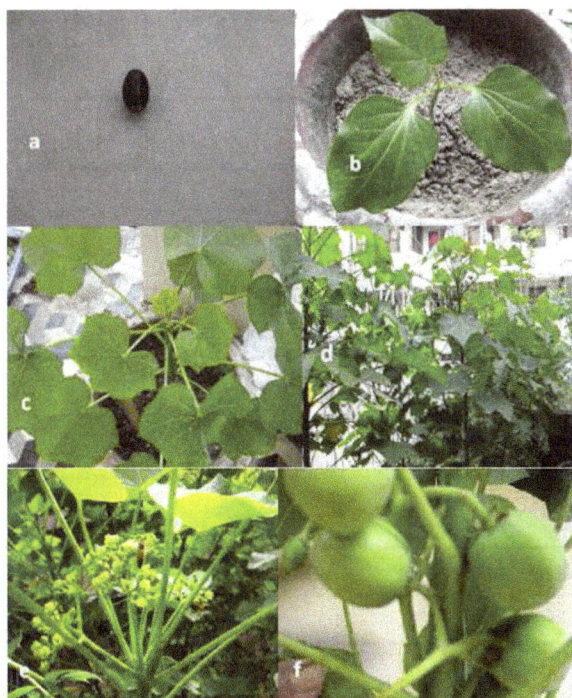

Figure 1. Different stages of plant growth. a-b. Seed to seedling, c-d. Plant at different growth stages, e-f. Multiple peduncles, Inflorescence initiation with blooming flowers and fruit formations (Flowering and fruiting after 2 years).

Figure 2. *Seed to Seed Cake: a-b. Seeds with and without shell, c-d. Jatropha oil and burning lamp, e-f. Seed, kernel, husk and seed cake.*

After harvesting, the samples were then dried for 48 hours at 47 degree Celsius to receive the dry wt of samples. The samples were than grind for chemical analysis and used for determination of nitrogen, crude protein and other nutrients content in plants. At maturity of the plant, the oil seeds were extracted and analyzed for their chemical and physical properties such as acid value, percentage free fatty acids (% FFA), iodine value, saponification value as well as Specific gravity and density etc.

3. Results and Discussion

In the present study, the height and other morphological and physiological parameters of 1 and 2 years of the plants and seeds, oil properties were recorded. Results revealed that all the morphological parameter responded better in mature 2 years old plant compared to 1 year old plant (Table 1). But the physiological parameters like nutrient content showed variation at two growth stages. The percentage of Nitrogen and Sulfur were higher in 2 years old leaves but the concentration of potassium and phosphorus was lower compared to 1 year old leaves. But in case of petiole the percentage of all the nutrients were lower in mature plant except the Potassium content compared to young plant (Table 2-3).

The underground part, like root length increased with increasing age, which is a very important factor to consider in plant growth and development. The nutrient contents were found higher in the leaves than the petiole and increased with increasing age. These results also confirm the findings of Nahar and Hoque (2013), who observed better morphological response and nutrient contents in mature plant compare to young plant.

Table 1. *Morphological parameters of 1 year and 2 years Jatropha plant.*

Jatropha Plant	Plant height (cm)	Stem diameter (cm)	Leaf fresh wt (gm)	Petiole fresh wt (gm)	Leaf dry wt (gm)	Petiole dry wt (gm)	Leaf petiole length (cm)	Leaf length (cm)	Leaf width (cm)	Root Length(cm)
1 year old	35.47	1.68	2.68	1.52	0.58	0.016	14.88	13.56	12.44	18.39
2 years old	75.63	2.74	2.97	1.98	0.79	0.019	18.96	16.77	17.59	76.53

Table 2. *Physiological parameters of 1 year and 2 years Jatropha plant.*

Jatropha Plant	Plant height (cm)	Stemdiameter (cm)	Leaf fresh wt (gm)	Petiole fresh wt (gm)	Leaf dry wt (gm)
1 year old	35.47	1.68	2.68	1.52	0.58
2 years old	75.63	2.74	2.97	1.98	0.79

Table 3. *Physiological parameter of 2 year Jatropha plants.*

Samples	% N	% Crude protein	% P	% K	% S
Jatropha leaf	1.96	11.89	0.19	0.38	0.19
Jatropha Petiole	0.84	4.73	0.09	0.99	0.14

Table 4. *Fatty acid composition (%) of Jatropha seed oil.*

Unsaturated fatty acid	Saturated fatty acid	Oleic acid	Linoleic acid	Stearic acid	Palmitic acid
78.74 %	21.26%	42.17%	33.27%	7.25%	16.13%

At maturity, after 2 years, the plants provide seeds and each plant produces about 1 kg of seeds which produces about 250-300 ml of oil. When dried seeds are crushed, the resulting Jatropha oil produced can be used in lighting lamp, while the residue seed cake can be used as fertilizer. The composition of the seed shown in Table 5, with the important byproduct seed cake.

Table 5. Jatropha Seed Kernel, Husk and Cake composition.

Samples	%N	% Crude Protein	%P	%K	%S
Jatropha Seed Kernel	6.97	43.15	0.994	0.98	1.20
Jatropha Seed Husk	1.59	9.94	0.031	0.98	0.137
Jatropha Seed Cake	3.60	22.5	1.92	1.53	1.12

Table 6. Characteristics of Jatropha oil and comparison with fossil diesel.

Characteristic/variable	Jatropha Oil	Fossil Diesel
Density g/cc at 20°C	0.91992	0.8556
Specific Gravity 20°C	0.9222	0.852
Iodine Value mg/g	103.67	-
Flush Point °C	252	87.6
Ignition Point °C	325.8	215
Saponification value mg/g	197.88	-
Cetane no.	58.7	52
Sulfur %	0.128	1.22

The energy gain from Jatropha does not only rely on the production of crude oil but also on the use for organic fertilizer with higher nutrients content in seed cake residue as byproduct obtained after extraction of oil, which could be used as an excellent organic fertilizer, with nutrients value, N: 3.6%, P_2O_5: 1.9% and K_2O: 1.5%. These results also confirm the findings of Achten (2008), Patolia (2007) and Wani (2006), who observed higher nutrient contents in the cake of Jatropha seed. The nutrients contained in the seed kernel, husk and seed cake are also shown in the Table 5. Seed husks can also be used as a feedstock for a gasification plant.

The property of different fats and oils depends upon characterization of the degree of unsaturation or saturation. Hence different oils are less or more saturated according as they contain greater or lesser proportion of the saturation in fatty acids. In this experiment, the various number of test parameters like: saponification value (SV), iodine value (IV), Cetane number (CN) and other parameters had been estimated and the results are presented in Table 6. The iodine value is a measure of the average amount of unsaturation of fats and oils. The oil shows a high iodine value due to its high content of unsaturated fatty acids (Table 6). Jatropha seed oil extracted from the plant has higher iodine as well as saponification value, which were 103.67mg/g and 197.88 mg/g respectively. This result also confirm the finding of Email et al., (2009), who observed higher Saponication, Iodine and acid values in the extracted oil.

The physiochemical properties of Jatropha oil were studied and compared with fossil diesel (Table 6). The specific gravity of diesel was 0.855 and it was also reduced to a significant extent when compared with the specific gravity of raw Jatropha oil (0.9222).The cetane index of oil was 58.7and it was found higher than the range of fossil diesel.

Fuel from Jatropha causes less air pollution during engine operation because it contains lower S concentration than petroleum diesel. Also due to higher density and specific gravity than the diesel, the fuel is safer to store than petroleum diesel since it has a higher flash point. These observations are also in agreements with Brittaine and Lutaladio (2010), who postulated the same findings on fuel

characteristics of Jatropha oil.

Flash point (FP) and Ignition point (IP) are the important factors to consider in the handling, storage, and safety of fuels and flammable materials. In the current research, the flash point and ignition point determined and presented in Table 6, (FP 252°C, IP 325°C), and was higher than the fossil diesel fuel. These results also confirm the findings of Singh and Padhi (2009), who also observed about the similar results.

According to Table 4, the free fatty acid content of raw Jatropha oil obtained from the research work was found to be nearly the same as described by Joshi et al (2011). The major long chain fatty acids present in the oil, which are palmitic acid (16.13%), stearic acid (7.25%), oleic acid (42.17%) and linoleic acid (33.27%). Based on the physicochemical evaluation of the oil obtained from the seeds, contains high percentage of unsaturated fatty acid which is about 78.74%. The extracted oil can be classified as an unsaturated oil due to the presence of sufficient amounts of oleic and linoleic acids. Hence the oil has a great potential for various applications including low pour point biofuel. Therefore, it is necessary to have more research on Jatropha in the future to explore its potential for future industrial oilseeds crop. The presence of unsaturated fatty acids with high iodine value, allows it to remain fluid at lower temperatures. Also the low sulfur content in the oil (Table 6) indicates less harmful sulfur dioxide (SO_2) exhaust emissions when the oil is used as a fuel. These characteristics make the oil highly suitable for producing biodiesel and other byproducts.

4. Conclusion

Vegetative growth and other physiological parameter of Jatropha, based on the measurement and analysis of the above-ground part of the plant at two growth stages in the pots, concluded that 2 years old plant has higher growth, crude protein and nutrient concentrations compared to 1 year old plant as it is not well-developed. After extraction, seeds are providing oil and seedcake as organic fertilizer, could be used for industrial purposes. In conclusion mature plant shows better morphological parameters and sustainability. So sustainability is a main success factor for biofuel production. Since feedstock is the single largest cost component in biofuel production, considerably less expensive feedstock like Jatropha is the cheapest and the viable options. In the view of current energy crisis, this oil seems to be an attractive alternative source of energy, along with properties comparable to that of conventional diesel oil, so, we can consider the plant as a sustainable perennial bioenergy feedstock, provides a clean and renewable energy source where Environmental, social and economic matters have to be considered.

Soils of Bangladesh are suitable to cultivate the plant. The waste lands, and other upland, low land, lake/riverside can be easily taken consideration under Jatropha cultivation. Since the productivity remains variable over the globe, there is an urgent need for better and available data to guide investments and for that more research work is needed to provide for a sustainable future for Jatropha, which will lead to energy development in Bangladesh and elsewhere while saving the environment from conventional energy production.

Acknowledgements

The authors wish to thank Professor Tamim Ahmed for providing the necessary fuel analysis laboratory facilities at the Department of Petroleum and Mineral Resources Engineering, Bangladesh University of Science and Technology (BUET) and Dr. Sirajul Haque for plant, seed and soil sample analysis at the Laboratory of the Department of Soil, Water and Environment, University of Dhaka, Bangladesh.

References

[1] Achten, W.M.J., L. Verchot, Y.J. Franken, E. Mathijs, V.P. Singh, R. Aerts, and B. Muys. 2008. Jatropha bio-diesel production and use. *Biomass and Bioenergy* 32 (12):1063-1084

[2] Alkabbashi, A.N., M.Z. Alam, M.E.S. Mirghani and A.M.A.Al-Fusaiel. 2009. Biodiesel production from crude palm oil by transesterification process. *J. Applied. Sci.*, 9:3166-3170.

[3] Joshi, A., P. singhal and R. K. Bachheti. 2011. Physicochemical Characterization of Seed oil of *Jatropha curcas* l. collected from Dehradun (Uttarakhand) India. International Journal of Applied Biology and pharmaceutical technology. 2(2): 123-127.

[4] Becker, K. and G. Francis. 2010. Bio-diesel from Jatropha plantations on degraded land. Multifunctional Plants – Food, Feeds and Industrial Products Department of Aquaculture Systems and Animal Nutrition, University of Hohenheim.

[5] Brittaine, R., and N. Lutaladio. 2010. Jatropha: *A Smallholder Bioenergy Crop – The Potential for Pro-Poor Development*.Rome, Italy: Food and Agriculture Organization of the United Nations. Accessed August 25, 2011. http://www.fao.org/docrep/012/i1219e/i1219e.pdf.

[6] Chawla P. C. 2010. CSIR NEWS: Progress, Promise and Prospects. 60 (7) - (8):74. ISSN 0409-7467.

[7] Cerrate, S., F. Yan, Z. Wang, C. Coto, P. Sacakli and P.W. Waldroup. 2006. Evaluation of glycerine from biodiesel production as a feed ingredient for broilers. *Int. J. Poult. Sci.*, 5: 1001-1007.

[8] Deng, X., Z. Fang, Y. Liu. 2010. Ultrasonic transesterification of Jatropha curcas L. oil to biodiesel by a two-step process. *Energy Conversion Management*; 51: 2802-2807.

[9] Duke, J.A. 1983. Handbook of Energy Crops.Purdue University. Centre for New Crops and Plant Products. Unpublished.

[10] Dagar, J. C., O.S. Tomar, Y. Kumar, H. Bhagwan, R.K. Yadev and K. Tyagi. 2006. Performance of some under-explored crops under saline irrigation in a semiarid climate in Northwest India, *Land Degrad. Develop.* 17:285-299

[11] Email, A., Z. Yaakob., S.K.Kamarudin.,M. Ismail and J. Salimon. 2009. Characteristic and Composition of *Jatropha Curcas* Oil Seed from Malaysia and its Potential as Biodiesel Feedstock Feedstock. European Journal of Scientific Research. 29 (3):396-403. ISSN 1450-216X

[12] Grimm, C. 1996. The Jatropha project in Nicaragua.Bagani Tulu (Mali) 1: 10-14.

[13] Gubitz, G.M., M. Mittelbach, and M. Trabi. 1999. Exploitation of the tropical oil seed plant Jatropha curcas L. *Bio res Technol* 67:73-82.

[14] Ghosh, A., D. R. Chaudhary, M. P. Reddy, S. N. Rao, J.Chikara, and J. B. Pandya. 2007. "Prospects for *Jatropha*Methyl Ester (Biodiesel) in India." *Int. J Environ. Stud.* 64:659–674.

[15] Heller, J. 1996. Physic nut. Jatropha curcas L. Promoting the conservation and use of underutilized and neglected crops. 1. Institute of Plant Genetics and Crop Plant Research, Gatersleben, International Plant Genetic Resources Institute, Rome.

[16] Henning, R. 2002. Using the Indigenous Knowledge of Jatropha – The use of Jatropha curcas oil as raw material and fuel .IK Notes. No.47. August. World Bank

[17] Henning, R. K. 2010. Jatropha curcas in Africa – an Evaluation.Assessment of the impact of the dissemination of "the Jatropha System" on the ecology of the rural area and the social and economic situation of the rural population (target group) in selected countries in Africa.

[18] IEA- International Energy Agency (2007b): Bioenergy Project Development and biomass supply. IEA good practice guidelines. OECD/IEA, Paris.

[19] Jones, N., J.H. Miller. 1991. Jatropha Curcas- A multipurpose species for problematic sites. *Land resources series*. 1:40-43.

[20] Jepsen, J.K., Henning, R.K. and Nyati, B. 2006. Generative propagation of Jatropha curcas L. on Kalahari Sand.Environment Africa. Zimbabwe.

[21] Jongschaap, R.E.E., W.J.Corre, P.S Bindraban and W.A. Brandenburg. 2007. Claim and Facts on Jatropha curcas L. Global Jatropha curcas evaluation, breeding and propagation programme. Plant Research International. Wageningenur.

[22] Muok, B. 2008. Feasibility study of Jatropha curcas as a biofuel feedstock in Kenya. African Centre for Technology studies (ACTS).

[23] Makkar, H.P.S., Becker, K and B. Schmook. 2001. Edible provenances of Jatropha curcas from Quintna Roo state of Mexico and effect of roasting on antinutrient and toxic factors in seeds. Institute for Animal Production in the Tropics and Subtropics (480), University of Hohenheim, D-70593 Stuttgart, Germany.

[24] Nahar, K. 2011. Cultivation of Jatropha curcas L. in Bangladesh: A Sustainable Solution to the Energy, Environmental and Socioeconomic Crisis. *VDM Publisher.* ISBN 9783639365801.

[25] Nahar, K and S.A. Sunny.2011. Extraction of Biodiesel from a Second Generation Energy Crop (*Jatropha curcas* L.) by Transesterification Process. Journal of Environmental Science and Technology.4 (5): 498-503. DOI: 10.3923/jest.2011.498.503

[26] Nahar. K., S. A. Sunny and S. S. Shazi, 2011. Land Use requirement and urban growth Implications for the production of biofuel in Bangladesh. Canadian Journal on Scientific and Industrial Research. 2(6): 195-208.

[27] Nahar, K and Sirajul, H. 2013. A Morphological and Physiological Study of *Jatropha curcas* Linn.propagated from Seeds in Bangladesh. Middle-East Journal of Scientific Research 13 (8): 1115-1118.DOI: 10.5829/idosi.mejsr.2013.13.8.623

[28] Openshaw, K. 2000. A review of Jatropha curcas: An oil plant of unfulfilled promise. *Biomass Bioenergy*, 19: 1-15. Doi: 10.1016/s0961-9634(00)00019-2

[29] Rockefeller Foundation. 1998. The Potential of Jatropha curcas in Rural Development and Environment Protection – An Exploration.Concept paper.Rockerfeller Foundation and Scientific & Industrial Research & Development Centre, Harare, Zimbabwe 1998.

[30] Singh, R. K and S.K. Padhi. 2009. Characterization of Jatropha oil for the preparation of biodiesel.Natural Product Radiance. 8(2): 127-132

[31] Tiwari, A. k., Kumar, A., Raheman, H. 2007. Biodiesel production from Jatropha oil (Jatropha curcas) with high free fatty acids: An optimized process. *Biomass Bioenergy*, 31(8): 569-575.

[32] Patolia, J. S., A. Ghosh., J. Chikara, D. R. Chaudharry, D.R. Parmar, and H. M. Bhuva. 2007. "Response of *Jatropha curcas* L. Grown on Wasteland to N and P Fertilization."

[33] Paper presented at the FACT Seminar on *Jatropha curcas* L. Agronomy and Genetics, March 26–28, Wageningen. Article No.34.

[34] Wani, S. P., M. Osman, E. D'Siva, and T. K. Sreedevi. 2006."Improved Livelihoods and Environmental Protection through Biodiesel Plantations in Asia." *Asian Biotechnology and Development Review* 8 (2): 11–29.

29

Impact of extraction methods upon light absorbance of natural organic dyes for dye sensitized solar cells application

Barness Chirazo Mphande[*]**, Alexander Pogrebnoi**

Dept. of Materials Science and Engineering, The Nelson Mandela African Institution of Science and Technology, Arusha, Tanzania

Email address:

bcmphande@gmail.com (B. C. Mphande), alexander.pogrebnoi@nm-aist.ac.tz (A. Pogrebnoi)

Abstract: Aqueous extraction, cold ethanol, and Soxhlet hot ethanol extraction methods were used to study the general trend in performance of dyes as sensitisers for dye sensitized solar cells (DSSC) from different plants based on optical absorbance, and consequently light harvesting efficiency (LHE). *Spathodea campanulata*, *Thevetia peruviana*, *Hibiscus sabdariffa*, *Delonix regia* and *Acalypha wilkesiana 'Haleakala'* were used in this study. From the UV/Visible spectrophotometer with the recorded absorption measurements in the range between 300 – 700 nm, the cold ethanol and Soxhlet hot ethanol extracts exhibited LHE between 80 – 100% over 400 ~ 550 nm of visible range, and 40 – 99% for water extracts dyes between 400 – 700 nm. Ethanol extract of *Acalypha wilkesiana 'Haleakala'* had the highest LHE and a widely spread optical spectrum between 400 – 700 nm; it was earmarked as a potential sensitizer candidate for DSSC. The phytochemical screening was applied to detect the presence of anthocyanins, quinones, cuomarines and others in the extracts. Based on the phytochemical screening, there was no appreciable impact of the extraction methods on the presence of the organic compounds relative to individual samples; and also the optical absorption showed that no extraction method was found consistently better than the other in all extracts.

Keywords: Dye Sensitized Solar Cell, Extraction Method, Optical Absorbance, Light-Harvesting-Efficiency, Natural Organic Dye, Phytochemical Screening

1. Introduction

Dye sensitized solar cell (DSSC) is a solar cell that employs dyes to absorb light from the sun and convert it into electrical energy. DSSC, pioneered by Michael Grätzel and Brian O'Regan in 1991 [1, 2] are a third generation solar cell technologies that is attracting huge attention in the scientific research. Dyes are classified as synthetic and natural organic dyes of which the latter are the area of focus. Natural organic dyes are extracted from plants using various extraction methods. There are three most commonly used extraction methods, namely: Soxhlet hot ethanol, cold ethanol, and heating in water. Each extraction method is said to have an effect on the performance of the DSSC. Several studies have been done on extractions where one or two methods have been used to determine which one gives better results. Trial-and-error has been the

criterion for selection of the method. There is no report available that shows the effect of these methods of extraction upon the performance of the solar cells for individual plant species and if those effects also apply over a range of plants. It is hypothesized that there is a strong correlation between the method of extraction and optical absorbance or light harvesting efficiency (LHE) of the dye. The LHE is indicative of the incident photon-to-current conversion efficiency (IPCE) of the DSSC.

The main research objective was to investigate the effects of extraction methods of natural dyes upon LHE for DSSC applications for which five plant species were studied. Through the expected trend between the method of extraction and optical absorbance, the findings will contribute to a pool of knowledge for applications and further research in DSSC technologies.

2. Materials and Methods

2.1. Preparation

Five dyes were prepared from the flowers of *Spathodea campanulata*, *Thevetia peruviana*, *Hibiscus sabdariffa*, *Delonix regia* and leaves of *Acalypha wilkesiana 'Haleakala'*. Both flowers and leaves were air dried for some time until they became invariant in weight. Using the lab blender (WARING COMMERCIAL, Torrington Connecticut – USA), each specimen was ground to fine particles. Aqueous, cold ethanol and Soxhlet hot ethanol were the three extraction methods that were studied. Using the analytical scale (OHAUS Corporation – USA, made in China), 12.5 g of each sample and 125 ml of the solvent were measured. Distilled water was the solvent for aqueous extraction, and ethanol 96% v/v (AVONCHEM Ltd, Wellington House, Macclesfield Cheshire SKII 6PJ) for cold ethanol and Soxhlet hot ethanol extraction methods. For aqueous extraction method, each sample was heated at six different temperatures (40 – 90 °C with 10 °C step) for 30 – 50 min after which Whatman No. 41 filter paper was used to filter out solid particles. The optimum extraction temperature of each sample was determined based on the optical absorbance at uniform acidity (pH 3). It was this sample at optimum temperature that was used for comparison with other extraction methods. The ground *Hibiscus sabdariffa* formed a jelly-like stuff on either heating or soaking in cold water. The uncrushed dry

flowers were, instead, heated in distilled water for 15 min and filtered. For cold ethanol method, each sample was soaked in ethanol for 1 week, after which the filtrates were obtained. Using the Soxhlet method, the filtrates for each sample were also obtained after 5 to 6 hours of extraction process. The extracts were kept in the refrigerator at 4 °C until they were required for characterization.

2.2. Characterization

Each sample underwent the optical absorbance test using the UV/Vis spectrophotometer (SQ 2800 Single beam spectrophotometer, UNICO – USA); and LHE for each sample was calculated using the following formula [3]

$$(1-10^{-A\,(\lambda)})\times100 \qquad (1)$$

where $A(\lambda)$ is the absorbance at a specific wavelength.

To obtain the absorbance curve, the scanning for each sample was repeated for at least three times in the wavelength range between 300 – 700 nm, and the respective averaged curves were considered hereafter.

2.3. Phytochemical Screening

A qualitative screening was carried out to detect the presence of phytochemicals in plant extracts. Flavonoids, quinones, cuomarines, anthocyanins, anthraquinones, and carotenoids were the chemical compounds that are of practical relevance in DSSC applications. The phytochemical tests were performed on the liquid extracts using standard methods (Table 1).

Table 1. Phytochemical screening protocols

Phytochemicals	Test	Indicator	Reference
Phenol	1 ml extract + 2 ml of distilled water +10% $FeCl_3$.	Blue or green colour.	
Anthocyanin	a) Extract + 10% sodium hydroxide. a) Extract + conc. sulphuric acid. b) Small amount of extract + 2N NaOH.	a) Blue colour. b) Yellowish orange. c) Blue-green colour.	[4]
Quinones	1 ml of extract + 1 ml conc. H_2SO_4.	Red colour.	
Cuomarines	1 ml of 10% NaOH + 1 ml of extract.	Yellow colour.	
Anthraquinones	1ml extract + few drops of 10% NH_3.	Pink colour precipitate.	[5]
Flavonoids	1 ml extract + few drops of NaOH, 1 ml extract + few drops of NaOH + few drops of dilute acid.	An intense yellow colour, Colourless appearance.	[4]
Carotenoids	Extract + chloroform + drops of conc. H_2SO_4.	Deep-blue colored layers.	[6]

3. Results and Discussion

3.1. Dyes Spectral-Responses on the Extraction Methods

Many extraction methods for natural dyes from plants exist, but the most commonly used methods for DSSC applications are Soxhlet, aqueous method, and cold ethanol method. It is a common practice to use trial-and-error method to determine the appropriate extraction method for a particular plant(s) species. This study was aimed at

establishing the common trend of these methods relative to optical absorbance for different plant species.

For aqueous extraction method, the optimum temperatures for each type of extracts are as follows: *Acalypha wilkesiana 'Haleakala'* – 70 °C, *Hibiscus sabdariffa* – 50 °C, *Thevetia peruviana* – 60 °C, *Spathodea campanulata* – 60 °C, and *Delonix regia* – 50 °C.

Figure 1 depicts the results of the three extraction methods for five samples, and a set of observations were made. Within the near-ultraviolet region (NUV, approx. 300

– 400 nm), there are two occurrences common to all. All dyes have nearly the same absorbance except for *Spathodea campanulata* that has conspicuously shown lower absorbance for the water extract (Figure 1 (d)) and the water extract for *Acalypha wilkesiana 'Haleakala'* also drops further to 2.0 a.u. at 300 nm. But also, all plant species have the similar absorption spectra in the NUV region: their absorbance fall between 2.5 and 3.5 a.u. (Figures 1 and 2).

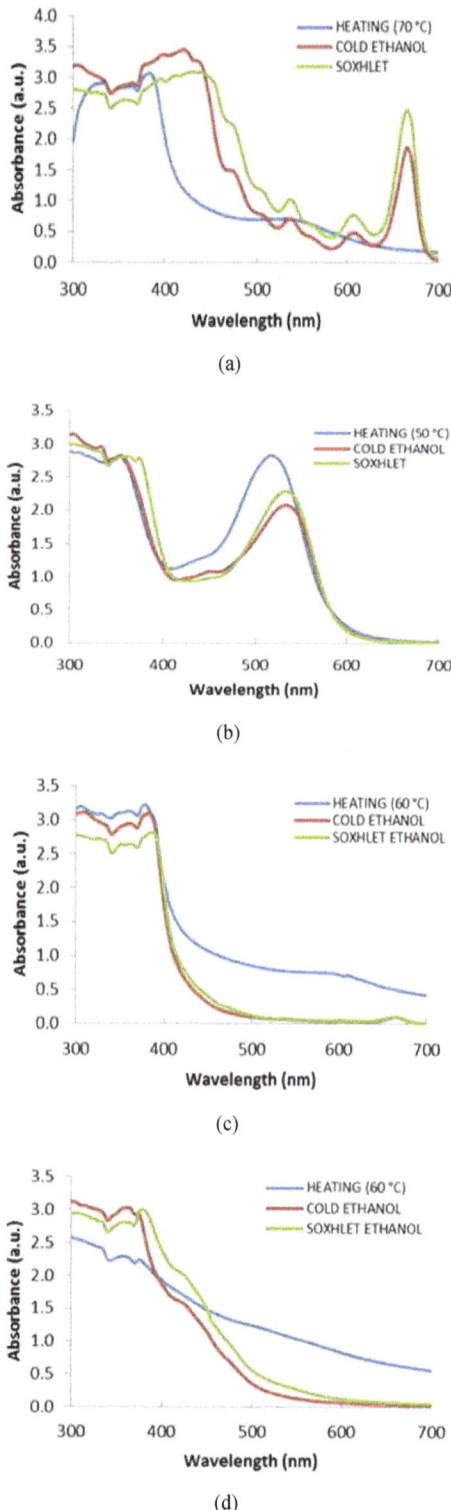

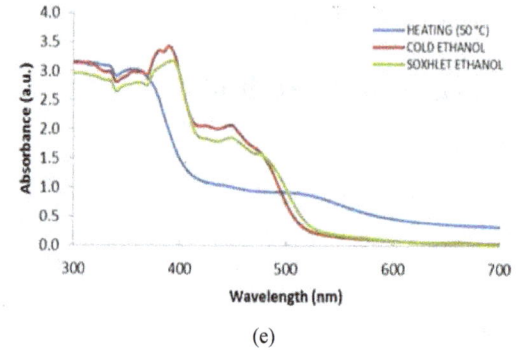

(a)

(b)

(c)

(d)

(e)

Figure 1. *Comparison of extraction methods for individual species, all at pH 3; (a) Acalypha wilkesiana 'Haleakala'; (b) Hibiscus sabdariffa; (c) Thevetia peruviana; (d) Spathodea campanulata; (e) Delonix regia*

This may be due to similar chemical composition of the constituents of the plants that are responsible for absorbance in the NUV range. In the visible region that ranges from 400 – 700 nm [7], the absorption peaks for both Soxhlet and cold ethanol extracts occur at the same wavelengths, and there is negligible difference (Figures 1(a), (b), (d) and (e)) or no difference at all (Figure 1(c)) in magnitudes of absorption. In a study by Boyo and co-workers [8] who used these two methods in dye sensitized solar cell with dyes extracted from *Lawsonia inermis* stem bark, this marginal difference was also reported. Extracts obtained by heating in water have not shown a predictable trend of absorbance as compared to ethanol extracts. For instance, the absorbance for water extract from *Acalypha wilkesiana 'Haleakala'* is low between 400 – 500 nm; also it is closely equal to that of Soxhlet and cold ethanol extracts between 500 – 650 nm (Figure 1(a)); for *Thevetia peruviana* and *Spathodea campanulata*, the absorbance of water extracts is higher than those of Soxhlet and cold ethanol extracts between 400 – 700 nm and 450 – 700 nm respectively. Similarly, *Delonix regia* water extract has shown lower absorbance between 400 – 500 nm, but higher thereafter to 700 nm. Besides, water extracts have shown to have wider and nearly consistent spectra between 400 – 700 nm wavelengths.

All the plant samples were studied for their responses relative to the extraction method. Figure 2 shows the spectra of the five samples for cold ethanol and aqueous extracts. Since the absorption profiles of Soxhlet and cold ethanol spectra are similar (Figure 1), Figure 2(a) is as well representative of Soxhlet extracts spectra. *Hibiscus sabdariffa* was found to have the maximum absorption peak at 535 nm for both Soxhlet and cold ethanol methods. In various studies, it has been reported to have the maximum absorbance peak between 500 and 550 nm [9-11] which is in agreement with the finding in this study. It was hereby used as a bench mark. In our study, the dye from *Delonix regia* was found to have an absorption maximum peak at 450 nm for cold ethanol and Soxhlet extracts, and an elevation at 506 nm in water extract (Figures 1(e), 2 (a) and (b)). *Delonix regia* was also used before in DSSC by

Kimpa et al. [12]. The dye was extracted by cold ethanol, and the absorption spectrum was found to be between 350 and 500 nm whose peak was at 415 nm. In another study by Adje et al. [13], the dye was extracted by heating in water; the phytochemical analysis showed that the dye had three different anthocyanins, namely; cyanidin 3-O-glucoside, cyanidin 3-O-rutinoside, and pelargonidin 3-O-rutinoside whose absorbance were at 516, 516 and 506 nm respectively, and this concurs with the finding about absorption peaks of *Delonix regia* by Godibo [14] who researched on *Amaranthus caudatus*, *Bougainvillea spectabilis*, *Delonix regia*, *Nerium oleanders*, and *Spathodea campanulata*. These results are in support of our findings for *Delonix regia* and *Spathodea campanulata*.

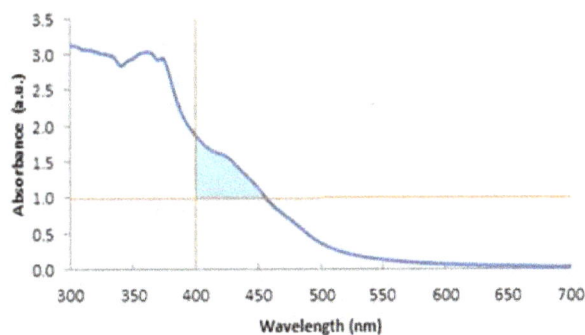

(a)

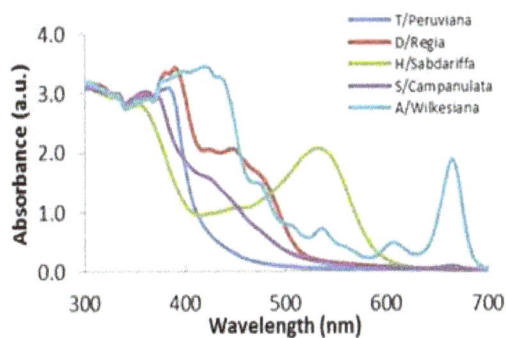

(b)

Figure 2. *Spectra of extracts of the five samples: (a) cold ethanol; (b) Aqueous (water).*

Spathodea campanulata was found to have no definite maximum absorption peak within the visible range for the water extract, and stretched over a wider area in the electromagnetic spectrum between 400 – 700 nm. For cold ethanol and Soxhlet extracts (Figure 1 (d)), there was a narrow area of absorption (i.e. between 400 and 450 nm) if 1.0 a.u. and 400 nm (Figure 3) would be arbitrarily taken as the minimum values of the curve for effective optical absorbance and wavelength respectively under consideration in the spectra above. The maximum peak was deemed to be at 400 nm for cold ethanol and Soxhlet hot ethanol methods. The results are also in full agreement with the previous study by Godibo [14]. In that study, acidified

water (0.1M HCl) was used as one of the extracting solvents. From the absorbance spectrum, it was also observed that *Spathodea campanulata* had no obvious maximum absorption peak [14]. In our study, both *Spathodea campanulata* and *Delonix regia* were also used to compare their performance relative to different extraction methods based on optical absorbance for their consideration in DSSC suitability and application.

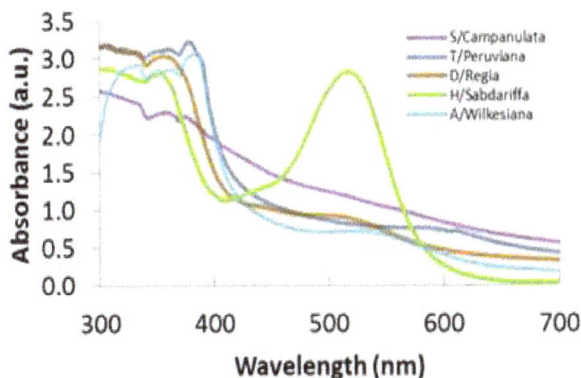

Figure 3. *An illustration of the effective area of optical absorption under the curve within the visible range for the sample.*

Thevetia peruviana and *Acalypha wilkesiana 'Haleakala'* have not been studied before for the application in DSSCs. *Thevetia peruviana*, in Figure 1 (c), shows that it has a very narrow absorption area (~400 – 450 nm) and negligible absorbance beyond 500 nm except for a minute shoulder at 663 nm regarding cold ethanol and Soxhlet methods in the visible range of the electromagnetic spectrum. The shoulder is indicative of the presence of chlorophyll a [14]. However, there was a considerable absorbance in aqueous extract and over a wider range between 400 and 700 nm for *Thevetia peruviana*.

Acalypha wilkesiana 'Haleakala' showed several peaks of absorbance over the visible range for cold ethanol and Soxhlet methods. The maximum absorption peak is at 420 nm with a corresponding peak at 663 nm which are indicative of the presence of chlorophyll a, but also a peak at 440 nm with the corresponding peak at 605 nm show the presence of chlorophyll b [14]. The two chlorophylls complement each other in absorbance of sunlight. The peaks at 515 and 535 nm are indicative of the presence of anthocyanins which may be responsible for its coppery appearance. Anthocyanins absorb sunlight more within this region. For this sample, cold ethanol and Soxhlet extracts have consistently shown higher absorbance than water extracts (Figure 1(a)).

In Figure 1, it is seen that Soxhlet extracts absorb slightly better than cold ethanol in the visible range, but the difference between the two is negligible. The advantage of Soxhlet hot ethanol method is that it has a short extraction time (3 – 6 hrs) as opposed to 12 hrs – 1 week for cold ethanol method, though it has a likelihood of the dye not performing long in DSSC due to possible thermal decomposition of the target compounds [15] caused by high extraction temperatures (> 78 °C, boiling point of ethanol).

On the other hand, the absorption trend for water extracts is closely similar for all plant species used (Figure 2 (b)) except for *Hibiscus sabdariffa*. The absorbance at 400 nm is high but drops almost uniformly towards 700 nm. The level of absorbance on the greater part of absorption spectra in the visible range is rather relatively low considering the suitability of water extracts for DSSC applications. The low absorbance contributes to the lower IPCE in DSSC as compared to ethanol extracts. However, water extracts have shown to be better in absorbance than other ethanol extract for some plant species, *Thevetia peruviana* and *Spathodea campanulata* for instance (Figure 1(c) and (d)). Prior tests are therefore necessary for water extracts. In this study, *Acalypha wilkesiana 'Haleakala'* proved to be the potential

candidate as a sensitizer in DSSC. Its absorption capability covers the visible range from 400 – 700 nm, and its absorbance pattern is similar to that of *Tectona grandis* (Teak) species whose IPCE was 37% in a study by Kushwaha and co-workers [16].

3.2. Light Harvesting Efficiencies

Besides the optical absorption spectra, LHEs were also calculated for the visible region ranging from 400 – 700 nm considering the fact that their absorbance was almost equal in the NUV region except for *Spathodea campanulata*'s water extract (Figure 1(d)) whose absorbance is slightly lower.

Table 2. Light harvesting efficiencies

Sample	Extraction method	Wavelength, λ (nm)	LHE (%)	Sample	Extraction method	Wavelength, λ (nm)	LHE (%)
Hibiscus sabdariffa	CE	400	92.5	*Thevetia peruviana*	CE	400	98.9
		535	99.2			450	61.1
	SHE	400	95.3		SHE	400	98.9
		535	99.5			450	61.1
	Aq.	400	94.0		Aq.	400	98.9
		520	99.9			500	85.9
Delonix regia	CE	400	99.9			600	81.3
		445	99.1	*Acalypha wilkesiana 'Haleakala'*	CE	400	100.0
		500	81.5			420	100.0
	SHE	400	99.9			470	96.7
		445	98.6			500	84.0
		500	84.0			535	81.2
	Aq.	400	96.8			665	98.7
		445	90.6		SHE	400	99.9
		500	89.6			420	99.9
		535	84.5			470	99.5
Spathodea campanulata	CE	400	98.6			500	93.9
		425	97.3			535	89.7
		500	72.5			665	99.7
	SHE	400	99.6		Aq.	400	99.0
		425	99.2			500	84.3
		500	72.5			535	77.5
	Aq.	400	96.6				
		500	94.2			665	39.7
		600	86.4				

Key: CE = cold ethanol; SHE = Soxhlet hot ethanol; Aq. = aqueous

Table 2 depicts the LHEs for each sample with regard to extraction method at specific wavelengths (λ) at which absorption peaks were seen. From the Table, the LHEs show clearly that there is no any remarkable difference between cold ethanol and Soxhlet extraction methods based on optical absorbance of the dyes since the values are almost equal for each sample. This, therefore, means that the two methods can be used interchangeably. However, caution has to be taken considering thermal degradation of the dyes due to hot ethanol in Soxhlet extraction method [8, 11]. Extraction by use of ethanol is not universally applicable for all species as evidenced by LHE for *Thevetia peruviana* as well as the spectra in Figure 1 (c). *Acalypha wilkesiana 'Haleakala'* showed both from the spectra and LHE its better absorption capability.

3.3. Phytochemical Screening

Frequently, dyes have been used as crude extracts (extracts without isolation) for DSSC applications. There are a large number of plant constituents available in the crude extracts that contribute to absorbance and consequently the impact on IPCE of the DSSC. It was for this reason that the phytochemical screening was conducted to investigate the impact of the extraction methods on the presence of such organic compounds relative to each extraction method. Phytochemicals such as quinones, flavonoids, anthraquinones, anthocyanins, and cuomarines were the target compounds because they have very important roles to play in DSSC. A brief overview of their functions in plants and importance in DSSC has been discussed in the following paragraphs.

Phenols (C_6H_6O) are a class of chemical compounds consisting of a hydroxyl group (-OH) bonded directly to an aromatic hydrocarbon group. Depending on the number of phenol units in the molecule, phenolic compounds are classified as simple phenols or polyphenols. Phenolic compounds are ubiquitous groups of secondary metabolites found throughout the plant kingdom [4]. They form a diverse group that includes the widely distributed hydroxybenzoic and hydroxycinnamic acids [17]. By virtue of having hydroxyl and carboxylic groups, plant phenolics make good metal chelators [18] – the most needful requirement in DSSC, besides playing key roles as major red, blue and purple pigments, as well as UV sunscreen.

Flavonoids comprise a large group of polyphenol compounds having a benzo-γ-pyrone structure and are ubiquitously present in plants; they serve as ultraviolet filters. Ultraviolet radiation is classified into UV-A (320 – 390 nm), UV-B (280 – 320 nm), and UV-C (< 280 nm) [19], UV-B radiation is the most damaging class against which the flavonoids shield by absorption [20].

Quinones occur as biological pigments that are available in plants. Various colour pigments such as chloranil and lawsone are quinone derivatives. Alizarin (1,2-dihydroxy-9,10-anthraquinone) extracted from the madder plant, for instance, was the first natural dye to be synthesized from coal tar [21]. Currently, there is an ongoing research study about the possible application of quinone (Alizarin and Lawsone) as a photosensitizer in DSSC that commenced in 2012 by Sreekala and Achuthan [22]. Quinones produce yellow, red or brown coloration in plants [23], and are very important in conversion of light into chemical energy [23]. Quinones in general are expected to absorb light between 420 – 430 nm [24].

Anthraquinone ($C_{14}H_8O_2$) is a polycyclic aromatic hydrocarbon containing two opposite carbonyl groups (C=O) at 9, 10 position, a subgroup of quinones. It occurs in plants, fungi, lichens, and insects as a parent material for colouring of yellow, orange, red, red-brown, or violet [25]. Anthraquinones and their derivatives are also used as photosensitizers in photovoltaic cells [26, 27]. Chaoyan Li

et al. [27] designed three anthraquinone dyes with carboxylic acid as anchoring group as sensitizers for DSSCs. They found out that anthraquinone dyes have very low performance on DSSC applications despite having broad and intense absorption spectra in the visible region up to 800 nm in the near infra-red (NIR) region.

Anthocyanins are a class of flavonoids responsible for the attractive bright red, purple, violet, and blue colours of most fruits, vegetables, flowers, leaves, roots and other plant storage organs [11]. The basic C6-C3-C6 anthocyanin structure is the source of diverse colours. Besides carotenoids and chlorophylls, anthocyanins in flavonoids are also the most important group of plant pigments [23], and the most applied dyes in DSSCs because of their excellent chelation capability, generally, to TiO_2 and ability to convert light energy to electrical energy. Their spectral absorbance lies within the range of 460 – 550 nm in the electromagnetic spectrum with the reported maximum absorbance at 520 nm [28].

Coumarin ($C_9H_6O_2$) is a colourless natural substance that is found in many plants that is a fragrant organic chemical compound in the benzopyrone class. It is used as a sensitizer [29, 30] in DSSC. The dyes are developed and applied as coumarin derivatives to improve IPCE [31]. As coumarin derivatives, the photon-to-current conversion efficiency of 7.4% has been achieved [32].

Carotenoids ($C_{40}H_{56}$ – the general molecular structure) comprise a large group of lipid soluble pigments that exhibit yellow, orange, and red colours, and are found in all kinds of plants. Other colours such as green, orange, or blue are exhibited when carotenoids are bound to proteins [6]. Carotenoids are involved in several aspects of photosynthesis, notably light absorption and energy transfer to the reaction centers (RC) complex and protection of the photosynthetic apparatus from damage by strong illumination [33]. Attempts have been made to apply carotenoids in DSSC. Yamazaki et al. [34] fabricated a DSSC using crocetin and crocin as photosensitizers from which crocetin performed the best with an IPCE of 0.56%. The absorption maximum depends on the number of conjugated bonds; it is around 450 nm for β-carotene [33].

Table 3. Results of the phytochemical preliminary screening of the crude plant extracts

Phytochemical	Acalypha wilkesiana 'Haleakala'			Hibiscus sabdariffa			Thevetia peruviana			Delonix regia			Spathodea campanulata		
	SHE	CE	Aq	SHE	CE	Aq	SHE	CE	Aq	SHE	CE	Aq	SHE	CE	Aq
Phenols	++	+	+++	+	+	++	+	+	+	+	+	+	+	+	+
Flavonoids	++	+	+	+	+	+	+	+	+	+	+	+	+	+	+
Quinones	-	-	+	+	+	+	+	+	+	++	++	++	+	+	+
Anthocyanins	+	+	+	++	++	+++	-			+	+	+	+	+	+
Anthraquinones	-		-	-	-	-	-	-	-	-	-	-	-	-	-
Cuomarines	+	+	+	+	+	+	+	+		-	-	-	+	+	+
Carotenoids	+	+	-	+	+	-	++	++	-	+	+	-	+	+	-

From the qualitative testing of the phytochemical constituents of all plant samples (Table 3), a number of observations were made. As per sample, *Acalypha wilkesiana 'Haleakala'* showed the absence of quinones in both ethanol extracts as opposed to water extract. This could have been the reason why ethanol extracts appeared green instead of reddish appearance as it was with water extract. By colour intensity, *Hibiscus sabdariffa* showed the presence of higher concentration of anthocyanins than all other samples, and even more in water extracts than in ethanol extracts. Higher light absorbance as shown in Figures 1(b), 2(a) and 2(b) are supported by this result; and better IPCE is expected from this sample. *Thevetia peruviana*, on the other hand, indicated to have a high concentration of carotenoids in ethanol extracts than all other samples, but anthocyanins were not present in all of these extracts' samples. The absence of anthocyanins is confirmed by the absorbance curves in Figure 1 (c) and Figure 2 (b) that do not show any peak or elevation in the region within 460 – 550 nm where anthocyanins absorb. The rather high absorbance for water extracts in this area may be attributed to chlorophyll as depicted by a small peak at around 620 nm. Despite having high concentration of carotenoids in ethanol extract, the absorbance is very low. This indicates that carotenoids are not good enough in absorbing light. The yellow colour of the extracts may be attributed to quinones, cuomarines and other constituents that may not have been detected. *Delonix regia* does not show the presence of cuomarines in all of its extracts, but *Thevetia peruviana* and *Spathodea campanulata* have them in ethanol extracts. In general, carotenoids are present in ethanol extracts of all samples except in water extracts. Carotenoids are generally insoluble in water [35], with the exception of crocin [4].

As per extraction method, the trend of the presence of phytochemicals in all plant extracts is similar in both cold ethanol and Soxhlet hot ethanol extracts. The reason may be due to the same extraction solvent used in both the extraction methods. This implies that performance in both optical absorbance (LHE) and IPCE are expected to be similar as Figure 1 ascertains the results. Anthraquinones were conspicuously not present in any of the plant extracts regardless of the extraction method. Flavonoids and phenols were found in every sample being the ubiquitous plant constituents [4]. Generally the impact of extraction methods on the phytochemical presence was slight with the exception of *Thevetia peruviana*.

4. Conclusion

The three methods of extracting natural organic dyes (i.e. cold ethanol, Soxhlet hot ethanol, and aqueous methods) have been applied and analysed. Based on light absorption of extracts, it has been found that there is a slight difference between cold ethanol and Soxhlet hot ethanol. These two methods can be used interchangeably though consideration

has to be made with regard to other factors such as thermal degradation, time and ease of extraction, and availability of the apparatus. On the other hand, the trend of absorbance of light for all water extracts used was closely similar except for *Hibiscus sabdariffa*. Compared to ethanol extracts, the area of effective absorbance of light for water extracts is, however, very narrow. Though the area of absorption stretches over the whole visible range, the magnitude of absorption is not high enough for practical applications in DSSC for commercial purposes. The phytochemical screening has shown that there is no appreciable impact of the extraction methods on the presence of organic compounds relative to individual samples. From the findings, no extraction method of the three is outstandingly better than the other in all plant samples. However, the difference in absorbance between water extracts and ethanol extracts may be due to differences in concentrations and nature of phytochemicals.

Acknowledgements

The authors would like to thank The Nelson Mandela African Institution of Science and Technology (NM-AIST) for the sponsorship. We are also pleased to acknowledge the valuable assistance and service by the School of Life Science and BioEngineering at NM-AIST.

References

[1]   Hedbor, S. and Klar, L., *Plant Extract Sensitised Nanoporous Titanium Dioxide Thin Film Photoelectrochemical Cells.* Examensarbete. 2005, p. 20

[2]   O'Regan, B. and Grätzel, M., *A Low-Cost, High Efficiency Solar Cell Based on Dye-Sensitized Colloidal TiO₂ Film.* Nature. 353(6346), 1991, p. 737-740.

[3]   Jasim, K.E., *Dye Sensitized Solar Cells - Working Principles, Challenges and Opportunities*, in *Solar Cells - Dye-Sensitized Devices*, P.L.A. Kosyachenko, Editor 2011, InTech: Kingdom of Bahrain. p. 35.

[4]   Harborne, J.B., *Phytochemical Methods: A guide to modern techniques of plant analysis.* 1973, New York: Chapman and Hall. 279.

[5]   Kapoor, L.D., Singh, A., Kapoor, S.L., and Srivastava, S.N., *Survey of Indian plants for saponins, alkaloids and flavonoids. I.* Lloydia. 32(3), 1969, p. 297-304.

[6]   Saluja, M.P., Kumar, R., and Agarwal, A., *Advanced Natural Products.* Carotenoids, ed. S. Rastogi. Vol. 1. 2008, Raj Printers, Meerut: Satyendra Rastogi Mitra (KRISHNA Prakashan Media (P) Ltd. 362.

[7]   Jones, A.Z. *The Visible Light Spectrum.* [cited 2014 25 May]; Available   from: http://physics.about.com/od/lightoptics/a/vislightspec.htm.

[8]   Boyo, A.O., Boyo, H.O., Abudusalam, I.T., and Adeola, S., *Dye-sensitised Nanocrystalline Titania Solar Cell using Laali Stem Bark (Lawsonia inermis).* Transnational Journal of Science and Technology. 2(4), 2012, p. 13.

[9]   Okonkwo and Nnaemeka, T.J., *Hibiscus Sabdariffa Anthocyanidins: A Potential Two-Colour End-Point Indicator in Acid-Base and Complexometric Titrations.* International Journal of Pharmaceutical Sciences Review and Research. 4(3), 2010, p. 123-128.

[10]   Selim, K.A., Khalil, K.E., Abdel-Bary, M.S., and Abdel-Azeim, N.A., *Extraction, Encapsulation and Utilization of Red Pigments from Roselle (Hibiscus sabdariffa L.) as Natural Food Colourants,* in *5th Alex. Conference of food & dairy science and technology* 2008, Alexandria Journal of Food Science and Technology: Alexandria, Egypt. p. 7 - 20.

[11]   Wongcharee, K., Meeyoo, V., and Chavadej, S., *Dye-sensitized solar cell using natural dyes extracted from rosella and blue pea flowers.* Solar Energy Materials and Solar Cells. 91(7), 2007, p. 566-571.

[12]   Kimpa, M.I., Momoh, M., Isah, K.U., Yahya, H.N., and Ndamitso, M.M., *Photoelectric Characterization of Dye Sensitized Solar Cells Using Natural Dye from Pawpaw Leaf and Flame Tree Flower as Sensitizers.* Materials Sciences and Applications. 3(5), 2012, p. 281-286.

[13]   Adje, F., Lozano, Y.F., Meudec, E., Lozano, P., Adima, A., Agbo N'zi, G., and Gaydou, E.M., *Anthocyanin Characterization of Pilot Plant Water Extracts of Delonix regia Flowers.* Molecules. (13), 2008, p. 7.

[14]   Godibo, D.J., *Screening of Natural Dyes for Use in Dye Sensitized Solar Cells,* in *Materials Science*2012, Addis Ababa: Addis Ababa, Ethiopia. p. 70.

[15]   Handa, S.S., Khanuja, S.P.S., Longo, G., and Rakesh, D.D., *Extraction Technologies for Medicinal and Aromatic Plants.* Decoction and Hot Continuous Extraction Techniques, ed. S. Tandon and S. Rane. 2008, Triesta, Italy: International Centre for Science and technology. 266.

[16]   Kushwaha, R., Srivastava, P., and Bahadur, L., *Natural Pigments from Plants Used as Sensitizers for TiO$_2$ Based Dye-Sensitized Solar Cells.* Journal of Energy. 2013, 2013, p. 8

[17]   Ray Sahelian, M.D. *Phenolic Compounds and Acids, benefit of phenols.* 2014 [cited 2014 20 May]; Available from: http://www.raysahelian.com/phenolic.html.

[18]   Lattanzio, V., Lattanzio, V.M.T., and Cardinali, A., *Role of phenolics in the resistance mechanisms of plants against fungal pathogens and insects,* in *Phytochemistry: Advances in Research,* F. Imperato, Editor. 2006, Research Signpost: Fort P.O., Trivandrum-695 023, Kerala, India. p. 23-67

[19]   Burger, J. and Edwards, G.E., *Photosynthetic Efficiency, and Photodamage by UV and Visible Radiation, in Red versus Green Leaf Coleus Varieties.* Plant Cell Physiol. 37(3), 1996, p. 395-399

[20]   Kootstra, A., *Protection from UV-B-induced DNA damage by flavonoids.* Plant Molecular Biology. 26(2), 1994, p. 771-774.

[21]   Wikimedia. *Quinone.* 2014 [cited 2014 16 May]; Available from: http://en.wikipedia.org/wiki/Quinone#Dyes.

[22]   Sreekala, C.O. and Achuthan, K. *Natural Dyes as Efficient Candidate for Enhancing the Photovoltaic Properties of Dye Sensitized Solar Cells* 2012 [cited 2014 12 May]; Available from: http://biotech.amrita.edu/research/30.html.

[23]   Delgado-Vargas, F., Jiménez, A.R., and Paredes-López, O., *Natural Pigments: Carotenoids, Anthocyanins, and Betalains — Characteristics, Biosynthesis, Processing, and Stability.* Critical Reviews in Food Science and Nutrition. 40(3), 2000, p. 173-289.

[24]   Reusch, W. *Visible and Ultraviolet Spectroscopy.* 2013 [cited 2014 20 May]; Available from: http://www2.chemistry.msu.edu/faculty/reusch/virttxtjml/spectrpy/uv-vis/spectrum.htm.

[25]   Chemicalland21. *Anthraquinone.* [cited 2014 14 May]; Available from: http://www.chemicalland21.com/specialtychem/finechem/ANTHRAQUINONE.htm.

[26]   Lehr, F., *Anthraquinone dyes as photosensitizers in photovoltaic cells,* 2009, Google Patents.

[27]   Li, C., Yang, X., Chen, R., Pan, J., Tian, H., Zhu, H., Wang, X., Hagfeldt, A., and Sun, L., *Anthraquinone dyes as photosensitizers for dye-sensitized solar cells.* Solar Energy Materials and Solar Cells. 91(19), 2007, p. 1863-1871.

[28]   *Anthocyanin Absorption Spectrum.* [cited 2014 15 May, 2014]; Available from: http://cellbiologyolm.stevegallik.org/anthocyanin/page17.

[29]   Edward W. Castner, J., Kennedy, D., and Cave, R.J., *Solvent as Electron Donor: Donor/Acceptor Electronic Coupling Is a Dynamical Variable.* J. Phys. Chem. A 104, 2000, p. 2869-2885.

[30]   Faber, C., Duchemin, I., Deutsch, T., and Blase, X., *Many-body Green's function study of coumarins for dye-sensitized solar cells.* Physical Review B: Condens. Matter Mater. 86 (155315), 2012, p. 1-7.

[31]   Hara, K., Sayama, K., Ohga, Y., Shinpo, A., Sugab, S., and Arakawa, H., *A coumarin-derivative dye sensitized nanocrystalline TiO$_2$ solar cell having a high solar-energy conversion efficiency up to 5.6%.* The Royal Society of Chemistry. 2001, p. 569-570.

[32]   Hara, K., Wang, Z.-S., Sato, T., Furube, A., Katoh, R., Sugihara, H., Dan-oh, Y., Kasada, C., Shinpo, A., and Suga, S., *Oligothiophene-Containing Coumarin Dyes for Efficient Dye-Sensitized Solar Cells.* J. Phys. Chem. B 109, 2005, p. 15476-15482.

[33]   Mimuro, M. and Katoh, T., *Carotenoids in photosynthesis: absorption, transfer and dissipation of light energy.* Pure & Appl. Chern. 63(1), 1991, p. 123-130.

[34]   Yamazaki, E., Murayama, M., Nishikawa, N., Hashimoto, N., Shoyama, M., and Kurita, O., *Utilization of natural carotenoids as photosensitizers for dye-sensitized solar cells.* Solar Energy. 81(4), 2007, p. 512-516.

[35]   Rahiman, R., Ali, M.A.M., and Ab-Rahman, M.S., *Carotenoids Concentration Detection Investigation: A Review of Current Status and Future Trend.* International Journal of Bioscience, Biochemistry and Bioinformatics. 3(5), 2013, p. 466-472.

30

Tree species diversity and dominance in Gelai Forest Reserve, Tanzania

Noah Sitati, Nathan Gichohi, Philip Lenaiyasa, Peter Millanga, Michael Maina, Fiesta Warinwa, Philip Muruthi

African Wildlife Foundation, P.O Box 20 00207, Namanga, Kenya

Email address:

nsitati@awf.org (N. Sitati)

Abstract: Tree species diversity and dominance of Gelai Forest Reserve, an isolated montane forest located in an arid area of Northern Tanzania remains unknown. A systematic grid of 390 m x 780 m between 100 plots of 0.02 ha, along nine transects was used during the forest survey. The tree species present, location, diameter above breast height (dbh) and botanical names were recorded including regenerants of tree species and key shrub species. These parameters were then used to determine species diversity index, dominance index, number of tree species regenerants, number of stems per ha and tree basal area per ha. A total of 39 tree species were recorded. The tree species with the highest importance values were *Nuxia conjesta* (70.7), *Olea europaea* (44.4) and *Crotalaria stulhmanii* (40.4). The Simpson index value ranged between 0.0 and 0.034; with *Crotalaria stulhmanii* having the highest (0.034) index. The tree species diversity index ranged between 0.016 and 0.313. Forest stocking was 377 stems per ha while species basal area ranged between 0.098 m^2 and 439 m^2 per ha, with *Nuxia congesta* occupying the highest (439.07 m^2 per ha) area and *Acacia rovumae* the lowest (0.098 m^2 per ha), respectively. Seventy nine regenerants were recorded on 9% of the plots. Shrubs, herbs and grasses were found on 55% of the plots mainly without trees dominated by *Vernonia galamensis*, *Leonatis leonorus*, *Ocimum suave* and *Solonum incanum*. In conclusion, the forest has high tree species diversity which is a good stand characteristic of a natural forest. This survey established a baseline for future monitoring of the forest performance after mitigation of human activities.

Keywords: Baseline, Diversity, Dominance, Gelai Forest, Regenerants, Shrubs, Tanzania, Transects

1. Introduction

Forests on small protruding hills in dry areas are usually little known in East Africa and hence less studied. Most of the studies are on the Eastern Arc mountains (Burgess et al., 1998; Burgess, Doggart and Lovett, 2002; CEPF, 2003; Burgess et al., 2004; Burgess, et al., 2006) and the coastal forests (Burgess and Clarke, 2000; CEPF, 2003) describing the biodiversity and associated vegetation composition. In Tanzania, relatively very few reports exist about vegetation on these special forests. Forests on isolated hills are found in most parts on Tanzania. These forests are naturally occurring on landforms which are visible over a flat area. Such forests on the hills have a distinct diversity of micro-habitats and are rich in flora and fauna. Depending on the nature of the soils and the rock, trees or shrubs vary in number, but herbaceous angiosperms, algae, mosses, ferns and lichens are generally abundant. Equally, many of the

endemic ephemerals, herbaceous angiosperms, pteridophytes and lichens are restricted to these hilly areas. Species composition patterns are influenced by multiple environmental factors like soil type, elevation, rock aspect and micro-environments (Burke, 2005a). On the contrast, the savannah type of vegetation that usually surrounds the forests on the hills is usually extensively studies (Field and Ross, 1976; van Essen et al., 2002). Complete plant diversity on the isolated hills is not yet revealed satisfactorily (Burke, 2003, Burke, 2005a, Burke, 2005b).

It is known that major changes have taken place in the woodland communities in areas inhabited by the Masai community over the last 25 years (Sitati, Ucakuwun & Wishitemi, 2008). These changes include a loss of tree cover in the tall height classes put pressure on the remnant of the existing forests. According to Dublin (1986), some areas have experienced as much as a 95% decrease in tree cover since 1950 This decline which has largely been attributed to the impact of fire and elephants on the trees, and of their current continued negative influence on seedling regeneration (Dublin, 1986) has now been overtaken by increasing cultivation and demand for food (Sitati, Ucaku-won & Wishitemi, 2008). This impact is exacerbated by other factors such as increasing demand for timber and immigration of other communities (Sitati 2003).

Though the hill forests are considered to be isolated from the surrounding landscape, they are always surrounded by some vegetation or ecological niches. These surrounding areas and biotypes on it are indispensable factors influencing the hill forests biota as well as their ecological conditions. In this paper we present the findings of a study on the tree species diversity and abundance on the isolated hill outcrop the Gelai Forest Reserve in the northern Tanzania. A comprehensive study with respect to tree species diversity and abundance across the forest is reported for the first time from this unique isolated forest on a hill.

2. Study Area

Gelai forest is a Local government reserve that was es-

tablished in 1955 and covers about 2,341 ha of isolated peak of Gelai Hill with elevation of 2,942 m (http://www.fao.org/in-action/tanzania-forest-inventory-pro vides-critical-baseline-data/en/). About 452.7 ha of the forest has been encroached and settled by the local people. Gelai Forest Reserve (GFR) is one of the important dry montane forests that are water catchment in Longido District, in Tanzania. Located at 2°40' S, 36° 5' E (Figure 1) on volcanic soil, the area receives mean annual rainfall of between 500 – 750 mm and mean daily minimum and maximum temperature of about 17°C and 22°C, respectively. Shrubs, herbs and grass dominate main part of the forest, with dry montane forest at higher altitudes, but with a closed canopy only in riverine. The forest is surrounded by five villages, namely Alalilai, Lumbwa, Meirugoi, Magadini and Loon-dolou Esirwa. Inhabited by the Masai community, the main socio-economic activities in these villages are livestock keeping, agriculture and small microbusinesses. However, the forest which is managed by the Longido District Council is threatened by human activities including logging, charcoal burning, livestock grassing and cultivation.

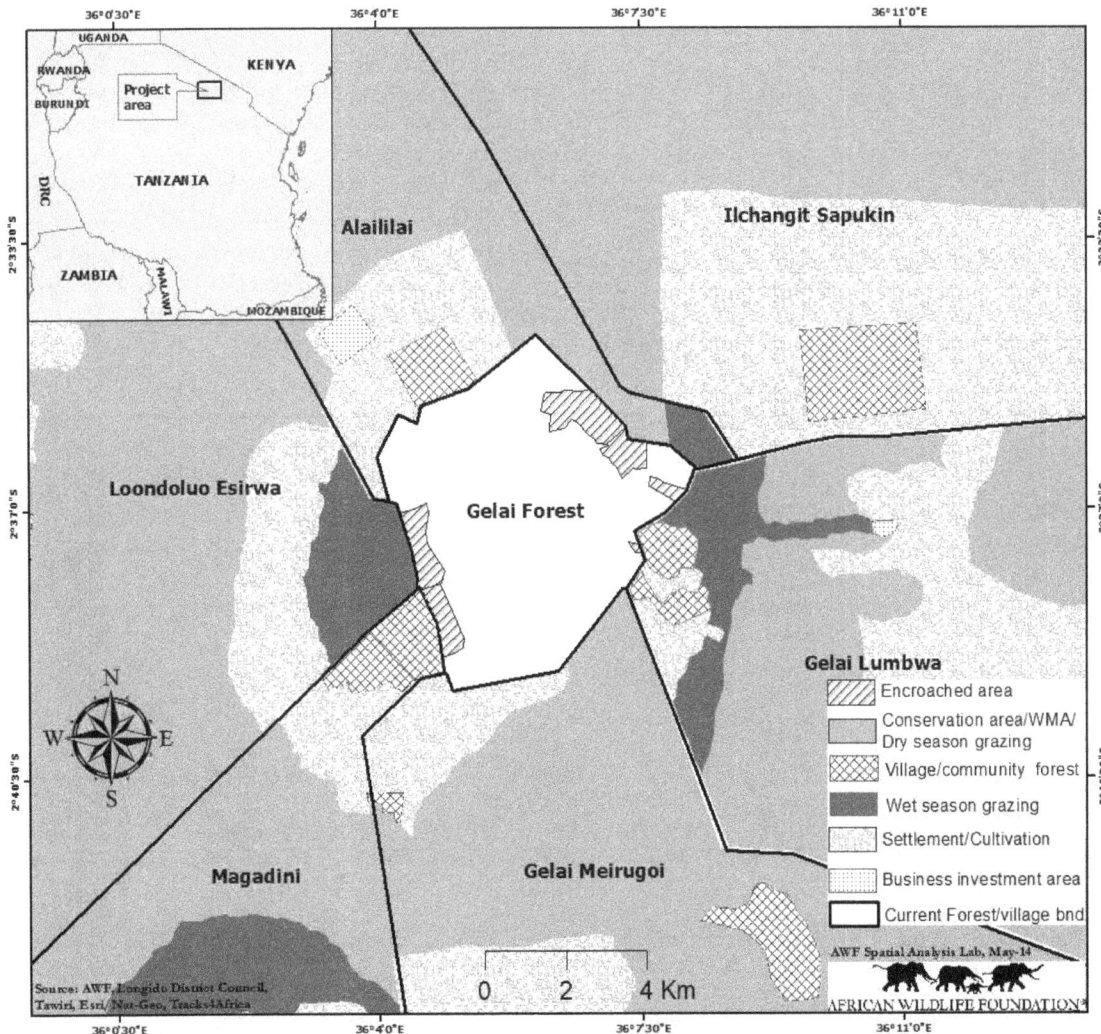

Figure 1. Map showing the location of the study area and status of land use in Gelai forest

3. Materials and Methods

Data on tree species were collected from 100 plots each of 0.02 ha located along nine transects (Figure 2). These plots were at distance of 390 m from each and 790 m between transects. In each plot, data on GPS readings, diameter at breast height (dbh) and botanical names of all trees with dbh ≥ 5 cm were recorded among other variables. Tree regeneration (seedlings and samplings) with > 10 cm tall and dbh<10 cm was assessed by counting them by species in two subplots of 1 m radius each established in South and North of each plot. Both the genus and species names of the trees were recorded. However, in cases where only the genus name was known, only the genus name followed with sp. was used (URT, 2010). Shrub species and other associated vegetation were also recorded during the survey.

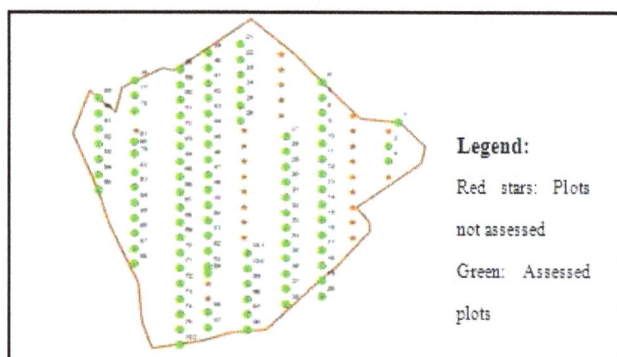

Figure 2. A sketch map of Gelai Forest Reserve showing the location of transects and the plots where the surveys were undertaken. Only 100 plots out of 127 were surveyed as shown by red and green colors.

Unknown collected plant specimens were processed for

the herbarium following standard techniques (Jain & Rao 1977). The herbarium specimens were carefully checked in the laboratory at the National Museums of Tanzania and their identity was confirmed with help of the floras, relevant monographs and published literature in scientific journals. Doubtful specimens were checked and confirmed using online database of IPNI (2013).

Using Statistical Package for Social Sciences (SPSS, 2010) version 20.0 (SPSS Inc., Chicago, Illinois, U.S.A) Simpson index (SI) and Shannon-Weiner index were used to determine the tree species dominance and diversity.

4. Results

4.1. Tree Species Dominance and Diversity

A total of 39 tree species were identified and recorded during the survey (Table I). Trees with the highest importance values (IVI) were those that exist in the greatest number. Three tree species with highest IVI on a scale of 300.0 were *Nuxia conjesta* (70.7), *Olea europaea* (44.4) and *Crotalaria stulhmanii* (40.4). The Simpson index (SI) which is a measure of diversity showed that tree specific values ranged between 0.0 and 0.034; with *Crotalaria stulhmanii* having the highest. The overall SI value for the forest was low (0.091) implying that the chance of picking two plant species being of the same species in the forest is low, due to high species diversity.

Tree species diversity indices (Shannon-Weiner index –H') ranged between 0.016 and 0.313 (Table I), with the overall H' being 2.848. Moreover, tree species that existed in greatest number (i.e. highest IVI values) had also the highest species diversity index and vice versa.

Table 1. Tree species dominance, diversity, importance values and diversity indices in Gelai Forest Reserve

Specie name	Frequency	Species basal area per ha	Relative density	Relative frequency	Relative basal area	Important Value Index	Simpson index	Shannon index
Acacia gerradii	1	0.181	0.265	0.265	0.015	0.546	0.000	0.016
Acacia nilotica	2	1.506	0.531	0.531	0.125	1.186	0.000	0.028
Acacia rovumae	1	0.098	0.265	0.265	0.008	0.539	0.000	0.016
Budulea sp.	1	0.565	0.265	0.265	0.047	0.577	0.000	0.016
Cassipourea gummiflua	4	9.237	1.061	1.061	0.764	2.886	0.000	0.048
Catha edulis	14	12.488	3.714	3.714	1.033	8.460	0.001	0.122
Celtis gerradii	4	9.873	1.061	1.061	0.817	2.939	0.000	0.048
Celtis milbradii	12	29.298	3.183	3.183	2.423	8.789	0.001	0.110
Chrysophyllum mannii	1	1.134	0.265	0.265	0.094	0.624	0.000	0.016
Clausena anisata	4	1.039	1.061	1.061	0.086	2.208	0.000	0.048
Crotalaria stulhmanii	70	39.577	18.568	18.568	3.273	40.409	0.034	0.313
Cussonia arborea	1	8.523	0.265	0.265	0.705	1.235	0.000	0.016
Cussonia spicata	25	26.087	6.631	6.631	2.158	15.420	0.004	0.180
Cynodenium sp.	2	1.021	0.531	0.531	0.084	1.145	0.000	0.028

Specie name	Frequency	Species basal area per ha	Relative density	Relative frequency	Relative basal area	Important Value Index	Simpson index	Shannon index
Dichapetalum deflexum	2	0.687	0.531	0.531	0.057	1.118	0.000	0.028
Dombeya rotundifolia	13	16.468	3.448	3.448	1.362	8.259	0.001	0.116
Erythrina abyssinica	2	7.136	0.531	0.531	0.590	1.651	0.000	0.028
Erythrococca fischeri	8	32.636	2.122	2.122	2.699	6.943	0.000	0.082
Euclea divinorum	9	6.780	2.387	2.387	0.561	5.335	0.001	0.089
Fagaropsis angolensis	7	5.814	1.857	1.857	0.481	4.194	0.000	0.074
Flacourtia indica	18	14.132	4.775	4.775	1.169	10.718	0.002	0.145
Grewia mollis	1	3.143	0.265	0.265	0.260	0.791	0.000	0.016
Juniperus procera	7	156.529	1.857	1.857	12.947	16.660	0.000	0.074
Maytenus ilicifolia	1	0.203	0.265	0.265	0.017	0.547	0.000	0.016
Maytenus lancifolia	3	0.915	0.796	0.796	0.076	1.667	0.000	0.038
Maytenus senegalensis	5	0.927	1.326	1.326	0.077	2.729	0.000	0.057
Maytenus sp.	15	7.559	3.979	3.979	0.625	8.583	0.002	0.128
Mystroxylon aethiopicum	4	5.217	1.061	1.061	0.432	2.554	0.000	0.048
Nuxia conjesta	65	439.073	17.241	17.241	36.316	70.799	0.030	0.303
Nuxia floribunda	2	3.343	0.531	0.531	0.276	1.337	0.000	0.028
Olea europaea	33	325.715	8.753	8.753	26.940	44.447	0.008	0.213
Pittosporum viridiflorum	2	1.711	0.531	0.531	0.142	1.203	0.000	0.028
Protea rubiobracteate	6	13.365	1.592	1.592	1.105	4.288	0.000	0.066
Synadenium sp.	1	0.318	0.265	0.265	0.026	0.557	0.000	0.016
Triscalria myritifolia	1	0.433	0.265	0.265	0.036	0.566	0.000	0.016
Turraea stulhmanii	1	2.990	0.265	0.265	0.247	0.778	0.000	0.016
Turrea robusta	1	1.134	0.265	0.265	0.094	0.624	0.000	0.016
Umbellifera sp.	1	0.565	0.265	0.265	0.047	0.577	0.000	0.016
Vepris nobilis	27	21.609	7.162	7.162	1.787	16.111	0.005	0.189

4.2. Regeneration/Forest Recruitment for Tree Species

The number of regenerants of tree species in this forest was low (Table 2). Only 9% of the plots had tree regenerants where 79 were recorded. The tree species with relatively higher number of regenerants were *Vepris nobilis* and *Abutilon sp.* The low number of tree regenerants could be attributed to human activities, particularly grazing.

4.3. Forest Stand Parameters

The number of tree stems per ha in the forest was 377 while species basal area per ha ranged between 0.098 m² and 439 m². Tree species with the highest and lowest basal area per ha were *Nuxia congesta* (439.07 m²) and *Acacia rovumae* (0.098 m²), respectively. The stand mean basal area per ha was 26.87 m².

The distribution of tree dbh (n=124) in the forest had a reversed *J*- shape with mature trees with dbh >46 cm being in low numbers, hence high SE of 40.607 (Figure 2, Table 4).

The mean ± SE of the Dbh was 17.32 ± 2.414 (Table 4).

Table 2. Regenerants of tree species recorded in Gelai Forest Reserve

Tree species	Number of regenerants counted
Abutilon sp.	10
Cassipourea gummiflua	7
Clausena anisata	7
Euclea divinorum	3
Fagaropsis angolensis	1
Flacourtia indica	2
Olea europaea	2
Umbellifera sp.	6
Vepris nobilis	41
Total	79

4.4. Key Shrub Species and Vegetation Associations

Shrubs, herbs and grasses were recorded on 55% of the surveyed plots majorly on treeless plots (Table 3). The most abundant four shrubs and herbs were *Vernonia galamensis* (14.05%) *Leonatis leonorus* (12.08%), *Ocimum suave* (10.44%) and *Solanum incanum* (9.2%). The three dominant grass species recorded were *Eleusine jaegeri, Panicum sp.* and *Paspalum sp.*

Table 3. List of shrubs and herb species recorded in Gelai Forest Reserve

Species	Frequency	Species	Frequency
Abutilon sp.	9	Lippia javanica	11
Allophylus camptostachys	2	Maytenus senegalensis	2
Artemisia afra	12	Ocimum suave	26
Asparagus sp.	2	Rhamnus sp.	3
Asplenium mosambicensis	17	Rhus longipes	4
Justicia sp.	17	Rhus vulgaris	1
Buddleia sp.	1	Rubus sp.	2
Caparis prinoides	1	Solanum incanum	23
Cassia floribunda	1	Sonchus sp.	1
Cyathula officinalis	1	Synadenium sp.	1
Diospyros fischeri	2	Trumpheta annua	2
Hibiscus ludwigii	1	Trumpheta sp.	5
Erythrococca fischeri	1	Turraea sp.	1
Grewia tembensis	2	Umbellifera sp.	1
Grewia tenax	1	Urtica massaica	16
Heliotropium sp.	2	Vernonia colorata	3
Hibiscus sp.	8	Vernonia galamensis	35
Leonatis leonorus	30	Vernonia sp.	2

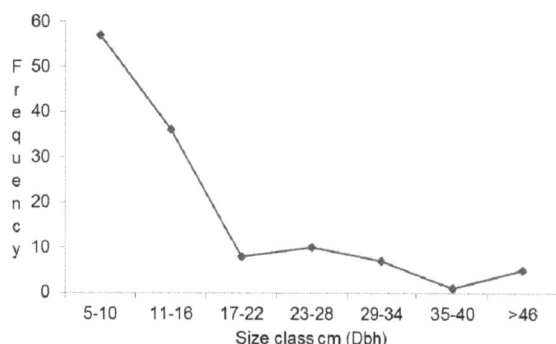

Figure 2. Size class distribution (Dbh in cm) of all tree species in the Gelai Forest Reserve showing a reversed J- shape

Table 4. Mean, Standard Deviation and Standard Error of Size class distribution (Dbh in cm) of all tree species in the Gelai Forest Reserve

Dbh (cm)	Mean	Std. Deviation	SE of Mean
5-10	7.6175	1.93586	0.25641
11-16	13.0222	1.46387	0.24398
17-22	19.85	1.91162	0.67586
23-28	26.22	2.02803	0.64132
29-34	31	1.49778	0.56611
35-40	35	.	.
>46	114.34	90.80004	40.60701
Total	17.3202	26.88437	2.41429

5. Discussion

The present study brings a significant finding about the species composition and abundance of often forgotten forests on small isolated hills in dry areas. The poorly managed forest is experiencing human disturbance and plant shifts related to various forms of forest disturbance were observed in cultivated areas where the original vegetation, mostly shrubs and herbs and closed forest were cleared for agriculture. Where closed forest was converted into agricultural land and later on abandoned, these areas were found to be dominated by bushes and shrubs of *Ocimum suave* and *Solanum incanum*. According to Mligo (2011), *Solanum incanum* is one of the early colonizer and invasive alien species that is mostly found in open areas disturbed by fires, grazing and agriculture. In spite of that, tree species diversity was high according to the Shannon diversity Index which usually ranges between 1.5 and 3.5 and rarely it exceeds 4.5 (Nagendra, 2002; Bhatt and Purohit, 2009). Forest disturbance is known to affect wildlife species that depends on the forest for habitat and food (Rabinowitz, 1997; Marshall, 2007).

According to studies by Bhatt and Purohit (2009), species diversity (richness) and dominance are inversely related to each other. However, selective harvesting of specific tree species may alter the species diversity index. For instance, *Juniperus procera* is highly preferred for building houses, making livestock fences and log beehives. *Ole capensis* is a good source of timber while *Olea europaea* provides good firewood and straight branches for cultural ceremonies. Another heavily used tree species is *Vepris nobilis* for making walking sticks, clubs and spear shafts. However, despite the disturbance, the reversed *J*- shape shows a good stand characteristic in a natural forest. This ensures good forest succession, whereby old trees when die then after a time young ones fill the gaps. This calls for sustainable management of natural resources which requires integration of protective, productive, social and environmental aspects of natural resources (Okali and Eyog-Matig, 2004).

The Tanzania Forest Act No.14 of 2002 recognizes all

forests under different management categories in the country including; national forest reserves, local authority forest reserves, village land forest reserves, community forest reserves, private forests, forest on general land and sacred and traditional forests. The law also recognizes partnerships in forest management, whereby the partnerships could be between state and community, or private and state or private and communities (URT, 2002). Ideally, through partnership arrangement, Gelai forest could be best managed with the local community by developing a Joint Forest Management plan.

Therefore, the inventory data collected in this survey can be used effectively as a benchmark from which to evaluate trends in tree species especially with the community involvement in conservation of the forest. The survey also demonstrated the importance of doing similar surveys to other small isolated forests of Tanzania to act as baselines for future monitoring. However, some parts of the forest were not systematically covered but with information collected from 100 plots (about 80%) out of 127 plots and low variation of vegetation types within this forest, the survey data will still meet the intended purposes of providing baselines information for planning, monitoring and measuring impacts. Ideally, the aim of the study, which was to estimate the tree species diversity and abundance following increased human encroachment and activities in the forest and to establish a baseline which can be used in follow-up studies to determine change, particularly after mitigation measures have been put in place. Indeed, the Gelai Forest Reserve is an area of exceptional importance for biodiversity conservation in the dry northern part of Tanzania. The majority of the forest is still intact although poorly managed and is a water catchment and a source of rivers that drains into Lake Natron. The forest reserve boundary, even where it is clearly defined, is encroached by the local people in some areas. However, the limited human resource available to manage the forest and the distance from the administrative unit makes management problematic and allows significant illegal activities to take place within the forests. The future of the forest resource will depend on the Joint Forest Management with the local people from the five villages who border the forest boundaries.

Acknowledgements

We thank EU for funding African Wildlife Foundation (AWF) through its project "Enhancing livelihoods through PFM in Northern Tanzania" (10th EDF NSA-ENV 2012/304-812). We also wish to appreciate the support of members of five Village Natural Resource Committees (VNRCs) who were involved in data collection, Village governments and Longido District Council Facilitation team. Finally, we also thank the Masaai morans who provided protection against dangerous wild animals. This work was undertaken under the aegis of AWF with permission of the Government of Tanzania to whom we are very grateful.

References

[1] Bhatt, V. and Purohit, V. K. (2009). Floristic structure and phytodiversity along an elevational gradient in Peepalkoti-Joshimath area of Garhwal Himalaya, India. Nature and Science, 7 (9) 63-74.

[2] Burgess, N.D., Butynski, T.M., Cordeiro, N.J., Doggart, N.H., Fjeldsa J., Howell, K.M., Kilahama, F.B., Loader, S.P., Lovett, J.C., Mbilinyi, B., Menegon, M., Moyer, D.C., Nashanda, E., Perkin, A., Rovero, F., Stanley, W.T., Stuart, S.N., (2007). The biological importance of the Eastern Arc Mountains of Tanzania and Kenya. Biological Conservation, 134, 209 – 239. doi:10.1016/j.biocon.2006.08.015.

[3] Burgess, N.D., Lovett, J., Rodgers, A., Kilahama, F., Nashanda, E.,Davenport, T., Butynski, T., (2004). Eastern Arc Mountains and Southern Rift. In: Mittermeier, R.A., Robles-Gil, P., Hoffmann,M., Pilgrim, J.D., Brooks, T.M., Mittermeier, C.G., Lamoreux, J.L.,Fonseca, G.A.B. (Eds.), Hotspots Revisited: Earth's Biologically Richest and Most Endangered Ecoregions, second ed. Cemex, Mexico, pp. 245–255.

[4] Burgess, N.D., Doggart, N.H., Lovett, J.C., (2002). The Uluguru Mountains of eastern Tanzania: the effect of forest loss on biodiversity. Oryx 36, 140–152.

[5] Burgess, N.D., Clarke, G.P. (Eds.), (2000). The Coastal Forests of Eastern Africa. IUCN Forest Conservation Programme, Gland and Cambridge.

[6] Burgess, N.D., Nummelin, M., Fjeldsa° , J., Howell, K.M., Lukumbyzya, K., Mhando, L., Phillipson, P., Vanden Berghe, E. (Eds.), (1998). Biodiversity and conservation of the Eastern Arc Mountains of Tanzania and Kenya. Journal of East African Natural History 87, 1–367 pp.

[7] Burke, A. (2003). Inselbergs in a changing world - global trends. Diversity and Distributions 9: 375–383.

[8] Burke, A. (2005a). Vegetation types of mountain tops in Damaraland, Namibia. Biodiversity and Conservation 14: 1487–1506.

[9] Burke, A. (2005b). Biodiversity patterns in arid, variable environments. Mountain Research and Development 25(3): 228–234.

[10] CEPF, (2003). Ecosystem Profile: Eastern Arc Mountains and Coastal Forests of Tanzania and Kenya Biodiversity Hotspot. Critical Ecosystem Partnership Fund, Washington, DC. Available from: <http://www.cepf.net>.

[11] Dublin, H.T. (1986). Decline of the Mara Woodlands: The Role of Fire and Elephants. PhD thesis, University of British Columbia.

[12] Field, C.R. & Ross, I.C. (1976). The savanna ecology of Kidepo Valley Park. Part II: Feeding ecology of elephant and giraffe. E. Afr.Wildl. J. 14,1 - 15.

[13] IPNI (2013). <http://www.ipni.org> On-line version dated 29 April 2013.

[14] Jain, S.K. &. Rao, R.R. (1977). Field & Herbarium Methods. Today & Tomorrow's. Printers & Publishers, Delhi, 157pp.

[15] Marshall, A.R. (2007). Disturbance in the Udzungwas: Responses on Monkeys and Trees to Forest degradation. PhD Thesis. The University of York. Pp 151.

[16] Mligo, C. (2011). Anthropogenic disturbance on the vegetation in Makurunge woodland, Bagamoyo district, Tanzania. Tanz. J. Sci. Vol. 37, 95-108.

[17] Nagendra, H. (2002). Opposite trends in response for the Shannon and Simpson indices of landscape diversity. Applied Geography 22, 175–186.

[18] Okali, D., Eyog-Matig, O. (2004). Lessons learnt on sustainable forest management for Africa: Rain forest management for wood production in West and Central Africa. A report prepared for the project KSLA/AFORNET/FAO project. pp 79.

[19] Rabinowitz, A.R. (1997). Wildlife Field Research and Conservation Training Manual. Wildlife Conservation Society, 185th Street and Southern Blvd. Bronx, New York. Pp 281.

[20] Sitati, N.W. (2003). Human-elephant conflict in Transmara District adjacent to Masai Mara National Reserve, Kenya. PhD Thesis, DICE, University of Kent, UK.

[21] Sitati, N.W., Ucakuwun, E.K., Wishitemi, B.E.L. (2008). Spatio-temporal analysis of land use types in the Masai Mara dispersal areas, Kenya. East African Journal of Pure and Applied Science. vol. 4. Pp. 24-31.

[22] SPSS (2011). Statistics for Windows, Version 20.0. Armonk, NY: IBM Corp.

[23] United Republic of Tanzania (URT). (2002). The Forest Act No. 14 of 2002. Ministry of Natural Resources and Tourism, Dar es Salaam, Tanzania, Government Printer. pp. 1281.

[24] van Essen, L.D., J. du P. B othma, N. van Rooyen and W. S.W. Trollope (2002). Assessment of the woody vegetation of Ol Choro Oiroua, Masai Mara, Kenya. Afr. J. Ecol., 40, 76 - 83.

31

Analysis of wind energy potential in north east Nigeria

A. Ahmed[1], A. A. Bello[2], D. Habou[2]

[1]Department of Mechanical Engineering, Kano University of Science and Technology, Wudil, Nigeria
[2]Department of Mechanical Engineering, Abubakar Tafawa Balewa University, Bauchi, Nigeria

Email address:
abdula2k2@yahoo.com (A. Ahmed), biieeyz@yahoo.com (A. A. Bello), hdandakuta@gmail.com (D. Habou)

Abstract: This research reports wind energy potential evaluation of two locations in the north east Nigeria (Bauchi and Borno). The evaluation is based on Weibull and Rayleigh models using 17 years mean monthly wind speed data covering the period (1990-2006). The result shows that Rayleigh is best fit model that describes the wind speed data at 10 m height. Reference mean power density (based on the measured probability distribution) was compared with those obtained from the Weibull and Rayleigh models. In calculating the percentage error, results shows that Weibull provided better power density estimation in all 12 months than the Rayleigh model. From this research work, it was found that Borno has high wind power density 273.16 W/m^2 for Weibull and 365.77 W/m^2 for Rayleigh in the month of June as compared Bauchi with highest power density of 31.45 W/m^2 for Weibull and 37.06 W/m^2 for Rayleigh in the month of May.

Keywords: Wind Energy Potential, Nigeria, Generation, Weibull, Rayleigh, Probability Density Function

1. Introduction

Wind energy is currently the most economic renewable energy apart from hydropower, its usage versatility and ability to use it as a decentralized energy form make its applications possible in rural areas where it is technically and economically feasible in the country. The major challenge to using wind as a source of electricity generation is that wind is intermittent and it does not always blow when electricity is needed. However, wind power is one of the most the promising and cost – effective renewable.

In the 1980's, California purchased large quantities of wind power invest on operating experience needed to bring the cost of wind power down to a power installed in California, and another 1000W installed in other parts to generate electricity for over 750,000 homes.

The number of wind farms in US has increased substantially in the wind farms installed. The US Department of Energy projects that by the power, enough to generate electricity for 1.7 million homes, due to the power will fall. Currently, wind power costs between 3 and 6 cents making it one of the cheapest resources available [1].

Nigeria is subject to the seasonal rain – bearing south – westerlies, which blow strongly from April to October and to the dry and dusty North- East trade wind which blow strongly from November to March every year. Most areas sometimes experience some periods of doldrums in between these periods. In Nigeria, wind energy reserves at 10 m height shows that some sites have wind regime for between 1.0 to 5.1 ms^{-1}.

Energy supply in Nigeria is a major problem for both large and small scale purposes. Highly centralized production and distribution units have not been equally distributed thus becomes inadequate in meeting the economic needs of both urban and rural populace in Nigeria. With respect to this problem, solar and wind energy are some of the alternative sources of energy that can be exploited to meet some of the populace needs. It is therefore necessary to evaluate the wind regimes in the country and assess the potential of wind, installing wind energy conversion system for the generation of electricity.

In this context, over the years researchers have carried out a number of studies in order to assess the wind energy potential in some parts of the world. Shata [1] worked on the potential of electricity generation on the coast of Red Sea in Egypt. Celik [5] studied the distributional parameters used in assessment of suitability of wind speed probability density function. Ozoptal, et al [8] studied the regional wind energy potential of Turkey.

In this presentation, 17 years (1990 - 2006) monthly mean wind speed data are obtained from Nigeria Meteorological Agency (NIMET) Abuja for some selected locations in the North east Nigeria (Bauchi and Borno), were statistically analyzed to evaluate wind power density based on the Weibull and Rayleigh models.

2. Data Collection and Wind Speed Characteristics

Table 1. Geographical data of the locations

Locations	State	Latitude (N)	Longitude (E)	Altitude
Bauchi	Bauchi	10°18'57	9°50'39	615
Maiduguri	Borno	11°50'47	13°9'37	299

Table 2. Summary of average wind speed V_m (ms^{-1}) and standard deviation σ (ms^{-1}).

	Bauchi		Borno	
Months	V_m(ms^{-1})	σ	V_m(ms^{-1})	σ
Jan	2.43	0.40	4.94	1.35
Feb	2.09	0.62	5.85	1.71
Mar	1.99	0.68	6.27	1.31
Apr	2.62	0.57	6.02	1.03
May	3.16	0.77	6.27	1.10
Jun	2.91	0.71	6.88	1.15
Jul	2.82	0.57	6.19	1.23
Aug	2.44	0.43	5.03	0.98
Sep	2.54	0.55	4.33	0.98
Oct	2.17	0.57	4.07	0.93
Nov	1.68	0.59	4.52	0.89
Dec	1.47	0.63	5.06	1.39

Table 3. Summary of average scale factor c (ms^{-1}), shape factor k and gamma function

	Bauchi			Borno		
Months	c(m/s)	k	Γ	c(m/s)	k	Γ
Jan	2.58	7.21	0.826	5.45	4.11	0.729
Feb	2.31	3.77	0.713	6.48	3.81	0.715
Mar	2.22	3.23	0.683	6.79	5.53	0.782
Apr	2.85	5.27	0.774	6.42	6.85	0.818
May	3.46	4.67	0.753	6.71	6.65	0.813
Jun	3.19	4.68	0.753	7.24	6.91	0.819
Jul	3.06	5.75	0.789	6.68	5.81	0.790
Aug	2.61	6.66	0.813	5.42	5.92	0.794
Sep	2.75	5.34	0.776	4.71	5.06	0.767
Oct	2.39	4.27	0.736	4.43	5.03	0.766
Nov	1.88	3.12	0.677	4.87	5.90	0.793
Dec	1.66	2.54	0.637	5.59	4.11	0.729

Table 4. Maximum and minimum values of wind speed (ms^{-1})

Locations	Max. Vel. (m/s)	Month	Min. Vel. (m/s)	Month
Bauchi	3.16	May	1.47	Dec
Borno	6.88	Jun	4.07	Oct

In this study, statistical analyses of wind speed and power density available in selected states of north central Nigeria are investigated. Seventeen years monthly mean speed (1990-2006) from Nigeria Meteorological Agency

(NIMET) Abuja at the height of 10m was used. Table 1 shows the geographical locations of the two locations in the north east of Nigeria (Bauchi and Borno). From the above Table it can be seen that Bauchi is located at longitude 9°50'39 East and latitude 10°18'57 North with a land scale slope of 615 meters. Borno is located at longitude 13°9'37 East and latitude 11°50'47 North with a land scale slope of 299 meters.

The summary of average wind speed V_m (ms^{-1}) and standard deviation σ (ms^{-1}) for the monthly distributional parameters for all the sites are presented in Table 2 below.

The summary of average scale factor c (ms^{-1}), shape factor k and gamma function for the monthly distributional parameters of the locations considered are presented in Table 3 below while Table 4 presents the maximum and minimum values of wind speed (ms^{-1}) for the two locations and months of their occurrences.

The frequency probability distribution for Bauchi in the month of January is presented in Table 5; the same pattern of table was computed for the two locations.

Table 5. Frequency probability distribution for January for Bauchi.

Vj	Vmj	fj	f(vj)	fw(vj)	fR(vj)
0 - 0.9	0.45	0	0.000	5.45262E-05	2.12698E-09
1 - 1.9	1.45	2	0.118	0.076813476	0.199965388
2 - 2.9	2.45	14	0.824	1.017947398	0.293669502
3 - 3.9	3.45	1	0.059	0.005020743	0.217221871
4 - 4.9	4.45	0	0.000	6.33475E-21	0.152524121
5 - 5.9	5.45	0	0.000	1.23105E-93	0.10994378
6 - 6.9	6.45	0	0.000	0.000	0.082081098
7 - 7.9	7.45	0	0.000	0.000	0.063266859
8 - 8.9	8.45	0	0.000	0.000	0.05010226
9 - 9.9	9.45	0	0.000	0.000	0.040584663
10 - 10.9	10.45	0	0.000	0.000	0.03350453
11 - 11.9	11.45	0	0.000	0.000	0.028106483
12 - 12.9	12.45	0	0.000	0.000	0.023902753
13 - 13.9	13.45	0	0.000	0.000	0.020568503
14 - 14.9	14.45	0	0.000	0.000	0.017881314

2.1. Monthly Average Wind Speed and Standard Deviation

The monthly average wind speed and the standard deviation can be obtained using equation 1 and 2 below.

$$V_m = N^{-1}\left[\sum_{i=1}^{N} Vi\right] \quad (1)$$

$$\sigma = \left[\frac{1}{N-1}\sum_{i=1}^{N}(Vi-Vm)^2\right]^{1/2} \quad (2)$$

2.2. Wind Speed Probability Distributions

The wind speed data in time series format is usually arranged in the frequency distribution format since it is more convenient for statistical analysis. Therefore, the available time series data were translated into frequency distribution format. This process is illustrated for an example for the month of January for Bauchi as presented in Table 5.

The wind speed is grouped into classes (bins) as given in the first column of Table 5.The mean wind speeds are calculated for each speed class intervals are in second column. The probability density distribution is presented in the third column. The probability density obtained from Weibull and Rayleigh parameters are presented in the fourth and fifth columns.

The wind speed distributions and the functions representing them mathematically are the main tools used in the wind related literature. Their use include a wide range of applications, from the techniques used to identify the parameters of the distribution function to the use of such functions for analyzing the wind speed data and wind energy economics. Two of the commonly used functions for fitting a measured wind probability distribution in a given location over a certain period of time are the Weibull and Rayleigh. The probability density function of the Weibull distribution is given by;

$$f_w(v) = (k/c)\,(v/c)^{k-1} \exp\left[-(v/c)^k\right] \qquad (3)$$

The corresponding cumulative probability function of the Weibull distribution is,

$$F_w(v) = 1 - \exp\left[-(v/c)^k\right] \qquad (4)$$

$$V_m = c\,\Gamma\left(1 + \frac{1}{k}\right) \qquad (5)$$

$$c = \left[\frac{k^{2.6674}}{0.184 + \left(0.816\,k^{2.73859}\right)}\right] \qquad (6)$$

$$k = \left(\frac{\sigma}{v}\right)^{-1.090} \qquad (7)$$

The Rayleigh model is a special and simplified case of the Weibull model. It is obtained when the shape factor k of the Weibull model is assumed to be = 2. The probability density and the cumulative functions of the Rayleigh model are given by,

$$f_R(V) = \frac{\pi}{2}\frac{v}{v^2_m}\exp\left[-\frac{\pi}{4}\left(\frac{v}{v_m}\right)^2\right] \qquad (8)$$

$$F_R(V) = 1 - \exp\left[-\frac{\pi}{4}\left(\frac{v}{v_m}\right)^2\right] \qquad (9)$$

One of the most distinct advantages of the Rayleigh distribution is that the probability density and the cumulative distribution functions could be obtained from the mean value of the wind speed. The Rayleigh model has

also widely been used to fit the measured probability density distribution.

2.3. Power Density Distribution & Mean Power Density

The power of the wind per unit area is given by;

$$P(V) = \frac{1}{2}PV^3 \qquad (10)$$

Where ρ is assume to be 1.225 kg/m^3 in this paper.

The wind power density for the measured probability density distribution which serves as the reference mean power density as shown below;

$$P_{m.R} = \sum_{j=1}^{n}\left(\frac{1}{2}PV^3{}_{mj}\,f(vj)\right) \qquad (11)$$

$$P_m(V) = \frac{1}{2}\rho.(V^3)_m \qquad (12)$$

$$P_w = \frac{1}{2}\rho.c^3\,\Gamma\left[1 + \frac{3}{k}\right] \qquad (13)$$

$$P_R = \frac{3}{\pi}\rho.V_m{}^3 \qquad (14)$$

The yearly average error in calculating power densities using both Weibull and Rayleigh functions is obtained by using equation below;

$$\text{Error}(\%) = \frac{1}{12}\sum_{i=1}^{12}\left(\frac{P_{W,R} - P_{m,R}}{P_{m,R}}\right) \qquad (15)$$

3. Analysis of Wind Speed Data

The surface wind characteristics and stochastic analysis of the wind speed data in the two locations of the north east part of Nigeria were carried out. It is seen from Table 2 that the highest average wind speed occurred in June (6.88 ms^{-1}) in Borno while the lowest average wind speed occurred in December (1.47 ms^{-1}) in Bauchi. The average scale factor c (ms^{-1}) ranges from 1.66 ms^{-1} in Bauchi to 7.24 ms^{-1} in Borno, while the shape factor k ranges from 2.54 in Bauchi to 6.91 in Borno as presented in Table 3.

3.1. Best – Fit Probability Distribution Model

The average monthly values of the correlation coefficient for the two locations in north east Nigeria ranges between 0.130 and 0.539 for the Weibull and the Rayleigh model ranges from 0.387 to 0.811 as indicated in Table 6.

The month – to – month comparison shows that Rayleigh model returns higher coefficient values in all the twelve months for Bauchi and Borno. It can be seen from Table 7 that in Borno, Rayleigh model returns higher coefficient in all months while Bauchi also returns higher coefficient values for Rayleigh except in the month of November and December.

Table 6. *Correlation coefficient values for all the location.*

MONTHS	Bauchi		Borno	
	W	R	W	R
JAN	0.130	0.430	0.356	0.519
FEB	0.336	0.439	0.539	0.716
MAR	0.407	0.437	0.425	0.762
APR	0.219	0.416	0.309	0.736
MAY	0.219	0.387	0.360	0.762
JUN	0.228	0.394	0.386	0.811
JUL	0.202	0.400	0.393	0.754
AUG	0.140	0.430	0.247	0.604
SEP	0.205	0.422	0.227	0.497
OCT	0.278	0.440	0.213	0.459
NOV	0.424	0.418	0.204	0.526
DEC	0.541	0.405	0.374	0.609
AVE				

Table 7. *Summary of best – fit probability distribution models.*

Months	Bauchi	Borno
Jan	Rayleigh	Rayleigh
Feb	Rayleigh	Rayleigh
Mar	Rayleigh	Rayleigh
Apr	Rayleigh	Rayleigh
May	Rayleigh	Rayleigh
Jun	Rayleigh	Rayleigh
Jul	Rayleigh	Rayleigh
Aug	Rayleigh	Rayleigh
Sep	Rayleigh	Rayleigh
Oct	Rayleigh	Rayleigh
Nov	Weibull	Rayleigh
Dec	Weibull	Rayleigh

4. Results and Discussion

The mean power density shows a large month – to – month variation as shown in Figure 1 and 2. From Figure 1 the maximum power density occurs in month of May for Rayleigh (37.06 W/m^2) while for Weibull is (31.45 W/m^2). From Figure 2 the maximum power density occurs in the month of June for Rayleigh (365.77 W/m^2) while for Weibull is (273.16 W/m^2).

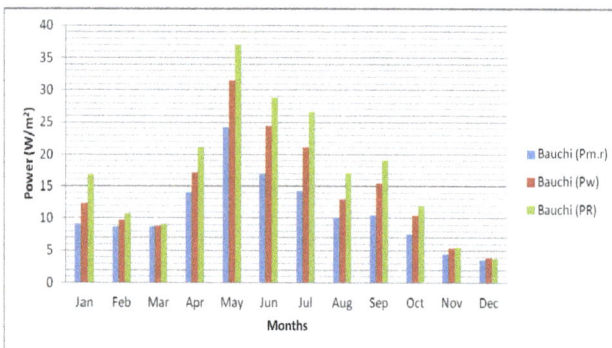

Fig 1. *Power density (W/m^2) for Bauchi ($P_{m.r}$, P_w & P_R)*

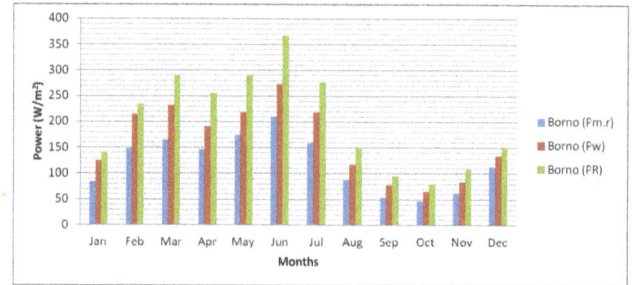

Fig 2. *Power density (W/m^2) for Borno ($P_{m.r}$, P_w & P_R)*

From Figure 3, it can be seen that Rayleigh returns a higher percentage error in almost all the month expect for the month of December where Weibull have a lower percentage error than Rayleigh model. It can be seen that from Figure 4, Rayleigh returns a higher percentage error in all the twelve months.

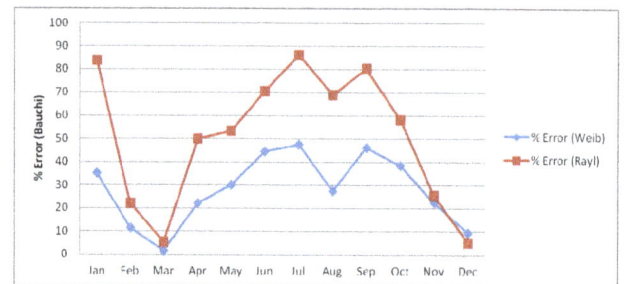

Fig 3. *% Error for Weibull and Rayleigh Bauchi*

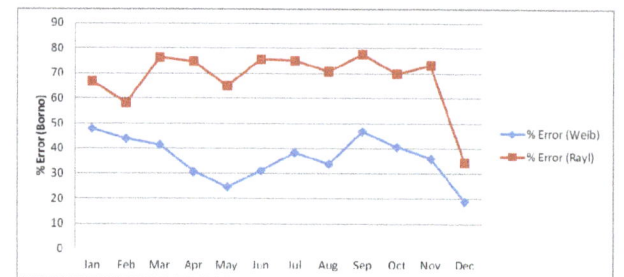

Fig 4. *% Error for Weibull and Rayleigh Borno*

5. Conclusion

In this study, it was found that Rayleigh returns a higher power density than Weibull and the highest power density is (365.77 W/m^2) for Rayleigh in Borno in the month of June, it was also found that Rayleigh returns the best fit probability distribution than Weibull in almost all the months in the two locations of north east part of Nigeria considered in this research work.

Finally, in calculating error in power density the Weibull model returns smaller error in calculating the power density compared to Rayleigh model in all locations. The power density is estimated by the Weibull model with a smallest error value of 1.5% for Bauchi in the month of March. From this research work, it was found that Borno in north east part of Nigeria has high wind power density for the generation wind energy with a maximum value of power density of 365.77 W/m^2 for Weibull. Borno can be classified under wind class II since the power density is greater than 100 W/m^2 while Bauchi can be classified under wind class I since the power density is less than 100 W/m^2.

References

[1] Ahmed A., Adisa AA, Habou D. An evaluation of wind energy potential in the northern and southern regions of Nigeria on the basis of Weibull and Rayleigh models. America Journal of Energy Engineering 2013; 1: 37 – 42.

[2] Ahmed Shata AS, Hanistsch R. The potential of electricity generation on the east coast of Red Sea in Egypt. Renewable Energy 2006; 31: 1597 – 625.

[3] Ahmed SA. Investigation of wind characteristics and wind energy potential at RAS Ghareb, Egypt. Renewable and Sustainable Energy Reviews 2011; 15: 2750 – 5.

[4] Akpinar EK, Akpinar S. Determination of the wind energy potential for Maden-Elazig, Turkey. Energy Conversion and Management 2004; 45: 2901-14.

[5] Brano VL, Orioli A, Ciulla G, Culotta S. Quality of wind speed fitting distributions for the urban area of Palermo, Italy. Renewable Energy 2011; 36:1026 – 39.

[6] Celik AN. On the distributional parameters used in assessment of the suitability of wind speed probability density functions. Energy Conversion and Management 2004; 45: 1735 – 47.

[7] Celik AN. A statistical analysis of wind power density based on the Weibull and Rayleigh Models at the southern region of Turkey. Renewable Energy 2004; 29:593 – 604.

[8] Celik AN. Assessing the suitability of wind speed probability distribution functions based on the wind power density. Renewable Energy 2003; 28:1563 -1574.

[9] Oztopal A, Sahin AD, Akgun N, Sen Z. On the regional wind energy potential of Turkey. Energy 2000; 25: 189 – 200.

[10] Sambo AS. The renewable energy for rural development. The Nigerian perspective "ISESCO" Science and Technology vision May, 2005; 1:16 -18.

[11] Salem AL. Characteristics of surface wind speed and direction over Egypt Solar Energy for sustainable development 2004; 4: 491 – 499.

[12] Seguro JV, Lambert TW. Modern estimation of the parameters of the Weibull wind speed distribution for wind speed distribution for wind energy analysis. J Wind Eng Ind Aerodyn 2000; 85: 75 – 84.

[13] Weisser D. Wind energy analysis of Grenada: an estimation using the Weibull density function. Renewable Energy 2003; 28: 1803 – 12.

[14] World Wind Energy Association (WWEA). World wind energy report 2011 website: http:// www.wwindea.org, Accessed August 2, 2012.

PERMISSIONS

The contributors of this book come from diverse backgrounds, making this book a truly international effort. This book will bring forth new frontiers with its revolutionizing research information and detailed analysis of the nascent developments around the world.

We would like to thank all the contributing authors for lending their expertise to make the book truly unique. They have played a crucial role in the development of this book. Without their invaluable contributions this book wouldn't have been possible. They have made vital efforts to compile up to date information on the varied aspects of this subject to make this book a valuable addition to the collection of many professionals and students.

This book was conceptualized with the vision of imparting up-to-date information and advanced data in this field. To ensure the same, a matchless editorial board was set up. Every individual on the board went through rigorous rounds of assessment to prove their worth. After which they invested a large part of their time researching and compiling the most relevant data for our readers.

The editorial board has been involved in producing this book since its inception. They have spent rigorous hours researching and exploring the diverse topics which have resulted in the successful publishing of this book. They have passed on their knowledge of decades through this book. To expedite this challenging task, the publisher supported the team at every step. A small team of assistant editors was also appointed to further simplify the editing procedure and attain best results for the readers.

Apart from the editorial board, the designing team has also invested a significant amount of their time in understanding the subject and creating the most relevant covers. They scrutinized every image to scout for the most suitable representation of the subject and create an appropriate cover for the book.

The publishing team has been an ardent support to the editorial, designing and production team. Their endless efforts to recruit the best for this project, has resulted in the accomplishment of this book. They are a veteran in the field of academics and their pool of knowledge is as vast as their experience in printing. Their expertise and guidance has proved useful at every step. Their uncompromising quality standards have made this book an exceptional effort. Their encouragement from time to time has been an inspiration for everyone.

The publisher and the editorial board hope that this book will prove to be a valuable piece of knowledge for researchers, students, practitioners and scholars across the globe.

LIST OF CONTRIBUTORS

Hamed H. H. Aly, M. E. El-Hawary
Department of Electrical and Computer Engineering, Dalhousie University, Halifax, Nova Scotia, Canada, B3H 4R2

Tayeb CHIHI
Laboratory for Elaboration of New Materials and Characterization (LENMC), University of Setif 1, 19000, Algeria

FATMI Messaoud
Research Unit on Emerging Materials (RUEM), University of Setif 1, 19000, Algeria

Laboratory of Physics and Mechanics of Metallic Materials (LP3M), University of Setif 1, 19000, Algeria

Molla Asmare
Centre of Competence for Sustainable Energy Engineering, Institute of Technology, Bahir Dar University, Ethiopia

Assefa Alena and Omprakash Sahu
Department of Chemical Engineering, Wollo University, Kombolcha, South Wollo Ethiopia

Muhammad Mukhtar, Chika Muhammad and Musa Usman Dabai
Department of Pure and Applied Chemistry, Usmanu Danfodiyo University, P. M. B, 2346, Sokoto, Nigeria

Muhammad Mamuda
Sokoto Energy Research Centre, Usmanu Danfodiyo University, P. M. B, 2346, Sokoto, Nigeria

Nuru Safarov, Gurban Axmedov and Sedreddin Axmedov
Department of Electronics, Telecommunications and Radio Engineering, Khazar University, Baku, Azerbaijan

Ahmed A. Rizk
Architectural Engineering Department, Faculty of Engineering, University of Tanta, Egypt

Visiting Professor of Architectural Engineering, University of Nebraska – Lincoln, South 67th Street, Peter Kiewit Institute, Omaha, NE

Gregor P. Henze
University of Colorado at Boulder, CEAE Department 428 UCB, Boulder, Colorado 80309-0428 U.S.A

Sadik Umar and Umar Kangiwa Muhammad
Department of Physics, Kebbi State University of Science and Technology Aliero, Aliero, Nigeria

Muhammad Mahmoud Garba and Hassan N. Yahya
Sokoto Energy Research Center, Usmanu Danfodiyo University Sokoto

Krishna Kumar and Omprakash Sahu
Department of Chemical Engineering, KIOT, Wollo University, Ethiopia

F. Ayaa and N. Banadda
Department of Agricultural and Bio-Systems Engineering, Makerere University, Kampala, Uganda

P. Mtui
Department of Mechanical Engineering University of Dar-es-Salaam, Dar-es-Salaam, Tanzania

J. Van Impe
Department of Chemical Engineering, Ku Leuven, Leuven, Belgium

Molla Asmare and Nigus Gabbiye
Centre of Competence for Sustainable Energy Engineering, Institute of Technology, Bahir Dar University, Bahir Dar, Ethiopia

Emmanuel Tete Okoh
Department of Furniture Design and Production, Accra Polytechnic, P O Box GP 561, Accra, Ghana

Anna Olsson, Robert Lundmark
Economics Unit, Luleå University of Technology, Luleå, Sweden

Ishmael Ackah
Department of Economics, Portsmouth Business School, University of Portsmouth, Portsmouth, UK

Dankwa Kankam
Opoku, Andoh& Co, Chartered Accountants and Management Consultants, Accra, Ghana

Kwaku Appiah- Adu
Central Business School, Central University College, Accra, Ghana

Emmanuel Tete Okoh
Department of Furniture Design and Production, Accra Polytechnic, P O Box GP 561, Accra, Ghana

Jyothilal Nayak Bharothu
Associate professor of Electrical & Electronics Engineering, Sri Vasavi Institute of Engineering & Technology, Nandamuru, A.P.; India

AbduL Arif
Assistant professor of Electrical & Electronics Engineering, Sri Vasavi Institute of Engineering & Technology, Nandamuru, A.P.; India

Abdullahi Ahmed
Department of Mechanical Engineering, Kano University of Science and Technology, Wudil, Nigeria

Adisa Ademola Bello and Dandakuta Habou
Department of Mechanical Engineering, Abubakar Tafawa Balewa University, Bauchi, Nigeria

Peter Mtui
College of Engineering and Technology, University of Dar es Salaam, ar es Salaam, Tanzania

Peter Mtui
College of Engineering and Technology, University of Dar es Salaam, P. O. Box 35131, Dar es Salaam, Tanzania

Kanhaiya Lal Meena, Vimala Dhaka and Prakash Chandra Ahir
Department of Botany, M.L.V. Government College, Bhilwara - 311001, Rajasthan, India

Mohammad Shuhrawardy and Kazi Tanvir Ahmmed
Department of Applied Physics, Electronics and Communication Engineering, University of Chittagong, Chittagong, Bangladesh

Hamed Aliyari and Ardavan Areyaei1
Electrical Engineering Department, Science and Research Alborz branch, Islamic Azad University, Alborz, Iran

Reza Effatnejad
Electrical Engineering Department, Karaj branch-Islamic Azad University, Alborz, Iran

Mohammad Shariful Islam and Enamul Basher
Electrical & Electronic Engineering, Bangladesh University of Engineering & Technology (BUET), Dhaka, Bangladesh

Asif Islam
Planning & Development Division, Power Grid Company of Bangladesh (PGCB) Ltd., Dhaka, Bangladesh

Dipendra Shah
NIS Department, Huawei Technologies Co. Ltd., Kathmandu, Nepal

Daniël Amrish Lachman
Institute for Graduate Studies and Research and Mechanical Engineering Discipline, Anton de Kom University of Suriname; FHR Lim A

Po Institute, Paramaribo, Suriname

Kamrun Nahar
Department of Environmental Science and Management, North South University, Bangladesh

Sanwar Azam Sunny
Department of Mechanical Engineering, University of Kansas, Lawrence, KS, United States

Barness Chirazo Mphande and Alexander Pogrebnoi
Dept. of Materials Science and Engineering, The Nelson Mandela African Institution of Science and Technology, Arusha, Tanzania

Noah Sitati, Nathan Gichohi, Philip Lenaiyasa, Peter Millanga, Michael Maina, Fiesta Warinwa and Philip Muruthi
African Wildlife Foundation, P.O Box 20 00207, Namanga, Kenya

A. Ahmed
Department of Mechanical Engineering, Kano University of Science and Technology, Wudil, Nigeria

A. A. Bello and D. Habou
Department of Mechanical Engineering, Abubakar Tafawa Balewa University, Bauchi, Nigeria

Index